Plant Abiotic Stress

Biological Sciences Series

A series which provides an accessible source of information at research and professional level in chosen sectors of the biological sciences.

Series Editor:
Professor J. A. Roberts, Plant Science Division, School of Biosciences, University of Nottingham, UK.

Titles in the series:

Biology of Farmed Fish
Edited by K.D. Black and A.D. Pickering

Stress Physiology in Animals
Edited by P.H.M. Balm

Seed Technology and its Biological Basis
Edited by M. Black and J.D. Bewley

Leaf Development and Canopy Growth
Edited by B. Marshall and J.A. Roberts

Environmental Impacts of Aquaculture
Edited by K.D. Black

Herbicides and their Mechanisms of Action
Edited by A.H. Cobb and R.C. Kirkwood

The Plant Cell Cycle and its Interfaces
Edited by D. Francis

Meristematic Tissues in Plant Growth and Development
Edited by M.T. McManus and B.E. Veit

Fruit Quality and its Biological Basis
Edited by M. Knee

Pectins and their Manipulation
Edited by G. B. Seymour and J. P. Knox

Wood Quality and its Biological Basis
Edited by J.R. Barnett and G. Jeronimidis

Plant Molecular Breeding
Edited by H.J. Newbury

Biogeochemistry of Marine Systems
Edited by K.D. Black and G. Shimmield

Programmed Cell Death in Plants
Edited by J. Gray

Water Use Efficiency in Plant Biology
Edited by M.A. Bacon

Plant Lipids – Biology, Utilisation and Manipulation
Edited by D.J. Murphy

Plant Nutritional Genomics
Edited by M.R. Broadley and P.J. White

Plant Abiotic Stress
Edited by M.A. Jenks and P.M. Hasegawa

Plant Abiotic Stress

Edited by

MATTHEW A. JENKS

Center for Plant Environmental Stress Physiology
Purdue University
Indiana, USA

and

PAUL M. HASEGAWA

Center for Plant Environmental Stress Physiology
Purdue University
Indiana, USA

Blackwell
Publishing

© 2005 by Blackwell Publishing Ltd

Editorial offices:
Blackwell Publishing Ltd, 9600 Garsington Road, Oxford OX4 2DQ, UK
Tel: +44 (0) 1865 776868
Blackwell Publishing Professional, 2121 State Avenue, Ames, Iowa, 50014-8300, USA
Tel: +1 515 292 0140
Blackwell Publishing Asia Pty Ltd, 550 Swanston Street, Carlton, Victoria 3053, Australia
Tel: +61 (0)3 8359 1011

First published 2005 by Blackwell Publishing Ltd

3 2007

ISBN: 978-1-4051-2238-2

Library of Congress Cataloging-in-Publication Data
Plant abiotic stress / edited by Matthew A. Jenks and Paul M. Hasegawa.—1st ed.
 p. cm.
 Includes bibliographical references and index.
 ISBN-10: 1-4051-2238-2 (hardback : alk. paper)
 ISBN-13: 978-1-4051-2238-2 (hardback : alk. paper)
 1. Crops-Effect of stress on. 2. Crops-Physiology. I. Jenks, Matthew A. II. Hasegawa, Paul M.
 SB112.5.P5 2005
 632′.1—dc22
 2004025753

A catalogue record for this title is available from the British Library

Set in 10 / 12pt Times New Roman
by Kolam Information Services Pvt. Ltd., Pondicherry, India
Printed and bound by Replika Press Pvt. Ltd., India

The publisher's policy is to use permanent paper from mills that operate a sustainable forestry policy, and which has been manufactured from pulp processed using acid-free and elementary chlorine-free practices. Furthermore, the publisher ensures that the text paper and cover board used have met acceptable environmental accreditation standards.

For further information on
Blackwell Publishing, visit our website:
www.blackwellpublishing.com

Contents

6 Adaptive responses in plants to nonoptimal soil pH   **145**
V. RAMÍREZ-RODRÍGUEZ, J. LÓPEZ-BUCIO and
L. HERRERA-ESTRELLA

Contributors

Manu Agarwal

Department of Botany and Plant Sciences, 2150 Batchelor Hall, Institute for Integrative Genome Biology, University of California, Riverside, California 92521, USA

Kenji Akiyama

Plant Mutation Exploration Team and Genomic Knowledge Base Research Team, RIKEN Genomic Sciences Center, Riken Yokohama Institute, 1-7-22 Suehiro-cho, Tsurumi-ku, Yokohama 230-0045, Japan

Miguel A. Botella

Dep. Biología Molecular y Bioquímica, Facultad de Ciencias, Universidad de Málaga, Campus de Teatinos s/n, Málaga, 29071 Spain

Ray A. Bressan

Center for Plant Environmental Stress Physiology, 625 Agriculture Mall Drive, Purdue University, West Lafayette, Indiana 47907–2010, USA

Daniel Cook

MSU-DOE Plant Research Laboratory, Michigan State University, East Lansing, MI 48824-1312, USA

William E. Dyer

Department of Plant Sciences, Montana State University, Bozeman, Montana 59717, USA

Akiko Enju

Plant Mutation Exploration Team, Plant Functional Genomics Research Group, RIKEN Genomic Sciences Center (GSC), RIKEN Yokohama Institute, 1-7-22 Suehiro-cho, Tsurumi-ku, Yokohama, Kanagawa 230-0045, Japan

Sarah Fowler

MSU-DOE Plant Research Laboratory, Michigan State University, East Lansing, MI 48824-1312, USA

Miki Fujita CREST, Japan Science and Technology Corporation (JST), Japan

S. Mark Goodwin Center for Plant Environmental Stress Physiology, 625 Agriculture Mall Drive, Purdue University, West Lafayette, Indiana 47907–2010, USA

Paul M. Hasegawa Center for Plant Environmental Stress Physiology, 625 Agriculture Mall Drive, Purdue University, West Lafayette, Indiana 47907–2010, USA

Luis Herrera-Estrella Director, Plant Biotechnology Unit, Centro de Investigacion y Estudios Avanzados, Km 9.6 Carretera Irapuato-Leon, 36500 Irapuato, Guanajuato, Mexico

Kei Iida Plant Mutation Exploration Team and Genomic Knowledge Base Research Team RIKEN Genomic Sciences Center, Riken Yokohama Institute, 1-7-22 Suehiro-cho, Tsurumi-ku, Yokohama 230-0045, Japan

Ishida Ishida Plant Mutation Exploration Team, Plant Functional Genomics Research Group, RIKEN Genomic Sciences Center (GSC), RIKEN Yokohama Institute, 1-7-22 Suehiro-cho, Tsurumi-ku, Yokohama 230-0045, Japan

Junko Ishida Plant Mutation Exploration Team, Plant Functional Genomics Research Group, RIKEN Genomic Sciences Center (GSC), RIKEN Yokohama Institute, 1-7-22 Suehiro-cho, Tsurumi-ku, Yokohama, Kanagawa 230-0045, Japan

Matthew A. Jenks Center for Plant Environmental Stress Physiology, 625 Agriculture Mall Drive, Purdue University, West Lafayette, Indiana 47907–2010, USA

Ayako Kamei Plant Mutation Exploration Team, Plant Functional Genomics Research Group, RIKEN Genomic Sciences Center (GSC), RIKEN Yokohama Institute, 1-7-22 Suehiro-cho, Tsurumi-ku, Yokohama, Kanagawa, 230-0045, Japan

Jane Larkindale Life Sciences South Building, Room 352, University of Arizona, P.O. Box 210106, Tucson, AZ 85721, USA

J. López-Bucio Plant Biotechnology Unit, Centro de Investigacion y Estudios Avanzados, Km 9.6 Carretera Irapuato-Leon 36500, Irapuato, Guanajuato, Mexico

Nakajima Maiko Plant Mutation Exploration Team, Plant Functional Genomics Research Group, RIKEN Genomic Sciences Center (GSC), RIKEN Yokohama Institute, 1-7-22 Suehiro-cho, Tsurumi-ku, Yokohama, Kanagawa 230-0045, Japan

Kyonoshin Maruyama Biological Sciences Division, Japan International Research Center for Agricultural Sciences (JIRCAS), Ministry of Agriculture, Forestry and Fisheries, 2-1 Ohwashi, Tsukuba 305-0074, Japan

Michael Mishkind Life Sciences South Building, Room 352, University of Arizona, P.O. Box 210106, Tucson, AZ 85721, USA

Saho Mizukado CREST, Japan Science and Technology Corporation (JST), Japan

Mari Narusaka Plant Mutation Exploration Team, Plant Functional Genomics Research Group, RIKEN Genomic Sciences Center (GSC), RIKEN Yokohama Institute, 1-7-22 Suehiro-cho, Tsurumi-ku, Yokohama, Kanagawa, 230-0045, Japan

Youko Oono Laboratory of Plant Molecular Biology, RIKEN Tsukuba Institute, 3-1-1 Koyadai, Tsukuba 305-0074, Japan; Master's Program in Biosystem Studies, University of Tsukuba, Tennoudai, Tsukuba, Ibaraki, 305-0074, Japan

V. Ramírez-Rodríguez Plant Biotechnology Unit, Centro de Investigacion y Estudios Avanzados, Km 9.6 Carretera Irapuato-Leon 36500, Irapuato, Guanajuato, Mexico

Abel Rosado Dep. Biología Molecular y Bioquímica, Facultad de Ciencias, Universidad de Málaga, Campus de Teatinos s/n, Málaga, 29071 Spain

Tetsuya Sakurai Plant Mutation Exploration Team and Genomic Knowledge Base Research Team, RIKEN Genomic Sciences Center, Riken Yokohama Institute, 1-7-22 Suchiro-cho, Tsurumi-ku, Yokohama 230-0045, Japan

Masakazu Satou Plant Mutation Exploration Team and Genomic Knowledge Base Research Team, RIKEN Genomic Sciences Center, Riken Yokohama Institute, 1-7-22 Suchiro-cho, Tsurumi-ku, Yokohama 230-0045, Japan

Motoaki Seki Plant Mutation Exploration Team, Plant Functional Genomics Research Group, RIKEN Genomic Sciences Center (GSC), RIKEN Yokohama Institute, 1-7-22 Suehiro-cho, Tsurumi-ku, Yokohama, Kanagawa, 230-0045, Japan
Laboratory of Plant Molecular Biology, RIKEN Tsukuba Institute, 3-1-1 Koyadai, Tsukuba Ibaraki 305-0074, Japan

Kazuo Shinozaki Plant Mutation Exploration Team, RIKEN Genomic Sciences Center, Riken Yokohama Institute, 1-7-22 Suchiro-cho, Tsurumi-ku, Yokohama 230-0045, Japan; Laboratory of Plant Molecular Biology, RIKEN Tsukuba Institute, 3-1-1 Koyadai, Tsukuba 305-0074, Japan; CREST, Japan Science and Technology Corporation (JST), Japan

Michael F. Thomashow MSU-DOE Plant Research Lab, 310 Plant Biology Building, Michigan State University, East Lansing, Michigan, USA

Taishi Umezawa Laboratory of Plant Molecular Biology, RIKEN Tsukuba Institute, 3-1-1 Koyadai, Tsukuba 305-0074, Japan

Elizabeth Vierling Department of Biochemistry & Molecular Biophysics, University of Arizona, 1007 E. Lowell Street, Tucson, Arizona 85721, USA

Stephen C. Weller   Center for Plant Environmental Stress Physiology,
                              625 Agriculture Mall Drive, Purdue University, West
                              Lafayette, Indiana 47907-2010, USA

Andrew J. Wood   Department of Plant Biology, Southern Illinois
                              University-Carbondale, Carbondale, IL 62901-6509,
                              USA

Kazuko   CREST, Japan Science and Technology Corporation
Yamaguchi-Shinozaki   (JST), Japan; Biological Sciences Division, Japan
                              International Research Center for Agricultural
                              Sciences (JIRCAS), Ministry of Agriculture, Forestry
                              and Fisheries, 2-1 Ohwashi, Tsukuba 305-0074, Japan

Jian-Kang Zhu   Department of Botany and Plant Sciences, 2150
                              Batchelor Hall, Institute for Integrative Genome
                              Biology, University of California, Riverside, CA
                              92521, USA

Preface

Over the past decade, our understanding of plant adaptation to environmental stress, including both constitutive and inducible determinants, has grown considerably. This book focuses on stress caused by the inanimate components of the environment associated with climatic, edaphic and physiographic factors that substantially limit plant growth and survival. Categorically these are abiotic stresses, which include drought, salinity, non-optimal temperatures and poor soil nutrition. Another stress, herbicides, is covered in this book to highlight how plants are impacted by abiotic stress originating from anthropogenic sources. Indeed, it is an important consideration that, to some degree, the impact of abiotic stress is influenced by human activities. The book also addresses the high degree to which plant responses to quite diverse forms of environmental stress are interconnected. Thus the final two chapters uniquely describe the ways in which the plant utilizes and integrates many common signals and subsequent pathways to cope with less favorable conditions. The many linkages between the diverse stress responses provide ample evidence that the environment impacts plant growth and development in a very fundamental way.

The unquestionable importance of abiotic stress to world agriculture is demonstrated by the fact that altogether abiotic factors provide the major limitation to crop production worldwide. For instance, Bray *et al.* (2000) estimates that 51–82% of the potential yield of annual crops is lost due to abiotic stress. Another example is the increasing use of aquifer-based irrigation by farmers worldwide, which poses a serious threat to the long-term sustainability of world agricultural systems. Over-utilization of these dwindling water supplies is leading to an ever-enlarging area in which productive farming itself has ceased or is threatened. With increasing irrigation worldwide comes the threat of increased salinization of field soils and, just as aquifer loss is shrinking crop yield, so soil salinization due to irrigation has, and will increasingly, reduce crop production in many parts of the world. Another major limitation to expansion of the production of traditional field crops is the problem of non-optimal temperatures, with conditions being either too cold for efficient crop production in the far northern and southern regions of the globe, or too warm in the more equatorial regions. Degradation of the soil by various factors (including anthropogenic) is also increasingly limiting crop yield, and so use of new crops with enhanced resistance to drought, salinity, sub- and supra-optimal temperatures, poor soil nutrient status and anthropogenic factors would benefit agriculture globally by reducing the use of groundwater resources and expanding the productivity of crops on existing and new lands.

The advent of new technologies for the efficient identification of genetic determinants involved in plant stress adaptation, fostered especially by the use of molecular genetics and high throughput transcriptome, proteome, metabolome and ionome profiling methods, has opened a door to exciting new approaches and applications for understanding the mechanisms by which plants adapt to abiotic stress, and should ultimately result in the production of new and improved stress-tolerant crops. This book seeks to summarize the large body of current knowledge about cellular and organismal mechanisms of tolerance to stress. Nine chapters written by leading scientists involved in plant abiotic stress research worldwide provide comprehensive coverage of the major factors impacting world crop production. While modifications to the environment (like increasing use of irrigation, agrichemicals or cultivation) or the expansion of farming into undisturbed lands poses an obvious risk to natural ecosystems, simple genetic changes to crops offer a relatively safe means of increasing yield at a minimal cost to the environment and the farmer. The material presented in this book emphasizes fundamental genetic, physiological, biochemical, and ecological knowledge of plant abiotic stress, which may lead to both traditional and biotechnological applications that result in improved crop performance in stressful environments.

We, the editors, would like to give a special thanks to the authors for their outstanding and timely work in producing such fine chapters. We would also like to thank Becky Fagan for her clerical assistance, and Blackwell Publishing's Graeme MacKintosh and David McDade for their advice and encouragement during the development of this important book.

Matthew A. Jenks and Paul M. Hasegawa

Cited above: Bray, E.A., Bailey-Serres, J. and Weretilnyk, E. (2000) Responses to abiotic stresses. In *Biochemistry and Molecular Biology of Plants*, B. Buchanan, W. Gruissem and R. Jones (eds), p 1160, American Society of Plant Physiologists.

1 Eco-physiological adaptations to limited water environments

Andrew J. Wood

1.1 Introduction

Water is the mother of the vine, The nurse and fountain of fecundity, The adorner and refresher of the world

– from *The Dionysia*, by Charles Mackay (1814–89)

Living organisms have two defining *essences* – a cellular organization and a requirement for liquid water. Although the cellular basis of life can be debated, particularly in introductory biology courses, the requirement for liquid water seems to be absolute. Organisms are able to exploit any ecological niche, no matter how extreme, if free water is available. In fact, we assume that no environment could be sterile if it contained liquid water. The search for extraterrestrial life, on Mars and elsewhere, has in reality become a search for water (Ganti *et al.*, 2003). Kramer and Boyer (1995) have summarized four functions of water in plants. Water is arguably the single most important constituent of a plant, comprising more than 90% of the fresh weight of most herbaceous plants. Water is also a solvent with unique biophysical properties – such as high heat of vaporization, high dielectric constant and high surface tension. These unique properties allow water to remain liquid over a broad temperature range and solvate a wide variety of ions, minerals and molecules. In addition to being the solvent, water acts as a reactant in a number of critical biochemical reactions including serving as the primary electron donor in photosynthesis. Finally, from a physiological perspective, water is the key component in maintaining cell turgor.

Water availability is one of the major limitations to plant productivity (Boyer, 1982), and is one of the major factors regulating the distribution of plant species. Over 35% of the world's land surface is considered to be arid or semiarid, experiencing precipitation that is inadequate for most agricultural uses (see Section 1.2). Regions that experience adequate precipitation can still be water-limiting environments. Precipitation is rarely uniform. All agricultural regions will experience drought (i.e. limited water availability of variable duration) – some areas experience a predictable 'dry season' while other areas experience unpredictable periods of drought. Agricultural regions affected by drought can experience yield loss up to 50% or more. Developing crops that are more tolerant to water deficits, while maintaining productivity, will become a critical requirement for enhancing agriculture in the twenty-first century. Understanding how plant cells tolerate water loss is a vital prerequisite for developing strategies that can impact agricultural and horticultural crop productivity and survival under these conditions of decreasing water availability. Significant work has been accomplished in this area. This chapter describes the arid regions of the world, compares strategies for water economy and provides some insight to the evolutionary aspects of adaptation to water-limiting environments.

1.2   Limited water environments

1.2.1   Arid and semiarid regions of the world

Temperature and precipitation are commonly used to determine climate. The Köppen Climate Classification System recognizes five major climatic types: A, Tropical Moist Climates; B, Dry Climates; C, Moist Mid-latitude Climates with Mild Winters; D, Moist Mid-latitude Climates with Cold Winters; and E, Polar Climates (Köppen, 1936). The Dry Climates are easily recognized (a desert is after all a desert) but water-limited environments can be difficult to classify precisely (see Table 1.1). Meigs (1953) developed a widely used system for classifying water-limited environments based upon mean precipitation (Figure 1.1). Extremely arid lands have at least 12 consecutive months without rainfall, arid lands have less than 250 mm of annual rainfall, and

Table 1.1   Definitions of key terms associated with adaptations to limited water environments

Drought: The limitation of water over a prolonged period of time. Refers to meterological events and should be reserved for field-grown plants. In plants denotes the loss of water from tissues and cells.

Dehydration: The loss of water from a cell. Plant cells dehydrate during drought or water deficit.

Desiccation: The extreme form of water loss from a cell. Denotes the process whereby all free water is lost from the protoplasm.

Homoiohydry: Water economy strategy whereby plants strive to maintain a high water potential under water limiting conditions.

Poikilohydry: Water economy strategy whereby plants lack the ability to control water loss to the environment.

Avoidance: The action of making void, or of having no effect.

Tolerance: The ability of any organism to withstand some particular environmental condition.

Drought avoidance: Avoiding the impact of drought upon an organism by utilizing adaptations that limit the perception of water deficit by the protoplasm.

Drought tolerance: The ability to withstand suboptimal water availability by utilizing adaptations that permit metabolism to occur at low water potential.

Desiccation tolerance: The ability to withstand severe water deficits. Denotes the ability for organisms to lose all free water from the protoplasm and recover normal metabolism upon rewatering.

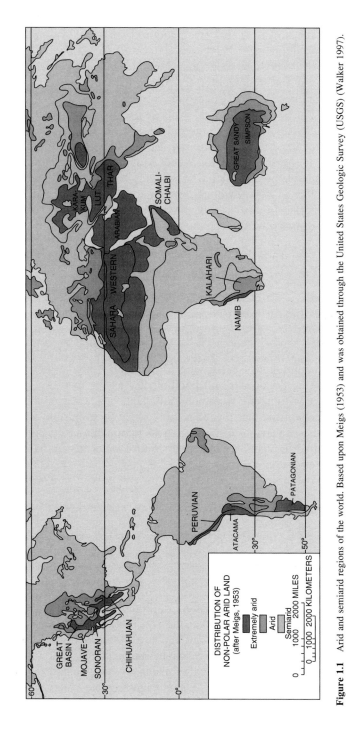

Figure 1.1 Arid and semiarid regions of the world. Based upon Meigs (1953) and was obtained through the United States Geologic Survey (USGS) (Walker 1997).

semiarid lands have a mean annual precipitation of between 250 and 500 mm. In this chapter, we define arid and extremely arid land as *desert*.

The majority of deserts are found near the equator between the Tropic of Cancer (30°N) and the Tropic of Capricorn (30°S) that encompasses the majority of Africa, the Arabian peninsula, the west coast of South America and nearly all of Australia (Figure 1.1). Notable exceptions are the Patagonian of southern South America, the Sonoran, Mojave and Great Basin of the western USA and the Lut, Karakum and Taklimakan of central Asia. The Atacama desert, in the shadow of the Andes Mountains, is the world's driest desert with measurable precipitation (i.e. >1 mm) observed once every decade on average (Walker, 1997). Although extremely dry, deserts and semiarid regions are not sterile. In this context, plants can be divided into two groups: *mesophytes* and *xerophytes*. Mesophytic plants grow in water-sufficient environments. Xerophytic plants are a unique flora generally restricted to a water-limited environment, while xeromorphic plants are 'desert adapted' plants that are not restricted to a water-limited environment.

1.2.2   *Plant strategies for water economy*

Water is one of the key components of life, and plants have evolved two major strategies for water economy: *homoiohydry* and *poikilohydry*. In this chapter, we will define homoiohydry as striving to maintain a high water potential under water-limiting conditions and poikilohydry as the inability to control water loss to the environment. Homoiohydric plants have evolved a hierarchy of protective mechanisms that maintain a favorable protoplasmic water content or mollify the deleterious effects of stress upon cellular constituents. In contrast, poikilohydric plants are unable to control water loss to the environment with the result that cellular water content fluctuates in concert with external water availability. The prefixes homo- and poikilo- are widely used in terminology related to eukaryotic physiology. For clarity they are defined, by the *Oxford English Dictionary* (http://dictionary.oed.com), as 'of the same kind' and 'variegated', respectively. The idea of homeostasis is central to physiology and the concept has been applied to a variety of characters. For example, homothermic organisms control their body temperature at some fixed value, while in poikilothermic organisms body temperature varies with the temperature of the environment (commonly known as 'cold-blooded'). However, despite extensive application of homoiohydry in the literature (Wickens, 1998), we think that neither the concept nor the terminology can be properly applied to water physiology in plants. It can be argued that plants strive to homoiohydry. However, we maintain that no plants are homoiohydric in the strict definition of the term because plants are incapable of maintaining their water content at a fixed value. Plants cannot create water where none exists, and ultimately all plants are unable to control water loss to the environment. Based upon these definitions, poikilohydric plants are limited to the *Bryophatea* (i.e. liverworts, hornworts and mosses) while homoiohydric plants would encompass the more derived tracheophytes (see Section 1.3 for greater detail).

1.2.3 Ability to survive in water-limited environments

Plants cannot maintain their cellular water content at a fixed point, and are therefore not truly homoiohydric. Both homoiohydric and poikilohydric plants (as defined earlier) are constantly losing water to the surrounding environment. In terms of water balance, land plants will always be in disequilibria with the atmosphere because the surrounding air is extremely 'dry' relative to the plant. Under fully hydrated conditions, leaf water potential (Ψ_w) for typical mesophytes, such as *Zea mays* or *Glycine max*, range from -0.3 to -0.5 MPa (Boyer, 1970). Air of 100% relative humidity (RH) is calculated to have a Ψ_{wv} of 0.00 MPa (see Table 1.2). Minor reductions in RH correspond to large reductions in Ψ_{wv}. While 99.6% RH corresponds to -0.54 MPa (in equilibrium with our typical mesophyte), 99% corresponds to -1.36 MPa and 90% corresponds to -14.2 MPa. Deserts and semiarid regions typically have RH below 50% with corresponding Ψ_{wv} significantly less than -93.6 MPa. As evident from these calculations, land plants cannot prevent water loss to the environment; and the problem is exacerbated since gas-exchange is required for other beneficial process, such as CO_2 fixation.

Table 1.2 Relative humidity and the corresponding values for Water vapor, water potential (Ψ_{wv})

Relative Humidity (%)	Ψ_{wv} (MPa)
100.0	0.00
99.6	-0.54
99.0	-1.36
90.0	-14.2
80.0	-30.1
70.0	-48.2
60.0	-70.0
50.0	-93.6
20.0	-217.3
10.0	-310.8
0.00	∞

Data are for 20°C, and are calculated using the formula $\Psi_{wv} = (135 \text{ MPa}) \ln\left(\frac{90}{100}\right) = -14.2 \text{ MPa}$. Adapted from Nobel (1983).

All land plants require free water for normal growth and development. One major strategy for surviving water-limited environments is to avoid them entirely. Mesophytic plants grow in water-sufficient environments. Some species of xerophytic and xeromorphic plants avoid drought-induced damage by maintaining a high cellular water contents (such as cacti) or by establishing extremely deep rooting systems. Other species of plants, including both mesophytic and xerophytic plants, escape drought-induced damage by rapidly completing their life cycles (i.e. ephemeral plants). For ephemeral plants all environments are water sufficient. Avoiding drought is an extremely important adaptation for survival in a water-limiting environment. However, plant physiologists are generally more interested in plants that are able to *tolerate* drought (i.e. plants that have evolved a number of anatomical, developmental, biochemical, physiological and molecular adaptations to limit the drying out of vegetative tissues) (see Section 1.3).

1.2.4   Surviving water deficit (drought) and severe water deficit (desiccation)

The vast majority of land plants, including all major crop and horticultural plants, would be classified as drought avoiders. Although vascular plants do produce specialized structures capable of withstanding severe stress (e.g. pollen, seeds and spores), few species can survive substantial loss of water from their vegetative tissues (Bewley & Krochko, 1982; Bewley & Oliver, 1992; Oliver *et al.*, 2000a). Tolerance is the ability to withstand a particular environmental condition. Under water-limiting conditions, plants will experience a net loss of water to the environment and cells will dehydrate (i.e. Ψ_w and relative water contents, RWC, will decline). Land plants can be classified based upon how they respond to this water deficit. Drought-avoiding plants strive to maintain elevated Ψ_w. Drought-tolerant plants are able to tolerate extended periods of water deficit. However, both drought-avoiding and drought-tolerant plants will reach a 'permanent wilting point' where Ψ_w has declined to such a degree that the plant cannot recover upon rewatering. Severe water deficit is defined as desiccation. It must be noted that dehydration and desiccation are not synonyms. Dehydration is the loss of water from protoplasm. Plants cannot partially desiccate – desiccation denotes the process whereby all free water is lost from the protoplasm. Desiccation-tolerant plants can survive severe water deficit (i.e. <25% RWC) and recover from the air-dried state (Bewley, 1979). Vegetative desiccation tolerance is uncommon; however, the phenotype is widely distributed and represented within the most major classes of plants (Oliver & Bewley, 1997; Oliver *et al.*, 2000a; Porembski & Barthlott, 2000). Approximately, 330 species of vascular plants (i.e. <0.15% of the total number) have been demonstrated as desiccation tolerant (Porembski & Barthlott, 2000; Proctor & Pence, 2002). Resurrection plants (i.e. desiccation-tolerant vascular plants) identified within ten angiosperm families and nine pteridophyte families. The majority of the Bryophatea, which represents approximately 30 000 species of mosses, liverworts and hornworts, are postulated to tolerate at least brief desiccation of low intensity (Proctor & Pence, 2002). Resurrection mosses include a number of extremely desiccation-tolerant species able to tolerate rapid desiccation (i.e. within minutes) and survive extended periods (i.e. >10 years) as desiccated material.

1.3   Adaptation to limited water environments

The world's deserts and arid regions are stressful environments. However, plants have been extremely successful in exploiting these ecological niches. Interestingly, many of the desert and arid regions described earlier (see Section 1.2.1) are relatively young geographic features – at least as compared to evolutionary time. In this section, we would like to explore several interesting questions: how did modern desert plants evolve? What adaptations enable modern plants to exploit water-limiting environments? What adaptations enabled embryophytes to invade the land – thereby exploiting a new environment? And, how do the adaptations of modern plants compare those with adaptations found in the earliest land plants?

1.3.1   Evolution of land plants

Terrestrial organisms are found in the fossil record for a fraction of the 3.5 billion years that life is thought to have existed on Earth. Fossils of land plants first appear in the upper Silurian and lower Devonian (Kenrick & Crane, 1997a; Wellman *et al.*, 2003). Embryophytes (i.e. land plants) are monophyletic (Kenrick & Crane, 1997b) and postulated to have fresh water rather than a marine origin (Karol *et al.*, 2001). Karol *et al.* (2001) suggest the common ancestor between algae (Charales) and land plants was a branched and filamentous organism with a haplontic life cycle. The 'invasion' of the land by freshwater algae is a key event in the evolution of life on earth. It is difficult to imagine the Earth before this event. The water, both fresh and marine, would have been teeming with life. The land, however, would have been completely devoid of vegetation and inhabited solely by bacteria and microorganisms. From the perspective of an aquatic organism, all land environments are water limiting. It seems probable that shallow pools, or seasonally wet areas may have served as intermediate environments for early land plants. But eventually plants left the water and became *land plants*. Some of the adaptations required for invasion of the land by aquatic plants are summarized in Table 1.3.

Table 1.3   Adaptations required for invasion of the land by aquatic plants

Primary (1°)
   Tolerance of vegetative tissues to desiccation

Secondary (2°)
   Reproductive and fertilization strategies in nonaqueous environments
   Gas exchange across a liquid–air interface
   Enhanced ion and metabolite transport

Tertiary (3°)
   Development of the cuticle
   Altered requirements for mechanical/physical support
   Tolerance of reproductive tissues to desiccation
   Development of vascular tissues

Primary and secondary adaptations would be required for the early land plants. Tertiary adaptations would be symptomatic for vascular land plants.

The first successful land-invading plant represents the common ancestor of bryophytes and tracheophytes (i.e. the embryophytes). We postulate that the successful 'land-invading' plant would have had three fundamental characteristics. The organism would have been:

- compact (i.e. <1 cm tall)
- poikilohydric
- desiccation tolerant

The first successful land-invading plant would have been an extreme form of life able to tolerate a suite of abiotic stresses – most notably high irradiation, temperature fluctuations and desiccation (see Figure 1.2). Recent synthetic phylogenetic analyses (Oliver *et al.*, 2000a) suggest that vegetative desiccation tolerance was primitively present in the bryophytes (the basal-most living clades of land plants), but was lost in the evolution of tracheophytes. Oliver *et al.* (2000a) postulate that vegetative desiccation tolerance was a crucial step required for the colonization of the land. To early land plants (and modern bryophytes), all terrestrial niches would have been water-limiting environments. As poikilohydric plants, the early land plants would have been unable to limit water loss to the environment because they had not yet evolved homiohydric phenotypes.

Early land plants were clearly poikilohydric plants. However, homoiohydric plants (i.e. the angiosperms) dominate modern terrestrial plants. Why? How were highly stress-tolerant plants supplanted? It is postulated that vegetative tolerance extracts a steep metabolic cost that depresses competitiveness (Oliver *et al.*, 2000a; Tuba & Proctor, 2002). In conjunction with the internalization of water economy, the loss of vegetative desiccation tolerance could have been favored as the vascular plants became more complex and exploited more

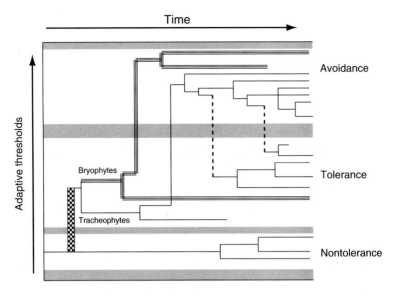

Figure 1.2 Evolution of water deficit-stress tolerance in plants. Grey bars indicate 'drought thresholds' that must be overcome by adaptation. The checkerboard line indicates the invasion of land by embryophytes. Triple stripped lines denote bryophyte lineages. Single lines indicate angiosperm lineages. Inspired by Simpson (1953).

competitive ecological niches. Gordon and Olson (1995) suggest four defining characteristics of homoiohydric land plants:

- lignified tracheids of water-conducting xylem
- protected reproductive cells
- cuticle
- cuticular pores and stomata

Vascular plants would have become less tolerant, as compared to both the ancestral land-invading plant and bryophytes (Figure 1.2). We postulate that early homoiohydric plants would have been mesophytic plants exploiting a wide variety of water-sufficient environments.

In the early land plants, water-limiting environments were 'caused' by the organism's inability to hold on to water and limit water loss to the environment. However for homoiohydric plants, water limitation becomes more a matter of geography and meteorology. Over time, we postulate that a series of adaptations would evolve creating plants able to thrive under water-limiting conditions (see Chapter 2 for review). Vegetative desiccation-tolerant vascular plants have independently evolved (or re-evolved) once in the ferns, and at least eight times in the angiosperms (Oliver *et al.*, 2000a). By evolving characters that are postulated to be similar to early land plants, resurrection plants are able to inhabit the deserts and semiarid regions of the world.

1.3.2 Tolerance to desiccation

Oliver *et al.* (2000a) have proposed that desiccation-tolerant plants can be divided into two main categories: cardinal and modified. Cardinal desiccation-tolerant plants that lose water to the environment at any rate (i.e. slowly or quickly) while modified desiccation-tolerant plants can only survive slow rates of water loss. Cardinal desiccation tolerance is apparently restricted to less complex plants (i.e. the bryophytes) and is postulated to apply to the earliest land plants. Modified desiccation tolerance is exhibited in the more complex, and derived, resurrection ferns and resurrection angiosperms. Regardless of the rates of water loss, or the category of desiccation tolerance, Bewley (1979) proposed three widely accepted criteria that must be met for plants to survive desiccation:

(1) damage must be limited to a repairable level;
(2) physiological integrity must be maintained in the dehydrated state; and
(3) repair mechanisms must exist, which function upon rehydration.

As the early land plants (i.e. small, poikilohydric and desiccation tolerant) evolved to exploit the numerous niches afforded by the terrestrial environment, vegetative desiccation tolerance was lost in favor of the adaptive advantages provided by increased metabolic rates, increased growth rates, phonological complexity and the evolution of vascular tissues. Vascular plants (i.e. homoiohydric plants) abandoned vegetative desiccation tolerance, but still retained the requirement for desiccation-reproductive tissues. The ability to produce large numbers of desiccation-tolerant seeds is a critical part of the angiosperm life cycle. It seems likely that constitutive protective and repair mechanisms found within the vegetative tissues were adapted by the reproductive tissues and placed under developmental control. Oliver *et al.* (2000a) have hypothesized that the mechanism of desiccation tolerance exhibited in seeds evolved from the most primitive form of vegetative desiccation tolerance. And, that modified desiccation-tolerant plants re-recruited tolerance mechanisms from reproductive tissues to the vegetative tissues under environmental rather than developmental control.

1.4 Refresher of the world – how to create more drought-tolerant crops

Plants have evolved a number of adaptations for surviving and thriving within the water-limited environments – and these adaptations are detailed in subsequent chapters. Plants must limit water loss to the environment. Fecundity must be maintained. Biochemical and metabolic processes must be fine-tuned. Protective mechanisms must be in place to limit damage to the protoplasm. Repair mechanisms must be able to repair stress-induced damage – and the list can be expanded. However, no single group of plants employs each adaptation. And most certainly they do not use the same genes to mediate each adaptation. To fully understand the suite of genes that control the adaptations to water-limiting environments, we must explore plants that span the spectrum of responses: poikilohydric plants, poikilohydric desiccation-tolerant plants (such as *Tortula ruralis*) (Oliver & Wood, 1997; Oliver *et al.*, 1997; Oliver *et al.*, 1998; Oliver *et al.*, 2000b; Wood *et al.*, 2000; Wood & Oliver, 2004), mesophytic homoiohydric plants (such as *Arabidopsis thaliana*) (Zhu, 2002), homoiohydric drought-tolerant plants (such as the grasses), homoiohydric drought-tolerant crop plants (such as *Sorghum bicolor*) (Wood *et al.*, 1996), xerophytic and xeromorphic plants and homoiohydric desiccation-tolerant plants such as *Craterostigma plantagineum* (Bartels & Salamini, 2001), *Myrothamnus flabellifolia* and *Xerophyta viscosa* (Farrant, 2000; Mundree & Farrant, 2000).

Drought and desiccation tolerance in plants is one of the most interesting phenomena in all of biology. And producing more drought-tolerant crops is one of the major challenges of the twenty-first century. In our opinion, these new drought-tolerant crops will be generated from the pyramiding of traits from all of these 'groups' of plants. *A. thaliana* has provided great insight to the genes associated with dehydration stress and the signal transduction mechanisms that mediate the response (Xiong *et al.*, 1999; Xiong & Zhu, 2002; Zhu, 2002) – however, it is not a tolerant plant. Mesophytic and drought-tolerant plants provide clear examples of the mechanisms used to delay water loss and/or tolerate modest dehydration (Ingram & Bartels, 1996; Bray, 2002). Poikilohydric desiccation-tolerant plants are more similar to the highly tolerant early land plants and offer novel genes potentially lost during angiosperm evolution. Desiccation-tolerant vascular plants demonstrate what tolerance mechanisms can effectively be exploited by crops (Ramanjulu & Bartels, 2002). The next five years will be an exciting time as comparative 'omics' (i.e. genomics, transcriptomics, proteomics and metabolomics) will define the suite of genes available in the 'tolerance to water deficit-stress' toolbox. The challenge lies in using classical genetics and physiology to use this information in order to build a better crop plant.

Acknowledgments

The author gratefully acknowledges Dr. Dale Vitt (Southern Illinois University) for helpful discussions.

References

Bartels, D. & Salamini, F. (2001) Desiccation tolerance in the resurrection plant *Craterostigma plantagineum*. A contribution to the study of drought tolerance at the molecular level. *Plant Physiology*, **127**, 1346–1353.

Bewley, J. D. (1979) Physiological aspects of desiccation-tolerance. *Annual Review of Plant Physiology*, **30**, 195–238.

Bewley, J. D. & Krochko, J. E. (1982) Desiccation-tolerance. In O. L. Lange, P. S. Nobel, C. B. Osmond & H. Ziegler (eds) *Encyclopedia of Plant Physiology*. Vol. 12B, *Physiological Ecology II*. Springer-Verlag, Berlin, pp. 325–378.

Bewley, J. D. & Oliver, M. J. (1992) Desiccation-tolerance in vegetative plant tissues and seeds: protein synthesis in relation to desiccation and a potential role for protection and repair mechanisms. In C. B. Osmond & G. Somero (eds) *Water and Life: A Comparative Analysis of Water Relationships at the Organismic, Cellular and Molecular Levels*. Springer-Verlag, Berlin, pp. 141–160.

Boyer, J. S. (1970) Leaf enlargement and metabolic rates in corn, soybean and sunflower at various leaf water potentials. *Plant Physiology*, **46**, 233–235.

Boyer, J. S. (1982) Plant productivity and the environment. *Science*, **218**, 443–448.

Bray, E. A. (2002) Abscisic acid regulation of gene expression during water-deficit stress in the era of the Arabidopsis genome. *Plant, Cell and Environment*, **25**, 153–161.

Farrant, J. M. (2000) Comparison of mechanisms of desiccation tolerance among three angiosperm resurrection plants. *Plant Ecology*, **151**, 29–39.

Ganti, T., Horvath, A., Berczi, S., Gesztesi, A. & Szathmary, E. (2003) Dark dune spots: possible biomarkers on Mars? *Origins of Life and Evolution of the Biosphere*, **33**, 515–557.

Gordon, M. S. & Olson, E. C. (1995) *Invasions of the Land*. Columbia University Press, New York.

Ingram, J. & Bartels, D. (1996) The molecular basis of dehydration tolerance in plants. *Annual Review of Plant Physiology and Plant Molecular Biology*, **47**, 377–403.

Karol, K. G., McCourt, R. M., Climino, M. T. & Delwiche, C. F. (2001) The closest living relatives of land plants. *Science*, **294**, 2351–2153.

Kenrick, P. & Crane, P. R. (1997a) The origin and early evolution of plants on land. *Nature*, **389**, 33–39.

Kenrick, P. & Crane, P. R. (1997b) *The Origin and Early Diversification of Land Plants*. Smithsonian Series in Comparative Evolutionary Biology. Smithsonian Institution Press, Washington, DC.

Köppen, W. (1936) Das geographische system der Klimate. In W. Köppen & R. Geiger (eds) *Handbuch der Klimatologie*, Vol. I. Borntraeger, Berlin.

Kramer, P. J. & Boyer, J. S. (1995) *Water Relations of Plants and Soils*. Academic Press, San Diego.

Meigs, P. (1953) World distribution of arid and semi-arid homoclimates. In *Reviews of Research on Arid Zone Hydrology*. Paris, United Nations Educational, Scientific, and Cultural Organization, Arid Zone Programme-1, pp. 203–209.

Mundree, S. G. & Farrant, J. M. (2000) Some physiological and molecular insights into the mechanisms of desiccation tolerance in the resurrection plant *Xerophyta viscosa* Baker. In J. Cherry (ed.) *Plant Tolerance to Abiotic Stresses in Agriculture: Role of Genetic Engineering*. Kluwer Academic Press, Dordrecht, The Netherlands, pp. 201–222.

Nobel, P. S. (1983) *Biophysical Plant Physiology and Ecology*. W.H. Freeman, New York.

Oliver, M. J. & Bewley, J. D. (1997) Desiccation-tolerance of plant tissues: a mechanistic overview. *Horticultural Reviews*, **18**, 171–214.

Oliver, M. J. & Wood, A. J. (1997) Desiccation-tolerance of mosses. In T. Koval (ed.) *Stress-inducible Processes in Higher Eukaryotic Cells*. Plenum Press, New York, pp. 1–26.

Oliver, M. J., Wood, A. J. & O'Mahony, P. (1997) How some plants recover from vegetative desiccation: a repair based strategy. *Acta Physiologiae Plantarum*, **19**, 419–425.

Oliver, M. J., Wood, A. J. & O'Mahony, P. (1998) "To dryness and beyond" – preparation for the dried state and rehydration in vegetative desiccation-tolerant plants. *Plant Growth Regulation*, **24**, 193–201.

Oliver, M. J., Tuba, Z. & Mishler, B. D. (2000a) The evolution of vegetative desiccation tolerance in land plants. *Plant Ecology*, **151**, 85–100.

Oliver, M. J., Velten, J. & Wood, A. J. (2000b) Bryophytes as experimental models for the study of environmental stress tolerance: desiccation-tolerance in mosses. *Plant Ecology*, **151**, 73–84.

Porembski, S. & Barthlott, W. (2000) Granitic and gneissic outcrops (inselbergs) as centers of diversity for desiccation tolerant vascular species. *Plant Ecology*, **151**, 19–28.

Proctor, M. C. F. & Pence, V. C. (2002) Vegetative tissues: bryophytes, vascular resurrection plants, and vegetative propagules. In M. Black & H. W. Pritchard (eds) *Desiccation and Survival in Plants: Drying Without Dying*. CABI Publishing, New York, pp. 207–237.

Ramanjulu, S. & Bartels, D. (2002) Drought- and desiccation-induced modulation of gene expression. *Plant, Cell and Environment*, **25**, 141–151.

Simpson, G. G. (1953) *The Major Features of Evolution*. Columbia University Press, New York.

Tuba, Z. & Proctor, M. C. F. (2002) Poikilohydry and homoiohydry: antithesis or spectrum of possibilities? *New Phytologist*, **156**, 327–349.

Walker, A. S. (1997) *Deserts: Geology and Resources*. United States Geologic Survey (USGS) (http://pubs.usgs.gov/gip/deserts/).

Wellman, C. H., Osterloff, P. L. & Mohiuddin, U. (2003) Fragments of the earliest land plants. *Nature*, **425**, 282–285.

Wickens, G. E. (1998) *Ecophysiology of Economic Plants in Arid and Semi-arid Lands*. Springer-Verlag, New York.

Wood, A. J. & Oliver, M. J. (2004) The molecular biology and genomics of the desiccation tolerant moss *Tortula ruralis*. In A. J. Wood, M. J. Oliver & D. J. Cove (eds) *New Frontiers in Bryology: Physiology, Molecular Biology and Applied Genomics*. Kluwer Academic Press, Dordrecht, The Netherlands, pp. 71–90.

Wood, A. J., Oliver, M. J. & Cove, D. J. (2000) Frontiers in bryological & lichenological research. I. Bryophytes as model systems. *The Bryologist*, **103**, 128–133.

Wood, A. J., Saneoka, H., Joly, R. J., Rhodes, D. & Goldsbrough, P. B. (1996) Betaine aldehyde dehydrogenase in *Sorghum bicolor*: molecular cloning and expression of two related genes. *Plant Physiology*, **110**, 1301–1308.

Xiong, L. & Zhu, J.-K. (2002) Molecular and genetic aspects of plant responses to osmotic stress. *Plant, Cell and Environment*, **25**, 131–139.

Xiong, L., Ishitani, M. & Zhu, J. K. (1999) Interaction of osmotic stress, temperature, and abscisic acid in the regulation of gene expression in Arabidopsis. *Plant Journal*, **17**, 363–372.

Zhu, J. K. (2002) Salt and drought stress signal transduction in plants. *Annual Review of Plant Biology*, **53**, 247–273.

2 Plant cuticle function as a barrier to water loss

S. Mark Goodwin and Matthew A. Jenks

2.1 Introduction

Terrestrial plants must adapt to an environment where the relative water content of the atmosphere is much lower than the internal water content of the plant. Fick's first law states that water will diffuse down a concentration gradient from high to low concentration. To prevent uncontrolled diffusion of water into the atmosphere, aerial tissues of terrestrial plants, including leaves, flowers, fruits and young stems, possess a very thin membrane called the cuticle that forms a protective covering over the outermost surface. In plants, the thickness of this membrane can vary from 0.05 μm in some mesophytic plants to as thick as 225 μm in xerophytic species.

Most angiosperms and gymnosperms have a limited ability to withstand dehydration. For example, irreversible leaf tissue damage occurs in most plants if roughly half the water content is lost (Burghardt & Riederer, 2003). To prevent this, plants close their stomata when internal water content falls below a certain threshhold. But even after stomatal closure, plants continue to lose water, albeit at a much lower rate, resulting in plant death if water-limiting conditions are sufficiently prolonged. It is the period after drought-induced stomatal closure when water loss is most influenced by the cuticle's resistance to water vapor flux and when the cuticle becomes most important to plant survival.

Recent studies have shed new light on the physicochemical characteristics of the cuticle and have provided a better understanding of the interrelationship between cuticle properties and cuticle permeability to water. These studies reveal that permeability of the cuticle is best understood at the submicroscopic level where water and lipid molecules interact. This chapter will discuss the significance of cuticular water loss to plant survival, cuticle properties involved in cuticle permeability and gene involvement in the development of these cuticle properties.

2.2 Cuticle structure and composition

To understand cuticle function in water loss, a clear understanding of cuticle morphology (Figure 2.1) and chemical composition is needed. The

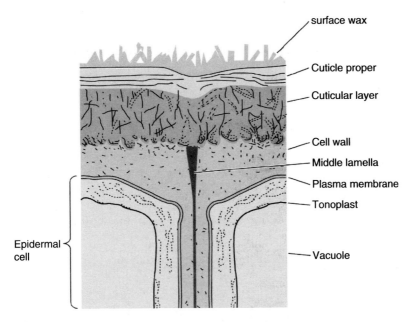

Figure 2.1 Schematic drawing of the plant cuticle ultrastructure as occurs covering a typical mature plant epidermis (Taiz & Zeiger, 2002). Reproduced, with permission, from Taiz and Zeiger (2002).

epicuticular wax layer visible on the surface of many plant species as a bluish-white colored coating called glaucousness or bloom is typically composed of *n*-alkanes, fatty acids, aldehydes, primary alcohols, secondary alcohols, ketones and esters (Walton, 1990; Jenks *et al.*, 2002). Studies using scanning electron microscopy (SEM) reveal that these waxes crystallize into many diverse structures specific to species, organ and environment (Jenks *et al.*, 1992; Beattie & Marcell, 2002). Recrystallization experiments (Jetter & Riederer, 1995) indicate that these crystalline structures are primarily determined by the corresponding wax composition (Rashotte & Feldman, 1998). Notwithstanding, some specific crystalline shapes have yet to be recreated *in vitro* (Meusel *et al.*, 1994), indicating that secretion site micromorphologies or leaf surface environments also play a role in determining wax crystallization patterns.

How hydrophobic waxes are transported across the aqueous outer cell walls and then the lipoidal cuticle layers is unknown. Wax-secreting microchannels (or pores) have been proposed, however microchannels traversing the cell wall or cuticle have not been observed by the use of electron microscopy. Lipid transfer proteins (LPT) constitute the major protein in the epicuticular wax layer, and LPTs have been localized to the cell wall, indicating that active transport mechanisms may be involved in wax secretion (Arondel *et al.*, 2000). However, the biological function of these specific LPTs remains unknown.

Instead of active transport, recent studies suggest that waxes may move together with water permeating through the cuticle (Neinhuis *et al.*, 2001). When a porous polyurethane film was pressed against a leaf surface, isolated cuticles or wax-loaded filter paper, wax crystals typical of the particular species were observed forming on the polyurethane. No wax crystals were formed, however, if water was not diffusing through the membranes, leading Neinhuis *et al.* (2001) to propose that wax was co-transported with water through the cuticle in a process akin to steam distillation. As such, variations in cuticle permeability to water could have significant effects on epicuticular wax deposition rates and crystallization patterns.

Although wax secretion is still a mystery, wax metabolism is starting to be revealed. Wax biosynthesis begins with elongases adding two carbons per elongation cycle to either C_{16} or C_{18} acyl-CoAs in a four-step process involving condensation between the acyl-CoA chain and malonyl CoA, β-keto reduction, dehydration and enoyl reduction (Kunst & Samuels, 2003). These elongated acyl-CoA chains are either shunted into the acyl-reduction pathway to form primary alcohols, or into the decarbonylation pathway leading to the formation of aldehydes, alkanes, secondary alcohols and ketones. Acyl-CoAs can also be cleaved by thioesterases to form free fatty acids and CoA, or condensed with primary alcohols to form esters. A model of the wax and cutin metabolic pathways is presented in Figure 2.2.

Below the epicuticular layer is the cuticle proper (Figure 2.1), which is made of a three-dimensional cutin and cutan matrix with embedded waxes. Whereas the waxes are soluble in organic solvents, the cutins and cutans forming the three-dimensional cuticle matrix are insoluble in those same solvents. It appears likely that the cuticle matrix of plants is composed of both cutin and cutan, with the exact ratios being species dependent. Cutin typically comprises between 40% and 80% by weight of the cuticle membrane (Nip *et al.*, 1986a, 1986b; Jeffree, 1996; Xiao *et al.*, 2004). The cuticular layer below the cuticle proper (Figure 2.1) has in addition to waxes and cutin/cutan, significant amounts of polysaccharide that extend as strands outward from the cell wall. Often, there is no sharp delineation between the cuticle proper above and the cell wall below, with the cutin/cutan matrix progressively expanding into the cell wall (Jeffree, 1996). Due to the hydrophilic nature of polysaccharides, the cuticular layer most likely provides little resistance to diffusion.

Intracuticular wax composition is apparently different from epicuticular wax composition. In *Prunus laurocerasus,* mechanical removal of epicuticular waxes produced entirely aliphatic waxes, whereas the remaining solvent-extracted intracuticular waxes comprised 63% triterpenoids (Jetter *et al.*, 2000). In similar studies, tomato fruit epicuticular waxes were exclusively aliphatic, while intracuticular waxes extracted with solvent had a sizable triterpenoid fraction (Vogg *et al.*, 2004). Others reported high proportions of intracuticular fatty acids in enzymatically isolated cuticles (Baker & Procopiou,

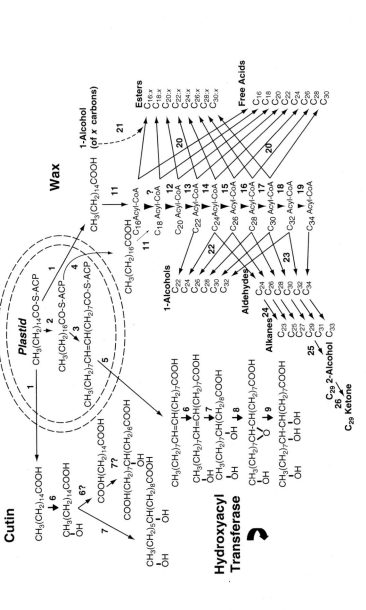

Figure 2.2 Model cuticular lipid biosynthesis pathways.

Plastidic reactions producing precursors for both waxes and cutin. (1) Palmitoyl-ACP thioesterase, (2) 3-ketoacyl-ACP synthetase II (KASII), (3) stearoyl-ACP desaturase, (4) stearoyl-ACP thioesterase and (5) oleoyl-ACP thioesterase (Kunst & Samuels, 2003).

Conversion of major constituents in the cutin monomer pathway. (6) ω-Hydroxylase, (7) mid-chain hydroxylase, (8) epoxidase and (9) epoxide hydrase. Monomers are linked together by a hydroxyacyl transferase (Le Bouquin *et al.*, 2001; Kolattukudy, 2002).

Conversion of major constituents in the cuticular wax pathway. (11) Acyl-CoA synthetase, (12–19) each a unique chain-length-specific acyl-CoA elongase complex comprising of β-ketoacyl synthase, β-keto reductase, dehydratase and enoyl reductase, (20) acyl-CoA thioesterase, (21) acyl-CoA:fatty alcohol transacylase, (22) acyl-CoA reductase (58 kDa), (23) acyl-CoA reductase (28 kDa), (24) aldehyde decarbonylase, (25) alkane oxidase and (26) secondary alcohol oxidase (Kolattukudy, 1996; Vioque & Kolattukudy, 1997; Kunst & Samuels, 2003). 1-Alcohols: primary alcohols; 2-Alcohols: secondary alcohols. Each wax constituent is represented by the number of carbons in the saturated chain, with full chemical formula omitted.

1975). These reports, however, must be viewed with caution since cuticles sorb internal fatty acids during incubation in the pectinase/cellulase solution (Schön-herr & Riederer, 1986), and it has not been determined whether triterpenoids from internal tissues were a source of triterpenoids found in the solvent-extracted waxes. Descriptions of the exact compositional difference between intracuticular and epicuticular waxes await improved analytical methods.

In cutin metabolism, feeding studies using ^{14}C-labeled precursors indicate that hexadecanoic acid is first hydroxylated at the ω-position to form the 16-hydroxy hexadecanoic acid followed by mid-chain hydroxylation to form the dihydroxy hexadecanoic acid (Figure 2.2; Kolattukudy, 2002). The P450 CYP94A5, isolated from *Nicotiana tabacum*, was able to catalyze the complete set of oxidation reactions from hydroxylation of the terminal methyl group to hydroxylation of the corresponding carbonyl group (Le Bouquin *et al.*, 2001). Potentially then, a single enzyme may be involved in reactions from the synthesis of hexadecanoic acid to the hexadecane-1,16-dioic acid (Figure 2.2). Octadecenoic acid, the precursor for C_{18} cutin monomers, is ω- and mid-chain hydroxylated, followed by epoxidation and then termination with the tri-hydroxy C_{18} monomer (Kolattu-kudy, 2002). Outside the cell wall, monomers are linked by transacylases to form a three-dimensional polymer with over 95% of primary hydroxyl groups ester-linked to other monomers, with about 40% of secondary hydroxyl groups ester-linked (Deas & Holloway, 1977). Besides ester-linked bonding between the monomers, cutin monomers can also be covalently bonded to polysaccharides, including arabinose, glucose and others found in the cuticular layer (Fang *et al.*, 2001). A detailed description of these linkages, and their role in the progressive cutinization of the outer cell walls, has yet to be made.

Cutan is the residue remaining after complete depolymerization of cutin. Use of Fourier transform infrared (FTIR) and ^{13}C-nuclear magnetic resonance (NMR) spectroscopy, X-ray diffraction and exhaustive ozonolysis indicates that this unsaponifiable polymeric material is an amorphous three-dimensional matrix of polymethylenic chains linked by ether bonds (Villena *et al.*, 1999). Cutan biosynthesis may involve a pathway where lipoxygenases and/or per-oxygenases act on polyunsaturated fatty acids to form fatty acid hydroper-oxides, a precursor for ether bonds (Blée & Schuber, 1993; Villena *et al.*, 1999). Data from *Clivia miniata* suggest that cutan is not the addition of a novel monomer, but rather modification *in situ* of previously deposited cutin since in *C. miniata* cutin content declines and cutan content increases over time (Reiderer & Schönherr, 1988).

2.3 Cuticle function as a barrier to plant water loss

According to Taiz and Zeiger (2002), for every gram of a typical plant, approximately 500 g of water is absorbed by the roots, transported to the leaf and then lost to the atmosphere. In fact, a leaf may exchange up to 100% of its internal water every hour. This continuous cycling of water through the plant is called transpiration, a process that can be divided into two main components: (1) stomatal transpiration, which is a gas-phase water diffusion through open stomata and (2) cuticular transpiration, which is a solid-phase water diffusion across the cuticle membrane itself. Water loss measurements that include both components are described by the term 'conductance' (g), with the maximum conductance (g_{max}) describing water loss when stomata are open, and minimum conductance (g_{min}) describing water loss when stomata are presumed closed. The term 'permeance' (P) is used to describe water movement by cuticular transpiration, as when measuring water movement through astomatous cuticles or water movement through stomatous cuticles using techniques that distinguish cuticular transpiration from stomatal transpiration (Kersteins, 1996).

Under water-sufficient conditions, the majority of water loss occurs through open stomata. In environmental conditions that cause stomata to close, such as during drought, the majority of water loss is via solid-phase cuticular transpiration. Körner (1993) compiled the conductance values of functionally related plants and demonstrated the effects of stomatal closure on water loss. In these studies, g_{min} was found to range from <1% of g_{max} in succulents to 5.6% of g_{max} in herbaceous shade plants. Thus, in well-watered environments, ~95–99% of all water loss occurs through the pores of stomata. By comparison, in conditions where plants close their stomata, such as during water insufficiency, water loss rates are determined primarily by the permeability of the plant cuticle. Interestingly, g_{max} varied only threefold between the diverse plant ecological groupings examined by Körner (1993), but there was nearly 50-fold variation in g_{min}, leading to speculation that variation in cuticle permeability could have important ecological consequences.

We suggest that cuticular water permeability plays a major role in plant adaptation to drought, especially if residual stomatal transpiration after stomatal closure is negligible. To better describe cuticular water loss researchers have sought to compare g_{min} with P. Until recently, however, P could be determined only for astomatous cuticles because of the inability to eliminate the possibility of residual stomatal water loss in stomatous cuticles. A novel approach examined species with hypostomatous leaves to compare g_{min} with P. In these studies, P values of enzymatically isolated, astomatous cuticles from the adaxial leaf surface were compared with g_{min} values of leaves of the same species under conditions where abaxial stomata were maximally closed. Measurements from *Acer campestre*, *Fagus sylvatica*, *Quercus petraea* and *Ilex aquifolium* found no significant difference between P and g_{min}, suggesting that residual stomatal transpiration from closed stomata made a negligible contribution to g_{min} (Burghardt & Riederer, 2003). However, the g_{min} of *Hedera helix* was threefold higher than P suggesting either the stomata were leaky or the abaxial cuticle itself was more permeable than adaxial cuticle, or both. This approach assumed that adaxial and abaxial cuticles on a single leaf have similar cuticular permeabilities, and this is unlikely true for all species. Recently, a technique was reported that permits the determination of cuticular permeance in enzymatically isolated stomatous cuticles that takes advantage of the condition that water vapor flux across a helium-filled pore is 3.6 times faster than water vapor flux across a nitrogen-filled pore, and that solid-phase flux across the cuticle is unchanged no matter the gas used. By using different gases, but with identical concentration gradients across the same pores, flux through pores could be calculated separately from flux through the cuticle. Using these methods, the abaxial cuticle of *H. helix* was found to have 11-fold higher permeability than the adaxial cuticle (Šantůcek *et al.*, 2004), thereby explaining the higher abaxial *P* in *H. helix* reported by Burghardt and Riederer (2003) earlier. This ability to distinguish the solid-phase permeability of cuticles from the gas-phase permeability of stomatal pores offers a new powerful tool for studies to elucidate the ecological significance of cuticles and stomata in the regulation of plant water loss.

Though often unconsidered, the cuticle also likely plays a role in determining the leakiness of stomatal pores. In most plants, the substomatal chamber is lined with cuticle (Osborn & Taylor, 1990), indicating that water movement from leaf mesophyll into the substomatal chamber must pass through a cuticle membrane. In *Cirsium horridulum*, for example, the substomatal chamber cuticle was one-third thinner than the epidermal cuticle (Pesacreta & Hasenstein, 1999). To what degree this substomatal chamber cuticle, however, provides resistance to water vapor flux is completely unknown. In addition, the outermost edge of the stomatal pore is lined with cuticular material that forms a ridge or lip. When the stomatal pore closes, these ridges come together in a lock and key fashion to seal the pore (Jenks, 2002). The location of these cuticle ridges provides evidence that the cuticle may play an important role in determining the degree to which closed stomata prevent water loss. Circumstantial evidence that these cuticle ridges function in drought adaptation comes from the fact that xerophytes usually possess more or larger cuticular ridges. Other studies indicate that most plants also possess an inner cuticle as a lining of the epidermal cell layer just above the inner mesophyll tissues (Pesacreta & Hasenstein, 1999). The degree to which the lower g_{min} of xerophytes is due to the efficiency in which their stomatal pores seal when closed, the permeability of the cuticle membranes lining their substomatal chambers or the diffusive resistance of the inner cuticle, is an important subject that needs further investigation.

Often, drought conditions occur in conjunction with either high temperatures or freezing temperatures. In high-temperature climates, leaf surface temperatures up to 50°C have been reported (Kuraishi & Nito, 1980). *Vinca major* and *H. helix* cuticle permeability to water increased 12- and 264-fold, respectively, as temperature increased from 10°C to 55°C (Schreiber, 2001). The basis for this increased permeability is that lipids found in plant cuticles undergo a phase transition at temperatures between 30°C and 39°C. It is difficult to reconcile this data and the ecophysiological consequences of such permeability increases with the many xerophytes that are successfully adapted to high-temperature climates. Moreover, many xerophytic plants, like CAM plants, close their stomata during the day, which limits evaporative cooling. To survive, the cuticles of plants adapted to growth in hot, dry environments must have unique lipid compositions and/or structures, or else they must utilize novel mechanisms to limit water loss during high-temperature that would be expected to increase cuticular permeability. While it is well known that cuticles of many desert plants are highly reflective of solar radiation, the ambient desert temperatures far exceed the temperatures at which phase transition and increased permeability has been shown to occur in mesophytes. Studies are needed to elucidate the composition and structure of xeromorphic cuticles, and their association with cuticle lipid-phase transitions and permeability.

In freezing environments, water becomes locked in the soil and unavailable for plant use. The cuticles of many broad-leaf evergreens in alpine and other cold winter climates are typically less permeable to water than mesophytic plants (Burghardt & Riederer, 2003), and it has been postulated that these modified cuticles provide high resistance to water flux during winter freezing-drought. Just as for hot climate plants, the role of cuticle composition and structure in preventing winter dehydration on evergreens needs further study.

Plants with the highest cuticle permeability to water are typically deciduous plants from temperate climates (Riederer & Schreiber, 2001), with xeromorphic plants of Mediterranean climates and evergreen epiphytic or climbing plants from tropical climates having much lower permeabilities (Figure 2.3A). The range of variation in permeability of these plants was over 2.5 orders of magnitude, and was closely associated with the water availability in the environment and the relative need to conserve water. Cuticle permeabilities can be up to one and one half orders of magnitude lower than synthetic polymers (Figure 2.3B). Water transport across the cuticle is by simple diffusion, however, no correlation between cuticle thickness and permeability was evident in a study of 57 species from diverse ecosystems (Riederer & Schreiber, 2001). This lack of correlation between cuticle lipid thickness and total amount is reported in many other studies as well (Lendzian & Kerstiens, 1991; Schreiber & Riederer, 1996). Recent studies suggest that the exact kind and location of cuticle lipid constituents within the cuticle membrane plays a primary role in determining cuticle permeability. Recent studies show, for example, that waxes located inside the cuticle membrane (intracuticular waxes) have a major influence on

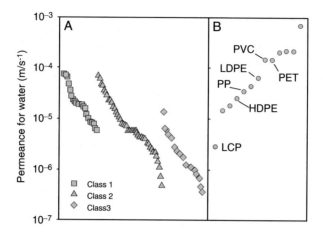

Figure 2.3 Range of permeances for water of leaf cuticular membranes (vapor-based driving force). Data are classified according to leaf anatomy and habitat; class 1 – deciduous species with mesomorphic leaves growing in temperate climates; class 2 – xeromorphic plant typically growing in a Mediterranean-type climate; class 3 – evergreen leaves from epiphytic or climbing plants naturally growing in a tropical climate (A). Permeances for water of synthetic polymer films 1 μm thick (B) (Riederer & Schreiber, 2001). HDPE, high density polyethylene; LCP, liquid crystal polymer; LDPE, low density polyethylene; PET, polyethylene terephthalate; PP, polypropene; PVC, polyvinyl chloride.

the rate of water movement across the cuticle membrane since removal of intracuticular waxes by solvent extraction greatly increased the cuticle permeability up to 550-fold (Knoche *et al.*, 2000; Vogg *et al.*, 2004). The mechanical removal of epicuticular waxes by comparison caused only a modest increase in water permeability of the cuticle.

To better understand how intracuticular waxes regulate diffusion of water, techniques like FTIR spectrophotometry, NMR, differential scanning calorimetry (DSC) and X-ray diffraction were used to shed light on the molecular arrangement of various cuticle lipid constituents within the cuticle membrane (Reiderer & Schreiber, 1995). These studies revealed that intracuticular waxes are arranged into crystalline and amorphous regions, with one study using FTIR spectrophotometry estimating that the aliphatic crystallinity of *H. helix* cuticular wax was ∼73% (Merk *et al.*, 1998). Crystalline regions within cuticles appear to have groups of aliphatic wax molecules closely packed together, with each molecule aligned perpendicularly to the cuticle surface in monolayers (Schreiber *et al.*, 1997; Casado & Heredia, 2001; Figure 2.4). These monolayers likely increase in size during the life of a leaf by extending laterally rather than by increasing in thickness. Many plants have wax profiles showing broad, skewed, or bimodal chain-length distributions, suggesting that some chain ends might protrude outside the top and bottom of these putative crystalline monolayers resulting in boundaries of irregular morphology. In the

amorphous regions that form between crystalline regions, chain ends project-
ing from crystalline regions have greater freedom of motion and greater
polarity due to the presence of free, and often charged, end groups having
carboxyl, hydroxyl or carbonyl terminal ends. A simple calculation of bond
lengths and angles reveals that individual wax constituent lengths range from
$\sim$2 nm for C_{16} acids to $\sim$5.8 nm for C_{46} esters (the longest detectable aliphatic
wax constituent in *Arabidopsis* for example). It is still unknown whether these
wax constituents associate randomly (as illustrated in Figure 2.4) or whether
they associate in some ordered way according to class or other affinity.
Depending on how these waxes associate, the hydrophilicity of the amorphous
regions could vary significantly. For example, amorphous regions could be
filled with the charged carboxylic terminal groups of fatty acids to be more
hydrophilic than amorphous regions bordering crystalline monolayers comprised of only alkanes. Notwithstanding, the lack of close packing in amorphous regions is expected to allow water molecules to diffuse more easily than in
crystalline regions. A broadening of the chain length distribution or an increase in the triterpenoid fraction would be expected to increase the proportion
of amorphous regions relative to crystalline regions. As the amorphous region

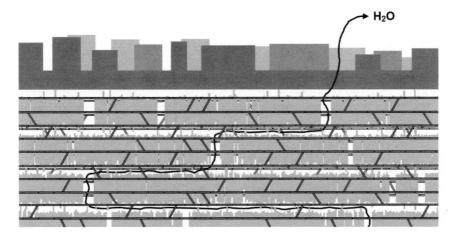

Figure 2.4   Model of the outer portion of a cuticle proper for a typical plant. Epicuticular wax layer (shades
of gray) is the outermost layer. The cuticle proper is comprised of crystalline wax regions (light gray) and
amorphous wax regions (white) embedded within a three-dimensional matrix of ester-linked cutin molecules
(dark gray). Crystalline areas are assumed to be random aggregates of wax molecules oriented perpendicularly
to the leaf surface resulting in irregular upper and lower surfaces due to variations in wax chain lengths. The
length of individual wax constituents range from $\sim$2 nm for C16 acids to $\sim$5.8 nm for C46 esters (drawn to
scale). The irregular surfaces of crystalline regions create amorphous wax areas between the crystalline
regions that are less densely packed and more polar due to the presence of free, and often charged, end groups.
Cutin monomers are pictured as cross-linked with dibasic acids, the largest class of cutin monomers found in
the cuticle of plants like Arabidopsis. Diffusion drives water molecules across the cuticle proper via these
tortuous polar pathways (Riederer & Schreiber, 1995).

fraction increases, water molecules are expected to have a shorter diffusion path to traverse the membrane (Figure 2.4) resulting in reduced diffusive resistance and increased cuticle permeability (Baur et al., 1999). What role the three-dimensional cutin matrix plays in establishing these diffusion pathways is still uncertain. Potentially, cutin provides a framework guiding the orientation of wax molecules within the cuticle, and in this way controls the shape of the diffusion pathways. Evidence from studies of plant cuticle membrane mutants discussed below supports this hypothesis. Regardless, previous studies reveal that the cuticle plays an important role in drought tolerance, with cuticle permeability as the primary determinant of water loss when stomata are closed during drought stress. Cuticles of drought-tolerant xerophytic plants with lower permeability to water support this premise, and also indicate that many xerophytes may have uniquely adapted cuticles. Understanding the role of cuticle in water loss, especially how cuticles help xerophytes survive in water-limiting environments, could lead to novel methods for the genetic modification of important crop plants to improve their drought tolerance.

2.4   Genetics of cuticle permeability

Plant cuticle mutants in numerous plant species have been isolated (Koorneef et al., 1989; Jenks et al., 1992; McNevin et al., 1993; Jenks et al., 1995), and chromosome walking and T-DNA tagging have been used to isolate the gene sequence altered in a few of these (Jenks et al., 2002; Chen et al., 2003; Xiao et al., 2004). Table 2.1 lists Arabidopsis genes characterized at the molecular level whose mutations caused a cuticle alteration phenotype, such as cutin monomer composition, cuticle ultrastructure and/or cuticular wax. Very little is known about the connection between the protein products of these genes and the cuticle lipid substrates or metabolic pathways they modify. Nevertheless, current findings demonstrate that cuticle permeability can be changed by alteration in the expression of certain 'cuticle' genes.

One approach that has been used to modify plant cuticle gene expression is chemical and insertion mutagenesis (Koorneef et al., 1989; McNevin et al., 1993; Jenks et al., 1994). The resulting plant cuticle mutants are beginning to elucidate the relationships between cuticle permeability and changes in wax and/or cutin. Measurement of water loss from fully hydrated stem segments in dark showed that certain eceriferum (cer) and other cuticle mutants, namely cer1, cer4, cer5, cer8, cer9, cer14, cer16, cer24, cer25, att1 and wax2 lost water at a significantly higher rate ($p < 0.01$) than an isogenic wild type (Goodwin & Jenks, unpublished). The highest water loss rates occurred in mutants att1, wax2 and cer25. The other cuticle mutants examined showed no effect on stem water loss using this assay. In-pot whole flowering plant transpiration rates of att1, wax2 and cer25 (Figure 2.5) show significantly

Table 2.1 Plant genes characterized at the molecular level having known and putative involvement in cuticle production

Loci Name	Predicted Protein	Genebank #/MIPs #	Reference
ACE1/HTH	Mandelonitrile lyase-like	*Arabidopsis* AAD55644/At1g72970	Nakatani *et al.* (1998), Krolikowski *et al.* (2003)
ALE1	Subtilisin-like serine protease	*Arabidopsis* AB060809/At1g62340	Tanaka *et al.* (2001)
CER1	Desaturase/receptor	*Arabidopsis* U40489/At1g02205	Aarts *et al.* (1995)
CER2	Transferase protein	*Arabidopsis* U40849/At4g24510	Xia *et al.* (1996, 1997)
CER3	E3 ligase (nuclear)	*Arabidopsis* X95962/At5g02310	Hannoufa *et al.* (1996)
CER6	Fatty acid synthase	*Arabidopsis* AF129511/At1g68530	Hooker *et al.* (2002)
CYP86A1	P450	*Arabidopsis* Z26358/At5g58860	Beneviste *et al.* (1998)
CYP86A2	P450 (*ATT1*)	*Arabidopsis* NM116260/At4g00360	Xiao *et al.* (2004)
CYP86A8	P450 (*LCR1*)	*Arabidopsis* AJ301678/At2g45970	Wellesen *et al.* (2001)
CYP94A1	P450	*Vicia* AF030260/At3g27670	Tijet *et al.* (1998), Pinot *et al.* (1999)
CYP94A2	P450	*Vicia* AF092917/At5g63450	Le Bouquin *et al.* (1999)
CYP94A5	P450	*Nicotiana* AF092916	Le Bouquin *et al.* (2001)
FAD2	Fatty acid desaturase	*Arabidopsis* L26296/At3g12120	Bonaventure *et al.* (2004) Okuley *et al.* (1994)
FATB	Thioesterase	*Arabidopsis* NM100724/At1g08510	Bonaventure *et al.* (2003, 2004)
FDH	β-ketoacyl-CoA synthase-like	*Arabidopsis* AF337910/At2g26250	Lolle *et al.* (1997), Pruitt *et al.* (2000)
GL1	Desaturase/receptor	*Zea mays* AAB87721	Hansen *et al.* (1997)
GL2	Transferase	*Zea mays* X88779/At3g23840	Tacke *et al.* (1995)
GL8	β-ketoacyl reductase	*Zea mays* U89509/At1g67730	Xu *et al.* (1997), Beaudoin *et al.* (2002)
HIC1	Fatty acid synthase	*Arabidopsis* AF188484/At2g46720	Gray *et al.* (2000)
KCS1	Fatty acid synthase	*Arabidopsis* AF053345/At1g01120	Todd *et al.* (1999)
LACS2	Acyl-CoA synthetase	*Arabidopsis* NM103833/At1g49430	Schnurr *et al.* (2004)
RST1	Unknown	*Arabidopsis* AY307371/At3g27670	Chen *et al.* (unpublished)
WAX2	Desaturase/receptor	*Arabidopsis* AY131334/At5g57800	Chen *et al.* (2003)
WIN1/SHN1	Transcription factor	*Arabidopsis* AY378101/At1g15360	Aharoni *et al.* (2004), Broun *et al.* (2004)
YBR159	β-ketoacyl reductase	*Arabidopsis* NM105441/At1g67730	Xu *et al.* (1997)

higher transpiration rates both in light and in dark (Chen *et al.*, 2003; Xiao *et al.*, 2004; Goodwin & Jenks, unpublished). These results indicated that increased cuticular permeability, not increased stomatal transpiration, was primarily responsible for the increased water loss rates measured in these

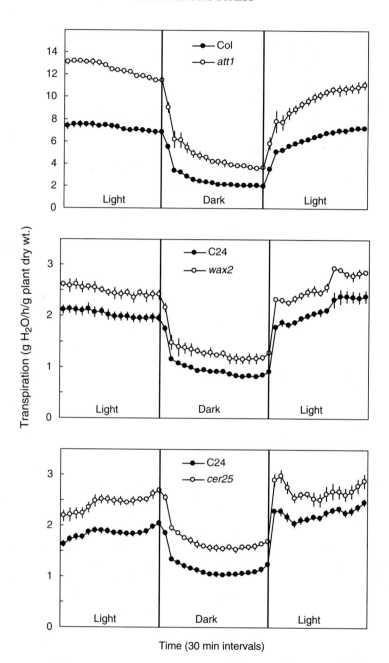

Figure 2.5 Whole plant in-pot transpiration rate (g H_2O/h/g plant dry wt ± s.e.) of *att1*, *wax2* and *cer25* mutants in both light and dark relative to their respective wild types, Col (Columbia) and C24 (Xiao *et al.*, 2004; Goodwin & Jenks, unpublished).

mutants. This conclusion is further strengthened in *wax2*, wherein stomatal densities were 86% and 70% on abaxial and adaxial leaf surfaces, respectively, of stomata densities in wild type (Chen *et al.*, 2003). Other evidence for cuticle involvement in whole plant water loss comes from studies of the *ale1* cuticle mutant whose seedlings wilt to the point of lethality in low humidity (Tanaka *et al.*, 2001). Phenotypically diverse mutants having alterations in many diverse genes result in similar elevations in cuticle permeability. Increased rates of chlorophyll extraction in ethanol were demonstrated for *ahd*, *bud*, *cod*, *ded*, *fdh*, *htd*, *thd*, *cer10*, *wax2*, *lacs2* and the overexpression mutant *win1/shn1* (Lolle *et al.*, 1998; Chen *et al.*, 2003; Aharoni *et al.*, 2004; Schnurr *et al.*, 2004). However, the degree to which the chlorophyll assay measures chlorophyll leaching through stomata pores has not been determined.

Many mutants showing evidence of altered cuticle permeability (via either elevated water loss or chlorophyll leaching), such as *ahd*, *ale1*, *bud*, *cod*, *ded*, *fdh*, *htd*, *thd*, *cer10* and *wax2*, exhibited a unique developmental program called post-genital fusion (Jenks *et al.*, 1996; Lolle *et al.*, 1998), wherein fusion occurs between leaves, stems and flower parts after primordial organs have differentiated. This fusion may be due to the altered secretion (or diffusion) of morphogenic factors that alter development coincident with contact between organs. Since *ahd*, *bud*, *cod*, *thd*, *htd* and *fdh* have no visible wax phenotype, it is possible that changes in nonwax lipids, like cutin or cutan, may have altered cuticle permeability to these factors. Further studies of cuticles on these fusion mutants using physical techniques like NMR, FTIR, DSC and X-ray diffraction could greatly advance our understanding of the cuticle characteristics that determine permeability.

Of the cuticle mutants examined by excised stem water loss, no correlations better than an $r^2 < 0.24$ could be found between the rates of water loss and either total cuticular wax, class or individual constituent amounts (Goodwin and Jenks, unpublished). Alterations in cutin are currently known for only five mutants, *att1*, *wax2*, *cer25*, *fad2* and *fatb* (Table 2.2; Bonaventure *et al.*, 2004). Interestingly, *att1*, *wax2* and *cer25* also exhibit reduced amounts of cuticle wax (data not shown). A recent study using $LiAlH_4$ to depolymerize total polyesters from whole epidermal tissues determined that octadecadiene-1, 18-dioate was a major polyester momoner in *Arabidopsis*, a constituent previously assumed restricted to suberin, and not reported in cutin except as a very minor constituent in cutins of only two plant species (see Bonaventure *et al.*, 2004). In an earlier study that employed BF_3 as the depolymerizing agent, cutin of isolated cuticle membranes from *Arabidopsis* was shown to possess only minor amounts of this polyunsaturated monomer (Xiao *et al.*, 2004). However, since BF_3 has the potential to degrade polyunsaturated monomers, additional studies are needed using $LiAlH_4$ instead of BF_3 as the cutin depolymerizing agent to determine whether isolated cuticle membranes of *Arabidopsis* might contain higher amounts of this unusual monomer. The

Table 2.2 Stem cutin monomer amount ($\mu g\ dm^{-2} \pm$ s.d.) examined using BF_3 depolymerization from three replicates of enzymatically isolated cuticles of Columbia and C24 ecotypes, with their isogenic mutants following. Depolymerization of the cutin polyester with BF_3 may degrade the polyunsaturated monomer octadecadien-1,18-dioic acid leading to an underestimate of the actual amount.

Cutin Monomer	Columbia	att1-1	C24	wax2	cer25
9-Hydroxy pentadecanoic acid	1.3 ± 0.4	0.7 ± 0.1	1.2 ± 0.2	1.4 ± 0.3	0.5 ± 0.0
10-Hydroxy heptadecanoic acid	1.9 ± 1.1	0.7 ± 0.3	0.7 ± 0.1	1.1 ± 0.3	1.0 ± 0.5
16-Hydroxy hexadecanoic acid	1.6 ± 0.3	1.1 ± 0.3	1.1 ± 0.1	1.4 ± 0.1	0.4 ± 0.2
10,16-Dihydroxy hexadecanoic acid	8.1 ± 2.6	3.5 ± 1.1	3.0 ± 0.6	3.0 ± 0.4	1.9 ± 0.2
Hexadecane-1,16-dioic acid	24.1 ± 3.5	5.4 ± 0.1	9.3 ± 1.4	9.8 ± 0.1	1.3 ± 0.1
7-Hydroxy hexadecane-1,16-dioic acid	5.5 ± 0.7	2.2 ± 0.3	1.3 ± 0.3	1.7 ± 0.5	0.3 ± 0.2
Octadecane-1,18-dioic acid	5.8 ± 1.1	1.8 ± 0.1	3.7 ± 0.8	3.8 ± 0.3	1.0 ± 0.2
Octadecadien-1,18-dioic acid	0.7 ± 0.1	0.8 ± 0.1	1.2 ± 0.4	1.4 ± 0.3	0.5 ± 0.1
Total	49.1 ± 8.2	16.1 ± 1.1	21.4 ± 2.2	23.6 ± 0.7	7.0 ± 0.2

Mutant att1-1 is in the Columbia background, whereas wax2 and cer25 are in C24. Position of the two double bonds in octadecadien-1,18-dioic acid not determined (Xiao et al., 2004; Goodwin & Jenks, unpublished).

cytochrome P450 mutant, att1, has a 70% decrease in monomer amount and altered cuticle membrane ultrastructure (Table 2.2; Figure 2.6). Synthesis of hexadecane-1,16-dioic acid was the primary defect of att1, being reduced to 75%. As such, ATT1 appears to have its primary function in the oxidation of 16-hydroxy hexadecanoic acid to hexadecane-1,16-dioic acid in the C_{16}-specific cutin pathway (Xiao et al., 2004). In wax2, the visible density of the cuticle membrane was reduced and membrane ultrastructure was highly disorganized, however, there was no change in cutin monomer composition (Table 2.2; Figure 2.6; Chen et al., 2003). The WAX2's function in cutin or cutan synthesis appears likely, but further studies are needed to confirm this. As one possibility, wax2's reduced visible density and structural organization of the cuticle membrane may be due to alteration in the polyester linkages that hold the cutin monomers together, rather than changes in the synthesis of

Figure 2.6 Arabidopsis stem pavement cell cuticle membranes visualized using transmission electron microscopy. Cuticle membrane bordered by arrowheads.

(A, B) The stem cuticle membrane of att1 is thicker, and less electron dense relative to Columbia (Xiao et al., 2004). Bar = 100nm.

(C, D) The stem cuticle membrane on wax2 is thicker, less electron dense and does not show the typical bilayer ultrastructure, lacking the dark-staining reticulum normally associated with the cuticular layer. The more translucent cuticle material bulges into the outer cell wall layer, revealing a disruption of the normal cutinization process (Chen et al., 2003). Bar = 100nm.

(E, F) The stem cuticle membrane on cer25 is thinner, and more electron dense, than C24 (Goodwin & Jenks, unpublished). Bar = 100nm.

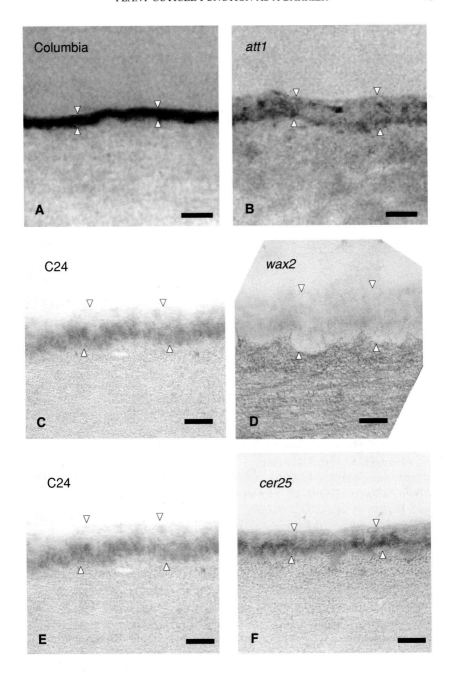

cutin monomer themselves. Conversely, the WAX2 protein has sequence homology with desaturases, a class of enzyme that function in synthesis of polyunsaturated acids. It is thus also possible that the polyunsaturated fatty acids that serve as precursors in cutan biosynthesis could be altered (Blée & Schuber, 1993; Villena *et al.*, 1999). The third cutin membrane mutant, *cer25*, has reduced membrane thickness, and a 67% reduction in cutin amount relative to wild type (Table 2.2; Figure 2.6). Like *att1* and *wax2*, these data suggest that the changes in cutin or cutan composition are playing a major role in cuticle permeability. Two other mutants, *fad2* and *fatb*, have been shown to have reduced epidermal polyester amounts (Bonaventure *et al.*, 2004), but cuticle permeability of these mutants has not been reported. Interestingly, mutants having the greatest increase in permeability were those having altered cutin, even though studies discussed above indicate that waxes are the primary hydrophobic barriers in cuticle membrane diffusion pathways. An important possibility exists then that cutin plays a role in forming the cuticle diffusion barrier by providing the framework in which wax molecules are arranged. Changes in the cutin monomer composition or structure could thereby change the size and proportion of crystalline regions, and ultimately the length and tortuosity of the diffusion pathways.

Changes in the expression of cuticle genes by plant exposure to drought or other abiotic stress suggest that these genes could play an important role in plant adaptability to stress. These findings are consistent with others that drought and high-temperature stress increase the accumulation of cuticular waxes on organs of numerous plants (Baker, 1980; Hadley, 1989; Jenks *et al.*, 2001). Microarray analysis of Columbia total RNA indicates that within 24 h of drought, cold and salt stress, the expression of the cutin associated P450, *CYP86A2*, was over twofold higher than in nonstressed plants (Seki *et al.*, 2002; Schuler & Werck-Reichhart, 2003). Elevated mannitol concentrations induced the P450's *CYP86A1*, *CYP86A2* and *CYP86A8* in roots by over twofold, whereas *CYP86A2* was repressed twofold in roots by salt and cold stress (Kreps *et al.*, 2002; Schuler & Werck-Reichhart, 2003). The *CER6* transcript accumulated with increasing osmotic stress (Hooker *et al.*, 2002). Drought-, cold- and light-inducible elements found in the *CER6* promoter explain in part the molecular basis for this induction. Since drought as well as other abiotic stress such as heat were typically transient phenomena, it is not surprising that plants have genetic-based mechanisms enabling adaptation to such stress. The induction responses for these genes indicate that the cuticle may play an important role in plant drought tolerance. Besides studies to elucidate the molecular basis for cuticle gene function in establishing cuticle permeability properties, it is also important that future studies examine cuticle permeability in response to the up- or down-regulation of these genes by abiotic stress as a means of determining whether cuticle response to environmental conditions has adaptive value.

The observed induction of cuticle genes by drought stress and related factors suggests that the use of biotechnology could be an effective strategy to modify cuticle permeability, and thereby improve drought tolerance of important crops. Clearly, however, further studies are needed to elucidate the impact of cuticle alteration on drought tolerance characteristics in crops. As an important consideration for biotech approaches with cuticle genes, recent studies suggest that alteration of cuticle gene expression could have major impacts on other agronomic traits. For example, cuticle gene overexpression has often produced unexpected plieotropic effects. Overexpression of *WIN1* resulted in shorter inflorescences, delayed development, delayed flowering and partial sterility (Broun *et al.*, 2004). A 35S:*FDH* construct resulted in plants with severe retardation of growth and development (Yephremov *et al.*, 1999). Overexpression of the 35S:*CER6* wax condensing enzyme construct had little effect on increasing total wax amounts (Hooker *et al.*, 2002), except in plants having multiple copies of the transgene expressed by the native promoter. Successful engineering of crop plant cuticles using biotechnology may require creative strategies involving the use of tissue specific or chimeric promoters to eliminate unwanted pleiotropic effects.

2.5   Conclusions

The cuticle is an essential plant adaptation that enhances plant survival in aerial environments. Variation in the permeability of plant cuticles also plays a major role in plant adaptation to different ecosystems, with plants adapted to arid zones having cuticle membranes one to two orders of magnitude less permeable than mesophytic plants. The uniqueness of the cuticle as a regulator of water movement across the epidermis is further underscored by the fact that the water barrier properties of plant cuticles can surpass the most impermeable synthetic polymer of equal thickness by over an order of magnitude. In changing environments, the exposure of plants to drought causes plants to close their stomatal pores and almost all water loss then occurs by diffusion across the cuticular membrane. With the additional evidence that prolonged drought induces increased expression of genes involved with cuticle synthesis, the evidence is strong that the cuticle is especially important for water conservation during drought.

The current state of cuticle research demonstrates that intracuticular waxes rather than epicuticular waxes are the primary determinants in cuticle permeability. Evidence suggests that intracuticular wax aliphatic constituents form crystalline regions that are water impermeable islands surrounded by regions of less densely packed and more polar amorphous regions that form the tortuous pathways in which water diffuses to the surrounding atmosphere. The basis for cuticle permeability is associated with the chain length distributions, constituent profiles and packing arrangements of the individual wax molecules. Cuticle membrane mutants reveal that the cutin matrix also plays an important role in cuticle permeability, perhaps providing a molecular framework that guides the arrangement of wax molecules. Further characterization of these and other yet unidentified cuticle loci is expected to reveal more about the critical role that cuticle plays in plant adaptation to arid and drought-prone environments. As the function of these genes is elucidated, researchers will be able to employ genetic engineering to manipulate the permeability of cuticles, enabling crop plants to use water more efficiently. Such crops with increased drought tolerance will be a very important addition to our world with an ever-increasing population to feed.

References

Aarts, M. G., Keijzer, C. J., Stiekema, W. J. & Pereira, A. (1995) Molecular characterization of the *CER1* gene of *Arabidopsis* involved in epicuticular wax biosynthesis and pollen fertility. *Plant Cell*, **7**, 2115–2127.

Aharoni, A., Dixit, S., Jetter, R., Thoenes, E., van Arkel, G. & Pereira, A. (2004) The SHINE clade of AP2 domain transcription factors activates wax biosynthesis, alters cuticle properties, and confers drought tolerance when overexpressed in *Arabidopsis*. *Plant Cell*, **16**, 2463–2480.

Arondel, V., Vergnolle, C., Cantrel, C. & Kader, J. (2000) Lipid transfer proteins are encoded by a small multigene family in *Arabidopsis thaliana*. *Plant Science*, **157**, 1–12.

Baker E. A. (1980) Chemistry and morphology of plant cuticular waxes. In D. F. Cutler, K. L. Alvin & C. E. Price (eds) *The Plant Cuticle*. Academic Press, New York, pp. 139–165.

Baker, E. A. & Procopiou, J. (1975) The cuticles of *Citrus* species. Composition of the intracuticular lipids of leaves and fruits. *Journal of the Science of Food and Agriculture*, **26**, 1347–1352.

Baur, P., Marzouk, H. & Schönherr, J. (1999) Estimation of path lengths for diffusion of organic compounds through leaf surfaces. *Plant, Cell and Environment*, **22**, 291–299.

Beattie, G. A. & Marcell, L. M. (2002) Effect of alterations in cuticular wax biosynthesis on the physicochemical properties and topography of maize leaf surfaces. *Plant, Cell and Environment*, **25**, 1–16.

Beaudoin, F., Gable, K., Sayanova, O., Dunn, T. & Napier, J. A. (2002) A *Saccharomyces cerevisiae* gene required for heterologous fatty acid elongase activity encodes a microsomal β-keto-reductase. *Journal of Biological Chemistry*, **277**, 11481–11488.

Benveniste, I., Tijet, N., Adas, F., Philipps, G., Salaun, J. P. & Durst, F. (1998) *CYP86A1* from *Arabidopsis thaliana* encodes a cytochrome P450-dependent fatty acid omega-hydroxylase. *Biochemical and Biophysical Research Communications*, **243**, 688–693.

Blée, E. & Schuber, F. (1993) Biosynthesis of cutin monomers: involvement of a lipoxygenase/peroxygenase pathway. *Plant Journal*, **4**, 113–123.

Bonaventure, G., Salas, J., Pollard, M. & Ohlrogge, J. (2003) Disruption of the *FATB* gene in *Arabidopsis* demonstrates an essential role of saturated fatty acids in plant growth. *Plant Cell*, **15**, 1020–1033.

Bonaventure, G., Beisson, F., Ohlrogge, J., & Pollard, M. (2004) Analysis of the aliphatic monomer composition of polyesters associated with Arabidopsis epidermis: occurence of octadeca-*cis*-6, *cis*-9-dioate as the major component. *Plant Journal*, **40**, 920–930.

Broun, P., Poindexter, P., Osborne, E., Jiang, C. & Riechmann, J. (2004) *WIN1*, a transcriptional activator of epidermal wax accumulation in *Arabidopsis*. *Proceedings of the National Academy of Sciences*, **101**, 4706–4711.

Burghardt, M. & Riederer, M. (2003) Ecophysiological relevance of cuticular transpiration of deciduous and evergreen plants in relation to stomatal closure and leaf water potential. *Journal of Experimental Botany*, **54**, 1941–1949.

Casado, C. G. & Heredia, A. (2001) Self-association of plant wax components: a thermodynamic analysis. *Biomacromolecules*, **2**, 407–409.

Chen, X., Goodwin, S. M., Boroff, V. L., Liu, X. & Jenks, M. A. (2003) Cloning and characterization of the *WAX2* gene of *Arabidopsis* involved in cuticle membrane and wax production. *Plant Cell*, **15**, 1170–1185.

Deas, A. H. B. & Holloway, P. J. (1977) The intermolecular structure of some plant cutins. In M. Tevini & H. K. Lichtenthaler (eds) *Lipids and Lipid Polymers in Higher Plants*. Springer-Verlag, Berlin, pp. 293–299.

Fang, X., Qui, F., Yan, B., Wang, H., Mort, A. J. & Stark, R. E. (2001) NMR studies of molecular structure in fruit cuticle polyesters. *Phytochemistry*, **57**, 1035–1042.

Gray, J. E., Holroyd, G. H., van der Lee, F. M., Bahrami, A. R., Sijmons, P. C., Woodward, F. I., Schuch, W. & Hetherington, A. M. (2000) The *HIC* signalling pathway links CO_2 perception to stomatal development. *Nature*, **408**, 713–716.

Hadley N. F. (1989) Lipid water barriers in biological systems. *Progress in Lipid Research*, **28**, 1–33.

Hannoufa, A., Negruk, V., Eisner, G. & Lemieux B. (1996) The *CER3* gene of *Arabidopsis thaliana* is expressed in leaves, stems, roots, flowers and apical meristems. *Plant Journal*, **10**, 459–467.

Hansen, J. D., Pyee, J., Xia, Y., Wen, T. J., Robertson, D. S., Kolattukudy, P. E., Nikolau, B. J. & Schnable, P. S. (1997) The *GLOSSY1* locus of maize and an epidermis-specific cDNA from *Kleinia odora* define class of receptor-like proteins required for the normal accumulation of cuticular waxes. *Plant Physiology*, **113**, 1091–1100.

Hooker, T. S., Millar, A. A. & Kunst, L. (2002) Significance of the expression of the CER6 condensing enzyme for cuticular wax production in *Arabidopsis*. *Plant Physiology*, **129**, 1568–1580.

Jeffree, C. E. (1996) Structure and ontogeny of plant cuticles. In G. Kerstiens (ed.) *Plant Cuticles: An Integrated Functional Approach*. BIOS Scientific Publishers, Oxford, pp. 33–82.

Jenks, M. A. (2002) Critical issues with the plant cuticle's function in drought tolerance. In A. J. Wood (ed.) *Biochemical and Molecular Responses of Plants to the Environment*. Research Signpost Press, Kerala, India, pp. 97–127.

Jenks, M. A., Eigenbrode, S. & Lemeiux, B. (2002) Cuticular waxes of *Arabidopsis*. In C. Somerville & E. Meyerowitz (eds) *The Arabidopsis Book*. American Society of Plant Biologists, Rockville, MD. DOI 10.1199/tab.0016. http://www.aspb.org/publications/arabidopsis/.

Jenks, M. A., Anderson, L., Teusink, R. & Williams, M. (2001) Leaf cuticular waxes of potted rose cultivars as effected by plant development, drought, and paclobutrazol treatments. *Physiologia Plantarum*, **112**, 62–70.

Jenks, M. A., Tuttle, H. A., Eigenbrode, S. D. & Feldmann, K. A. (1995) Leaf epicuticular waxes of the *eceriferum* mutants in *Arabidopsis*. *Plant Physiology*, **108**, 369–377.

Jenks, M. A., Tuttle, H. A., Rashotte, A. M. & Feldmann, K. A. (1996) Mutants in *Arabidopsis* altered in epicuticular waxes and leaf morphology. *Plant Physiology*, **110**, 377–385.

Jenks, M. A., Joly, R. A., Rich, P. J., Peters, P. J., Axtell, J. D. & Ashworth, E. N. (1994) Chemically-induced cuticle mutation affecting epidermal conductance to water vapor and disease susceptibility in *Sorghum bicolor* (L.) Moench. *Plant Physiology*, **105**, 1239–1245.

Jenks, M. A., Rich, P. J., Peters, P. J., Axtell, J. D. & Ashworth, E. N. (1992) Epicuticular wax morphology of *bloomless* (*bm*) mutants in *Sorghum bicolor*. *International Journal of Plant Science*, **153**, 311–319.

Jetter, R. & Riederer, M. (1995) In vitro reconstitution of epicuticular wax crystals: formation of tubular aggregates by long-chain secondary alkanediols. *Botanica Acta*, **108**, 111–120.

Jetter, R., Schäffer, S. & Riederer, M. (2000) Leaf cuticular waxes are arranged in chemically and mechanically distinct layers: evidence from *Prunus laurocerasus* L. *Plant, Cell and Environment*, **23**, 619–628.

Kerstiens, G. (1996) Cuticular water permeability and its physiological significance. *Journal of Experimental Botany*, **47**, 1813–1832.

Knoche, M., Peschel, S., Hinz, M. & Bukovac, M. J. (2000) Studies on water transport through the sweet cherry fruit surface: characterizing conductance of the cuticular membrane using pericarp segments. *Planta*, **212**, 127–135.

Kolattukudy, P. E. (1996) Biosynthetic pathways of cutin and waxes, their sensitivity to environmental stresses. In G. Kersteins (ed.) *Plant Cuticles: An Integrated Functional Approach*. BIOS Scientific Publishers, Oxford, pp. 83–108.

Kolattukudy, P. E. (2002) Cutin from Plants. In G. Steinbüchel (ed) *Biopolymers*. Wiley-VCH, Weinheim, pp. 1–40.

Koornneef, M., Hanhart, C. J. & Thiel, F. (1989) A genetic and phenotypic description of *eceriferum* (*cer*) mutants in *Arabidopsis thaliana*. *Journal of Heredity*, **80**, 118–122.

Körner, C. (1993) Scaling from species to vegetation: the usefulness of functional groups. In E. D. Schulze & H. A. Mooney (eds) *Biodiversity and Ecosystem Function*. Springer-Verlag, Berlin, pp. 117–140.

Kreps, J. A., Wu, Y., Chang, H., Zhu, T., Wang, X. & Harper, J. F. (2002) Transcriptome changes for *Arabidopsis* in response to salt, osmotic, and cold stress. *Plant Physiology*, **130**, 2129–2141.

Krolikowski, K., Victor, J. L., Wagler, T. N., Lolle, S. J. & Pruitt, R. E. (2003) Isolation and characterization of the *Arabidopsis* organ fusion gene *HOTHEAD*. *Plant Journal*, **35**, 501–511.

Kunst, L. & Samuels, A. L. (2003) Biosynthesis and secretion of plant cuticular wax. *Progress in Lipid Research*, **42**, 51–80.

Kuraishi, S. & Nito, N. (1980) The maximum leaf temperature of the higher plants observed in the inland sea area. *Botanical Magazine*, **93**, 209–220.

Le Bouquin, R., Pinot, F., Benveniste, I., Salaun, J. P. & Durst, F. (1999) Cloning and functional characterization of *CYP94A2*, a medium chain fatty acid hydroxylase from *Vicia sativa*. *Biochemical and Biophysical Research Communications*, **261**, 156–162.

Le Bouquin, R., Skrabs, M., Kahn, R., Benveniste, I., Salau, J., Schreiber, L., Durst, F. & Pinot, F. (2001) CYP94A5, a new cytochrome P450 from *Nicotiana tabacum* is able to catalyze the oxidation of fatty acids to the ω-alcohol and to the corresponding diacid. *European Journal of Biochemistry*, **268**, 3083–3090.

Lendzian, K. & Kerstiens, G. (1991) Sorption and transport of gases and vapors in plant cuticles. *Reviews of Environment Contamination and Toxicology*, **121**, 65–128.

Lolle, S. J., Hsu, W. & Pruitt, R. E. (1998) Genetic analysis of organ fusion in *Arabidopsis thaliana*. *Genetics*, **149**, 607–619.

Lolle, S. J., Berlyn, G. P., Engstrom, E. M., Krolikowski, K. A., Reiter, W. & Pruitt, R. E. (1997) Developmental regulation of cell interactions in the *Arabidopsis fiddlehead-1* mutant: a role for the epidermal cell wall and cuticle. *Developmental Biology*, **189**, 311–321.

McNevin, J. P., Woodward, W., Hannoufa, A., Feldmann, K. A. & Lemieux, B. (1993) Isolation and characterization of *eceriferum* (*cer*) mutants induced by T-DNA insertions in *Arabidopsis thaliana*. *Genome*, **36**, 610–618.

Merk, S., Blume, A. & Riederer, M. (1998) Phase behaviour and crystallinity of plant cuticular waxes studied by Fourier transform infrared spectroscopy. *Planta*, **204**, 44–53.

Meusel, I., Leistner E. & Barthlott W. (1994) Chemistry and micromorphology of compound epicuticular wax crystalloids (*Strelitzia* type). *Plant Systemetics and Evolution*, **193**, 115–123.

Nakatani, M., Araki, T. & Iwabuchi, M. (1998) *Adhesion of calyx edges*, a gene involved in the regulation of postgenital fusion in *Arabidopsis*. *Plant and Cell Physiology*, **39**, S64.

Neinhuis, C., Koch, K. & Barthlott, W. (2001) Movement and regeneration of epicuticular waxes through plant cuticles. *Planta*, **213**, 427–434.

Nip, M., Tegelaar, E. W. & de Leeuw, J. W. (1986a) A new non-saponifiable highly-aliphatic and resistant biopolymer in plant cuticles. Evidence from pyrolysis and ^{13}C-NMR analysis of present-day and fossil plants. *Naturwissenschaften*, **73**, 579–585.

Nip, M., Tegelaar, E. W. & Brinkhuis, H. (1986b) Analysis of modern and fossil plant cuticles by Curie-point py-gc-ms – recognition of a new, highly aliphatic and resistant bio-polymer, *Organic Geochemistry*, **10**, 769–778.

Okuley, J., Lightner, J., Feldman, K., Yadav, N., Lark, E., & Browse, J. (1994) *Arabidopsis FAD2* gene encodes the enzyme that is essential for polyunsaturated lipid synthesis. *Plant Cell*, **6**, 147–158.

Osborn, J. M. & Taylor, T. N. (1990) Morphological and ultrastructural studies of plant cuticular membranes. I. Sun and shade leaves of *Quercus velutina* (Fagaceae). *Botanical Gazette*, **151**, 465–476.

Pesacreta, T. C. & Hasenstein, K. H. (1999) The internal cuticle of *Cirsium horridulum* (Asteraceae) leaves. *American Journal of Botany*, **86**, 923–928.

Pinot, F., Benveniste, I., Salaün, J., Loreau, O., Noël, J., Schreiber, L. & Durst, F. (1999) Production in vitro by the cytochrome P450 CYP94A1 of major C_{18} cutin monomers and potential messengers in plant—pathogen interactions: enantioselectivity studies. *Biochemical Journal*, **342**, 27–32.

Pruitt, R. E., Vielle-Calzada, J. P., Ploense, S. E., Grossniklaus, U. & Lolle, S. J. (2000) *FIDDLEHEAD*, a gene required to suppress epidermal cell interactions in *Arabidopsis*, encodes a putative lipid biosynthetic enzyme. *Proceedings of the National Academy of Sciences*, **97**, 1311–1316.

Rashotte, A. M. & Feldmann, K. A. (1998) Correlations between epicuticular wax structures and chemical composition in *Arabidopsis thaliana*. *International Journal of Plant Science*, **159**, 773–779.

Riederer, M. & Schönherr, J. (1988) Development of plant cuticles fine structure and cutin composition of *Clivia miniata* Reg. leaves. *Planta*, **174**, 127–138.

Riederer, M. & Schreiber, L. (1995) Waxes: the transport barriers of plant cuticles. In R. J. Hamilton (ed.) *Waxes: Chemistry, Molecular Biology and Functions*. The Oily Press, Dundee, UK, pp. 131–156.

Riederer, M. & Schreiber, L. (2001) Protecting against water loss: analysis of the barrier properties of plant cuticles. *Journal of Experimental Botany*, **52**, 2023–2032.

Šantůček, J., Šimáňová, E., Karbulková, J., Šimková, M. & Schreiber, L. (2004) A new technique for measurement of water permeability of stomatous cuticular membranes isolated from *Hedera helix* leaves. *Journal of Experimental Botany*, **55**, 1411–1422.

Schnurr, J., Shockey, J. & Browse, J. (2004) The Acyl-CoA synthetase encoded by LACS2 is essential for normal cuticle development in *Arabidopsis*. *Plant Cell*, **16**, 629–642.

Schönherr, J. & Riederer, M. (1986) Plant cuticles sorb lipophilic compounds during enzymatic isolation. *Plant, Cell and Environment*, **9**, 459–466.

Schreiber, L. (2001) Effect of temperature on cuticular transpiration of isolated cuticular membranes and leaf discs. *Journal of Experimental Botany*, **52**, 1893–1900.

Schreiber, L. & Riederer, M. (1996) Ecophysiology of cuticular transpiration: comparative investigation of cuticular water permeability of plant species from different habitats. *Oecologia*, **107**, 426–432.

Schreiber, L., Schorn, K. & Heimburg, T. (1997) H^2 NMR study of cuticular wax isolated from *Hordeum vulgare* L. leaves: identification of amorphous and crystalline wax phases. *European Biophysics Journal with Biophysics Letters*, **26**, 371–380.

Schuler, M. A. & Werck-Reichhart, D. (2003) Functional genomics of P450s. *Annual Review of Plant Biology*, **54**, 629–667.

Seki, M., Narusaka, M., Ishida, J., Nanjo, T. & Fujita, M. (2002) Monitoring the expression profiles of 7000 *Arabidopsis* genes under drought, cold and high salinity stresses using a full-length cDNA microarray. *Plant Journal*, **31**, 279–292.

Tacke, E., Korfhage, C., Michel, D., Maddaloni, M., Motto, M., Lanzini, S., Salamini, F. & Doring, H. P. (1995) Transposon tagging of the maize *GLOSSY2* locus with the transposable element En/Spm. *Plant Journal*, **8**, 907–917.

Taiz, L. & Zeiger, E. (2002) *Plant Physiology, 3rd Edition*. Sinauer Associates, Sunderland, MA.

Tanaka, H., Onouchi, H., Kondo, M., Hara-Nishimura, I., Nishimura, M., Machida, C. & Machida, Y. (2001). A subtilisin-like serine protease is required for epidermal surface formation in *Arabidopsis* embryos and juvenile plants. *Development*, **128**, 4581–4689.

Tijet, N., Helvig, C., Pinot, F., LeBouquin, R., Lesot, A., Durst, F., Salaun, J. P. & Benveniste, I. (1998) Functional expression in yeast and characterization of a clofibrate-inducible plant cytochrome P-450 (*CYP94A1*) involved in cutin monomers synthesis. *Biochemical Journal*, **332**, 583–589.

Todd, J., Post-Beittenmiller, D. & Jaworski, J. G. (1999) *KCS1* encodes a fatty acid elongase 3-ketoacyl-CoA synthase affecting wax biosynthesis in *Arabidopsis thaliana*. *Plant Journal*, **17**, 119–126.

Villena, J. F., Domínguez, E., Stewart, D. & Heredia, A. (1999) Characterization and biosynthesis of non-degradable polymers in plant cuticles. *Planta*, **208**, 181–187.

Vioque, J. & Kolattukudy, P. E. (1997) Resolution and purification of an aldehyde-generating and an alcohol-generating fatty acyl-CoA reductase from pea leaves (*Pisum sativum* L.). *Archives of Biochemistry and Biophysics*, **340**, 64–67.

Vogg, G., Fischer, S., Leide, J., Emmanuel, E., Jetter, R., Levy, A. & Riederer, M. (2004) Tomato fruit cuticular waxes and their effects on transpiration barrier properties: functional characterization of a mutant deficient in a very-long-chain fatty acid β-ketoacyl-CoA synthase. *Journal of Experimental Botany*, **55**, 1401–1410.

Walton, T. J. (1990) Waxes, cutin, and suberin. In *Methods in Plant Biochemistry*. Academic Press, New York, pp. 4, 105–158.

Wellesen, K., Durst, F., Pinot, F., Benveniste, I., Nettesheim, K., Wisman, E., Steiner-Lange, S., Saedler, H & Yephremov, A. (2001). Functional analysis of the *LACERATA* gene of *Arabidopsis* provides evidence for different roles of fatty acid ω-hydroxylation in development. *Proceedings of the National Academy of Sciences*, **98**, 9694–9699.

Xia, Y., Nikolau, B. J. & Schnable, P. S. (1996) Cloning and characterization of *CER2*, an *Arabidopsis* gene that affects cuticular wax accumulation. *Plant Cell*, **8**, 1291–1304.

Xia, Y., Nikolau, B. J. & Schnable, P. S. (1997) Developmental and hormonal regulation of the *Arabidopsis* *CER2* gene that codes for a nuclear-localized protein required for the normal accumulation of cuticular waxes. *Plant Physiology*, **115**, 925–937.

Xiao, F., Goodwin, S. M., Xiao, Y., Sun, Z., Baker, D., Tang, X., Jenks, M. & Zhou, J. (2004) *Arabidopsis* *CYP86A2* negatively regulates *Pseudomonas syringae* type III genes and is required for cuticle development. *EMBO Journal*, **23**, 2903–2913.

Xu, X., Dietrich, C. R., Delledonne, M., Xia, Y., Wen, T. J., Robertson, D. S., Nikolau, B. J. & Schnable, P. S. (1997) Sequence analysis of the cloned *GLOSSY8* gene of maize suggests that it may code for a beta-ketoacyl reductase required for the biosynthesis of cuticular waxes. *Plant Physiology*, **115**, 501–510.

Yephremov, A., Wisman, E., Huijser, P., Huijser, C., Wellesen, K. & Saedler, H. (1999) Characterization of the *Fiddlehead* gene of *Arabidopsis* reveals a link between adhesion response and cell differentiation in the epidermis. *Plant Cell*, **11**, 2187–2201.

3 Plant adaptive responses to salinity stress

Miguel A. Botella, Abel Rosado, Ray A. Bressan
and Paul M. Hasegawa

3.1 Salt stress effects on plant survival, growth and development

Salt is so abundant on our planet that it is a major constraint to global food
crop production, jeopardizing the capacity of agriculture to sustain the bur-
geoning human population increase (Flowers, 2004). It is estimated that 20%
of all cultivated land and nearly half of irrigated land is salt-affected, greatly
reducing yield well below the genetic potential (van Schilfgaarde, 1994;
Munns, 2002; Flowers, 2004). Soil salinity is becoming a more acute problem,
primarily because of declining irrigation water quality (Rhoades & Loveday,
1990; Ghassemi et al., 1995; Flowers, 2004). Consequently, new strategies to
enhance crop yield stability on saline soils represent a major research priority.
Remediation of salinized soils is a possibility. However, this approach is
hypothetically based and unsubstantiated by compelling evidence of feasibility
(Tester & Davenport, 2003). A complementary strategy is to increase salt
tolerance of crop plants either by genetic introgression, using both traditional
as well as molecular marker-assisted breeding techniques, and biotechnology
(Hasegawa et al., 2000; Koyama et al., 2001; Borsani et al., 2003; Flowers,
2004).

Limited success in increasing the yield stability of crops grown on saline
soils is due, in some measure, to a minimal understanding about how salinity
and other abiotic stresses affect the most fundamental processes of cellular
function – including cell division, differentiation and expansion – which have
a substantial impact on plant growth and development (Hasegawa et al., 2000;
Zhu, 2001). In fact, many years ago we suggested that plant adaptation to
salinity stress most probably involves phenological responses that are important for fitness in a saline environment but may adversely affect crop yield
(Binzel et al., 1985; Bressan et al., 1990). Regardless, some degree of yield
stability is achievable in salt stress environments (Epstein et al., 1980; Yeo &
Flowers, 1986; Flowers & Yeo, 1995; Flowers, 2004). For example, wild
species and primitive cultivars of tomato are a repository of genetic variation
for salt adaptive capacity (Jones, 1986; Cuartero et al., 1992).

Whether sufficient genetic diversity for salt tolerance exists or is accessible
in cultivars commonly used in breeding programs is problematic (Yeo &
Flowers, 1986; Flowers, 2004). The narrow gene pool for salt tolerance in

modern crop varieties infers that domestication has selected against the capacity for salt adaptation. That is, negative linkages exist between loci responsible for high yield and those necessary for salt tolerance. Current research efforts directed at identification of salt tolerance determinants and dissection of the integration between salt tolerance and plant growth and development will provide substantial resources and insights for plant breeding and biotechnology strategies to improve crop yield stability. Molecular marker-based breeding strategies will identify loci responsible for salt tolerance and, to the extent possible, facilitate the separation of physically linked loci that negatively influence the yield (Foolad & Lin, 1997; Flowers et al., 2000; Foolad et al., 2001). Monogenic introgression of salt tolerance determinants directly into high-yielding modern crop varieties should simplify breeding efforts to improve yield stability and facilitate pyramiding of genes, which, in all probability will be required.

3.1.1   NaCl causes both ionic and osmotic stresses

Salt at higher concentrations in apoplast of cells generates primary and secondary effects that negatively affect survival, growth and development. Primary effects are ionic toxicity and disequilibrium, and hyperosmolality. Both Na^+ and Cl^- are inhibitory to cytosolic and organellar processes (Niu et al., 1995; Serrano et al., 1999; Zhu et al., 1998). Concentrations > 0.4 M inhibit most enzymes because of disturbances to hydrophobic–electrostatic balance that is necessary to maintain the protein structure (Wyn Jones & Pollard, 1983). However, Na^+ concentrations of 0.1 M are cytotoxic indicating that the ion directly affects specific biochemical and physiological processes (Serrano, 1996). High concentrations of salt also impose hyperosmotic shock by lowering the water potential causing turgor reduction or loss that restricts cell expansion (Munns & Termaat, 1986; Hasegawa et al., 2000; Zhu, 2002). Sufficiently negative apoplastic water potential can lead to cellular water loss simulating dehydration that occurs during the episodes of severe drought.

3.1.2   Secondary effects of salt stress

Principal secondary effects of NaCl stress include disturbance of K^+ acquisition, membrane dysfunction, impairment of photosynthesis and other biochemical processes, generation of reactive oxygen species (ROS) and programmed cell death (Serrano et al., 1999; Hasegawa et al., 2000; Rodriguez-Navarro, 2000; Zhu, 2003). K^+ is an essential element for plants and Na^+ competes with K^+ for uptake into cells, particularly when the external concentration is substantially greater than that of the nutrient (Niu et al., 1995; Rodriguez-Navarro, 2000). Ca^{2+} alleviates Na^+ toxicity through different mechanisms, including the control of K^+/Na^+ selective accumulation and others that are now being elucidated (Läuchli, 1990; Kinrade, 1998, 1999; Liu & Zhu, 1998; Rubio et al., 2004).

Ionic and hyperosmotic stresses lead to secondary metabolic effects that plants must alleviate in order to survive and to reestablish growth and development (Hasegawa et al., 2000; Zhu, 2002). For example, salt stress induces oxidative stress by increasing superoxide radicals (O_2^{-}), hydrogen peroxide (H_2O_2) and hydroxyl radicals (OH·) (Moran et al., 1994; Borsani et al., 2001a, 2001b). These molecules, known as ROS, are produced in aerobic cellular processes such as electron transport in the mitochondria or the chloroplast, or during photorespiration (Chinnusamy et al., 2004).

3.2 Plant genetic models for dissection of salt tolerance mechanisms and determinant function

Plants that are native to saline environments (halophytes) or not (glycophytes) use many conserved cellular and organismal processes to tolerate salt (Hasegawa et al., 2000; Zhu, 2003). Notwithstanding, some plants are salt avoiders, including those that use adaptive morphological structures such as glands or bladders on leaves as salt sinks (Adams et al., 1998; Tester & Davenport, 2003). However, even plants that possess such avoidance capabilities utilize conserved physiological processes once metabolically active cells are exposed to salt stress.

For some period, researchers have understood that crops have genetic variation for salt tolerance (Flowers et al., 1977; Epstein et al., 1980; Yeo & Flowers, 1986; Dvorak & Zhang, 1992; Flowers & Yeo, 1995; Munns, 2002; Flowers, 2004). Considerable emphasis over decades has been to dissect the critical mechanisms by which halophytes are able to tolerate extreme salinity (Flowers et al., 1977; Flowers & Yeo, 1995; Bohnert et al., 1995; Adams et al., 1998). This research characterized physiological, biochemical and molecular responses of several halophytes, including Salicornia, Sueada, Mesembryanthemum, etc., and determined many unique salt-adaptive features innate to these genotypes (Flowers et al., 1977, 1986; Bohnert et al., 1995; Adams et al., 1998; Bressan et al., 2001). Further insight came from the studies of wheat and tomato and their halophytic relatives (Dvorak & Zhang, 1992; Dubcovsky et al., 1994; Dvorak et al., 1994; Foolad et al., 2001, 2003). Research with tobacco cells and plants, and sorghum plants established that glycophytes could adapt to high levels of salinity, provided that stress imposition occurs in moderate increments (Amzallag et al., 1990; Hasegawa et al., 1994). Together, this research determined that all plants have some capacity to tolerate salt stress and many focal adaptation determinants are conserved in halophytes and glycophytes.

3.2.1   Arabidopsis thaliana as a model for glycophyte responses to salt stress

In the last decade there has been unprecedented progress in the identification of salt tolerance determinants, including components of signaling pathways that control plant responses to high salinity (Hasegawa *et al.*, 2000; Zhu, 2000, 2002, 2003; Bressan *et al.*, 2001; Xiong *et al.*, 2002; Xiong & Zhu, 2003; Shinozaki *et al.*, 2003). This progress is due primarily to the use of Arabidopsis as a molecular genetic model system in abiotic stress research, which has facilitated the identification of numerous salt adaptation determinants by loss- or gain-of-function experimental approaches (Kasuga *et al.*, 1999; Bressan *et al.*, 2001; Apse & Blumwald, 2002; Zhu, 2002, 2003). Arabidopsis will continue to be a model for dissection of plant salt stress responses being particularly pivotal for defining signal pathways networking and control node integration that coordinate determinant function.

Tomato is another glycophyte model that has been valuable for the dissection plant salt stress tolerance mechanisms because of its molecular genetic tractability and substantial genetic variation for adaptive capacity that is present in the primary gene pool (Borsani *et al.*, 2001a; Borsani *et al.*, 2002; Rubio *et al.*, 2004). Tomato is a widely distributed annual vegetable crop that is adapted to diverse climates. In spite of this, production is concentrated in a few warm and relatively dry areas where salinity is a serious problem (Szabolcs, 1994; Cuartero & Fernandez-Muñoz, 1999). For this reason, extensive physiological studies, particularly under field conditions, have been performed using tomato as a model crop plant to evaluate yield stability (Cuartero & Fernandez-Muñoz, 1999).

3.2.2   Thellungiella halophila (salt cress) – a halophyte molecular genetic model

Although all plants are able to tolerate some degree of salt stress, those indigenous to saline environments have substantially greater stress adaptation capacity. With the exception of specialized adaptations that are unique to some halophytes (Flowers *et al.*, 1977, 1986; Bohnert *et al.*, 1995; Adams *et al.*, 1998; Bressan *et al.*, 2001), there is only a rudimentary understanding of determinant function that mediates extreme salt tolerance of these plants. It is now time to dissect the integral processes of halophytism using a genetically facile plant model (Bressan *et al.*, 2001; Inan *et al.*, 2004; Taji *et al.*, 2004). Focal questions are: Do halophytes have unique salt tolerance determinants, 'better' alleles that encode more effective determinants, or regulatory cascades that endow these plants with greater capacity to mount a more effective defense in response to salt imposition?

Salt cress (*T. halophila*) is a halophyte – relative of Arabidopsis – that also exhibits a high degree of freezing tolerance (Bressan *et al.*, 2001; Inan *et al.*,

2004; Taji *et al.*, 2004). Salt cress is capable of reproduction after exposure to 500 mM NaCl or −15°C (Inan *et al.*, 2004). It is a tractable molecular genetic model because of its small plant size, prolific seed production, short life cycle, small genome (∼ 2X Arabidopsis) and ease of transformation (Bressan *et al.*, 2001). Furthermore, sequences in the genome (derived from cDNAs) average about 90% identity with Arabidopsis, which should facilitate orthology-based cloning of determinant genes (Inan *et al.*, 2004). Preliminary results also indicate that salt cress, like many halophytes, has substantially better water-use efficiency than crop plants (Lovelock & Ball, 2002). Therefore, it may be possible to use salt cress for the genetic dissection of water-use efficiency.

3.3 Plant adaptations to NaCl stress

Ion and osmotic homeostasis is necessary for plants to be salt tolerant. Intracellular ion homeostasis requires determinants that control toxic ion uptake and facilitate their compartmentalization into the vacuole (Niu *et al.*, 1995; Hasegawa *et al.*, 2000; Zhu, 2003). Since the vacuole is a focal compartment for cell expansion, ion accumulation in this organelle facilitates osmotic adjustment that drives growth with minimal deleterious impact on cytosolic and organellar machinery. Controlling ion uptake at the plasma membrane and vacuolar compartmentalization requires intracellular coordination of transport determinants, and the regulatory molecules and system(s) are beginning to be identified (Zhu, 2003). Osmotic homeostasis is accomplished by accumulation of compatible osmolytes in the cytosol for intracellular osmotic homeostasis (Hasegawa *et al.*, 2000). Processes that function to maintain ionic and osmotic homeostatic balance in a tissue and an organismal context are less deciphered. For example, root to shoot coordination that restricts ion movement to the aerial part of the plant minimizes salt load to the shoot meristem and metabolically active cells, but mechanisms involved are not well understood. Establishment of ionic and osmotic homeostasis after salt stress is essential not only to prevent cellular death but is also required for the physiological and biochemical steady states necessary for growth and completion of the life cycle (Bohnert *et al.*, 1995).

3.3.1 *Intracellular ion homeostatic processes*

Both Na^+ and Cl^- can be cytotoxic, however, primary focus has been the dissection of intracellular Na^+ homeostasis mechanisms (Blumwald *et al.*, 2000; Hasegawa *et al.*, 2000; Tester & Davenport, 2003; Zhu, 2003). Plant plasma membranes have an inside negative potential of between −120 and −200 mV (Niu *et al.*, 1995; Hirsch *et al.*, 1998; Borsani *et al.*, 2001a,

2001b; Rubio *et al.*, 2004). Thus, passive entry of Na^+ could concentrate the ion in the cytosol to $> 10^3$-fold relative to the apoplast (Niu *et al.*, 1995). Cytosolic Na^+ concentrations > 100 mM disturb the normal functioning of the cell (Serrano *et al.*, 1999), but plant cells can adapt to Na^+ concentrations < 100 mM in the cytosol (Flowers *et al.*, 1986; Binzel *et al.*, 1988). As indicated in Figure 3.1, intracellular homeostasis is achieved by tight coordinate control of Na^+ entry and compartmentalization of Na^+ and Cl^- into the vacuole, and there is some insight about how this regulation occurs (Niu *et al.*, 1995; Blumwald *et al.*, 2000; Hasegawa *et al.*, 2000; Tester & Davenport, 2003; Zhu, 2003).

3.3.1.1 Na^+ influx and efflux across the plasma membrane

Preventing or reducing net Na^+ entry into cells restricts accumulation of the ion in the cytosol, thereby increasing plant salt tolerance. Na^+ uptake across the plasma membrane has been attributed to low Na^+ permeability of systems that transport K^+ (Amtmann & Sanders, 1999; Maathuis & Amtmann, 1999; Blumwald *et al.*, 2000). Transport systems that have high affinity for K^+ but also low affinity for Na^+ include inward rectifying K^+ channels (KIRCs) like AKT1, outward rectifying K^+ channels (KORCs) and the KUP/HAK family of K^+/H^+ symporters (Maathuis & Amtmann, 1999; Blumwald *et al.*, 2000; Schachtman, 2000). However, there is no direct evidence of this function. The high-affinity K^+ transporter (HKT1), low-affinity cation transporter (LCT1) and nonselective cation channels are considered to be the most likely transport systems that mediate Na^+-specific cellular influx (Schachtman & Schroeder, 1994; Schachtman *et al.*, 1997; Amtmann & Sanders, 1999; Maathuis & Amtmann, 1999; Davenport & Tester, 2000; Amtmann *et al.*, 2001; Tester & Davenport, 2003; Zhu, 2003).

Energy-dependent Na^+ efflux is crucial to maintaining a low Na^+ concentration in the cytoplasm. Na^+ efflux occurs via a secondary active plasma membrane Na^+/H^+ antiporter that uses the H^+ motive force generated by the plasma membrane H^+-ATPase (Niu *et al.*, 1995; Blumwald *et al.*, 2000; Vitart *et al.*, 2001).

3.3.1.2 Na^+ and Cl^- compartmentalization into the vacuole

Compartmentalization of Na^+ and Cl^- into the vacuole is a salt-adaptation mechanism conserved in halophytes and glycophytes likely because vacuolar expansion is crucial for cell enlargement and it effectively reduces Na^+ and Cl^- toxicity in the cytosol (Blumwald *et al.*, 2000; Hasegawa *et al.*, 2000).

Na^+ and Cl^- are considered 'energy-efficient' osmolytes, as illustrated, with tomato seedlings for which growth is more sensitive to mannitol than to iso-osmotic concentrations of NaCl (Borsani *et al.*, 2002). Transport of Na^+ into the vacuole is mediated by cation/H^+ antiporters that are driven by the

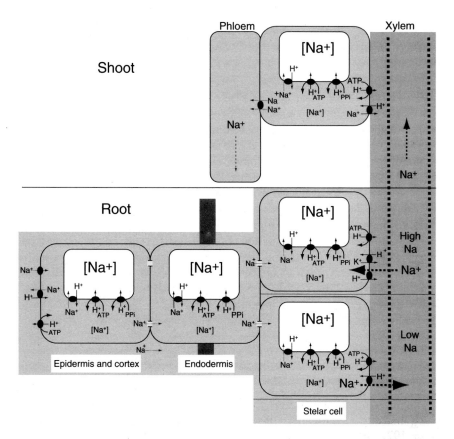

Figure 3.1 A model for Na$^+$ movement through the plant. Radial Na$^+$ transport from the soil solution across the root into the xylem has several control points that regulate the ion content that moved to the shoot as a byproduct of the transpirational flux required to maintain plant water status. Current evidence indicates that Na$^+$ enters the symplast of epidermal and other cells in the radial transport network through transport systems, like HKT1 and other yet unidentified proteins (multiple cortical layers cortex are represented as an individual cell). Continuation of symplastic transport is either via transport systems or plasmodesmata of interconnecting cells until the ion reaches the stele and is loaded into the xylem. Alternatively, movement of ions may occur through an extracellular matrix referred to as the apoplast. This pathway is restricted at the endodermis by the aqueous impermeable Casparian strip where movement occurs primarily through a symplastic mode. Cells that form the interconnecting network between the soil solution and the xylem reduce Na$^+$ load by functioning as sinks through vacuolar compartmentalization that is facilitated by *NHX*-type Na$^+$/H$^+$ antiporters; thereby reducing the total amount of Na$^+$ transported to the shoot. Depending on the apoplastic Na$^+$ concentration, the plasma membrane localized Na$^+$/H$^+$ antiporter SOS1 functions in loading the ion into (indicated as low Na$^+$) or retrieving it from (indicated as high Na$^+$) the xylem. Na$^+$ then moves into the shoot and eventually is transported across the plasma membrane of a leaf cell into the cytoplasm. Na$^+$ is then compartmentalized into the vacuole by NHX-type Na$^+$/H$^+$ antiporters or is transported back to the root by a phloem recirculation mechanism that is initiated by the action of HKT1 transport systems. Adapted from Epstein (1998).

H^+ electrochemical gradient generated by the vacuolar H^+-translocating enzymes, H^+-ATPase and H^+-pyrophosphatase (Blumwald *et al.*, 2000; Gaxiola *et al.*, 2001, 2002).

3.3.1.3 K^+/Na^+ selective accumulation

The capacity of plants to maintain intracellular K^+/Na^+ homeostasis is crucial for salt tolerance (Läuchli, 1990; Amtmann & Sanders, 1999; Maathuis & Amtmann, 1999; Hasegawa *et al.*, 2000; Zhu, 2002, 2003). How this is accomplished is still being determined but physiological and molecular genetic data reveal some of the focal mechanisms. Na^+, particularly at concentrations in saline soils, interferes with K^+ acquisition potentially creating a deficiency for this essential element. Na^+ competes with K^+ for intracellular uptake through both low- and high-affinity transport systems and Ca^{2+} facilitates K^+/Na^+ selective accumulation (Epstein, 1961, 1998; Läuchli, 1990; Niu *et al.*, 1995; Zhu, 2003; Rus *et al.*, 2004). Physiological evidence indicates that Ca^{2+} enhances K^+ over Na^+ uptake through a high-affinity system (Epstein, 1961, 1998). The Ca^{2+}-activated SOS signaling pathway regulates net Na^+ efflux across the plasma membrane by controlling the activity the SOS1 antiporter (Zhu, 2003).

3.3.2   Regulation of Na^+ homeostasis in roots and shoots

The fine-tuned control of net ion accumulation in the shoot involves precise *in planta* coordination between mechanisms that are intrinsically cellular with those that are operational at the intercellular, tissue or organ level (Flower *et al.*, 1977, 1986; Greenway & Munns, 1980; Hasegawa *et al.*, 2000; Munns, 2002; Munns *et al.*, 2002). Several processes are involved, including the regulation of Na^+ transport into the shoot, preferential Na^+ accumulation into the shoot cells that are metabolically not very active and the reduction of Na^+ content in the shoot by recirculation through the phloem back to the root (Munns, 2002; Berthomieu *et al.*, 2003). As previously mentioned, some halophytes use glands as salt sinks (Flowers *et al.*, 1977, 1986). These processes function to restrict cytotoxic ion accumulation into cells that are crucial for differentiation and development, or are necessary for carbon and nutrient assimilation (Johnson *et al.*, 1992; Läuchli, 1999).

Ions loaded into the root xylem are transported to the shoot largely by mass flow driven by the size of the transpirational sink (Läuchli, 1999; Flowers & Yeo, 1995; Munns, 2002; Munns *et al.*, 2002). A control response is to lower transpiration by a reduction in stomata aperture; however, this is only effective as a short-term response because plants need to maintain water status, carbon fixation and solute transport (Hasegawa *et al.*, 2000). Controlling ion load into the root xylem restricts accumulation in the shoot to a level where cells in this organ can be effective ion repositories by vacuolar compartmentalization (Hasegawa *et al.*, 2000; Munns, 2002; Munns *et al.*, 2002). Endodermal cells constitute a major control point in radial ion transport from the soil solution to the root xylem since the Casparian strip is an impermeable barrier to apoplastic solute movement (Azaizeh and Steudle, 1991; Cruz *et al.*, 1992; Niu *et al.*, 1995). However, bypass systems that function through 'leaks' in the Casparian strip barrier or movement through areas of the root where the specialized endodermal cells are not fully developed may be additional major entry points (Yadav *et al.*, 1996; Yeo *et al.*, 1999). Regardless, vacuolar compartmentalization in cells that form the interconnected network between the soil solution and the root xylem progressively lowers the content of ions that are entering the transpirational stream. It is presumed that *NHX*-like cation/H^+ transporters have a major function in this process (Blumwald *et al.*, 2000; Hasegawa *et al.*, 2000; Zhu, 2003).

Since Na^+ content is greater in older than in younger leaves, it is presumed that the former are ion sinks that restrict cytotoxic ion accumulation in cells that are crucial for development or are metabolically active (Greenway & Munns, 1980; Jeschke, 1984; Munns, 2002). Preferential accumulation of Na^+ in epidermal cells of barley leaves is attributed to increased activity of nonselective cation channels (Karley *et al.*, 2000). Despite the capacity of cells in the shoot to compartmentalize ions into the vacuole, this mechanism has finite capacity and is not operative in meristematic and other cells that are not fully expanded (Hasegawa *et al.*, 2000).

3.3.3   Sensing and regulatory pathways that control ion homeostasis

Osmotic and ionic stresses may be recognized by distinct sensors leading to the activation of signaling cascades that exert transcriptional and/or posttranscriptional control over determinants that mediate adaptation (Hasegawa *et al.*, 2000; Zhu, 2002, 2003; Shinozaki *et al.*, 2003). However, substantial complexity in operational function is envisaged because of overlapping stress effects, output diversity and pathway networking required to coordinate the processes necessary for an appropriate stress response (Hasegawa *et al.*, 2000; Zhu, 2002). Hyper- or hypo-osmolarity may cause ion channel gating, which leads to signal transduction pathway activation. Ca^{2+} has been long implicated as a decoder in stress signaling cascades (Knight *et al.*, 1997; Matsumoto *et al.*, 2002; Sanders *et al.*, 2002; Zhu, 2003; Hirschi, 2004). Alternatively, dissipation of the plasma membrane potential by electrophoretic ion flux could activate a signal cascade that controls osmotic or ion homeostasis (Navarre & Goffeau, 2000; Serrano & Rodriguez-Navarro, 2001). The two component histidine kinase AtHK1 functions as a phospho-relay system that controls ABA-independent or dependent activation stress tolerance determinants is a potential osmosensor (Shinozaki *et al.*, 2003).

3.3.4 Osmotic homeostasis: Compatible osmolytes

Osmotic balance in the cytoplasm is achieved by the accumulation of organic solutes that do not inhibit metabolic processes, called compatible osmolytes. These are sugars (mainly sucrose and fructose), sugars alcohols (glycerol, methylated inositols), complex sugars (trehalose, raffinose, fructans), ions (K^+), charged metabolites (glycine betaine) and amino acids such as proline (Hasegawa et al., 2000; Chen & Murata, 2002). The function of the compatible solutes is not limited to osmotic balance. Compatible solutes are typically hydrophilic, and may be able to replace water at the surface of proteins or membranes, thus acting as low molecular weight chaperones (Hasegawa et al., 2000). These solutes also function to protect cellular structures through scavenging ROS (Hasegawa et al., 2000; Zhu, 2001).

3.3.5 Damage response and antioxidant protection

Increased synthesis of several proteins in vegetative tissues occurs in response to salt stress. Most proteins are related to late embryogenesis-abundant (LEA) proteins (Ingram & Bartels, 1996), which are implicated to have properties similar to chaperones for the maintenance of protein structure during an osmotic stress episode (Ingram & Bartels 1996). Genetic evidence for the role of LEA proteins in salt and drought tolerance was provided from experiments based on the constitutive expression in rice of the barley *HVA1* gene (Xu et al., 1996). Transgenic rice plants showed increased tolerance to salt and mannitol relative to wild type (Xu et al., 1996).

In addition to protein miss-folding, salt stress causes an increase in ROS. Unless ROS are detoxified there is irreversible damage to membrane lipids, proteins and nucleic acids. Plants produce low molecular mass antioxidants such as ascorbic acid, reduced glutathione, tocopherol and carotenoids, and use detoxifying enzymes such as superoxide dismutases (SODs), catalases, ascorbate peroxidases, glutathione-*S*-transferases and glutathione peroxidases to scavenge ROS (Hasegawa et al., 2000; Apse & Blumwald, 2002; Chinnu-samy et al., 2004).

Evidence for the importance of ROS detoxification, first demonstrated based on gain-of-function experiments in tobacco (Tarczynski et al., 1993), was now confirmed by analysis of the Arabidopsis salt-tolerant mutant *pst1* (for *photoautotrophic salt tolerance1*) (Tsugane et al., 1999). The *pst1* plants are also more able to tolerate oxidative stress, however, tolerance is not attributable to proline, Na^+ or K^+ accumulation. Under salt stress, the *pst1* mutant has significantly higher SOD and ascorbate peroxidase activities than wild-type Arabidopsis, indicating that salt tolerance is linked to increased ROS scavenging. Additional support for the importance of ROS detoxification in salt tolerance comes from transgenic studies. Ectopic expression of a gene encoding a protein with both glutathione-S-transferase and glutathione peroxidase activity in tobacco plants improved salt stress tolerance due to enhanced ROS scavenging and prevention of membrane damage (Roxas *et al.* 1997, 2000). Likewise, salt tolerance of transgenic rice overexpressing the yeast Mn-SOD targeted to the chloroplast was increased by the reduction in O_2^- levels (Tanaka *et al.*, 1999). Transgenic tobacco plants expressing the *ANP1* gene were also tolerant to salt and other abiotic stresses (Kovtun *et al.*, 2000). *ANP1* encodes a MAPKKK, and transgenic plants with a constitutively active protein exhibited constitutively a signal through a MAPK cascade to increase the expression of the glutathione-S-transferase-6 gene. It is hypothesized that induction of antioxidant enzymes is controlled by MAPK signaling in response to osmotic and salt stress (Chinnusamy *et al.*, 2004).

3.4   Plant salt tolerance determinants identified by functional genetic approaches

One goal for many researchers has been the identification of genes responsible for salt tolerance, i.e. genetic determinants that can enhance yield stability of crops in saline conditions (Epstein *et al.*, 1980; Bohnert *et al.*, 1995; Hasegawa *et al.*, 2000; Apse & Blumwald, 2002). A common approach has been to identify genes based on expressional responses to salt stress. The great number of salt-regulated genes provides indication that plant salt adaptation is multi-faceted or, at least, the response to salt stress is very complex. However, even though many of these genes are regulated in response to salt imposition, direct functional evidence is required (Zhu *et al.*, 1997). Table 3.1 lists the genes that have been identified either by loss- or gain-of-function approaches as salt tolerance determinants. However, these approaches may not identify genes that are essential for growth and development that may be linked to yield stability, or have innate redundant function. Similarly, genes whose overexpression result in lethal or pleiotropic phenotypes may not be recognized as salt tolerance determinants.

Salt tolerance determinant genes can be categorized into two functional groups (Hasegawa *et al.*, 2000). The first includes genes that encode effectors, which are responsible for processes that are necessary for stress alleviation or adaptation. The second consists of regulatory genes that control expression and activity of these effectors (Hasegawa *et al.*, 2000). The effectors include proteins that function by protecting cells from high cytoplasmic Na^+, such as transporters, enzymes required for biosynthesis of various compatible osmolytes, LEA proteins, chaperones and detoxification enzymes. Regulatory molecules include transcription factors, and signaling intermediates like protein kinases and enzymes involved in phosphoinositide metabolism (Zhu, 2002).

Table 3.1 Salt tolerance determinants of plants identified by functional analysis

Name	Source species	Gene product	Function	Identification method	Harbourg species	Reference
Sodium influx						
AtHKT1	*A. thaliana*	Na^+ transporter	Na^+/K^+ homeostasis	Mutation	Arabidopsis	Rus *et al.* (2001)
HKT1	*T. aestivum*	Na^+/K^+ cotransporter	K^+/Na^+ homeostasis	Overexpression	Wheat	Laurie *et al.* (2002)
Sodium efflux						
SOS1	*A. thaliana*	Plasma membrane, Na^+/H^+ antiporter	Na^+ detoxification	Mutation, overexpression	Arabidopsis	Shi *et al.* (2000, 2003a, 2003b)
AHA4	*A. thaliana*	P-type H^+-ATPase	Plasma membrane energization	Mutation	Arabidopsis	Vitart *et al.* (2001)
Sodium compartmentation						
AtNHX1	*A. thaliana*	Vacuolar Na^+/H^+ antiporter	Na^+ vacuolar secuestration	Overexpression	Arabidopsis, cabbage, tomato	Apse *et al.* (1999), Zhang *et al.* (2001), Zhang and Blumwald (2001)
GhNHX1	*G. hirsutum*	Putative Na^+/H^+ vacuolar antiporter	Na^+ vacuolar secuestration	Overexpression	Tobacco	Wu *et al.* (2004)
AgNHX1	*A. gmelini*	Vacuolar Na^+/H^+ antiporter	Na^+ vacuolar secuestration	Overexpression	Rice	Ohta *et al.* (2002)
AVP1	*A. thaliana*	Vacuolar H^+-pyrophosphatase	H^+ transport, vacuolar acidification	Overexpression	Arabidopsis	Gaxiola *et al.* (2001)
Root/Shoot sodium distribution						
SOS1	*A. thaliana*	Plasma membrane, Na^+/H^+ antiporter	Xylem Na^+ loading/unloading	Mutation	Arabidopsis	Shi *et al.* (2002)
AtHKT1	*A. thaliana*	Na^+ transporter	Na^+ recirculation from shoot to root	Mutation	Arabidopsis	Mäser *et al.* (2002), Berthomieu *et al.* (2003)

Synthesis osmoprotectants

Gene	Species	Protein	Function	Method	Organism	Reference
P5CS	V. aconitifolia	Δ¹-pyrroline-5-carboxylate synthase	Proline synthesis	Overexpression	Tobacco, rice	Kishor et al. (1995)
ProDH	A. thaliana	Proline dehydrogenase	Proline degradation	Antisense	Arabidopsis	Nanjo et al. (1999)
BADH	S. oleracea, D. Carota	Betaine aldehyde dehydrogenase	Betaine synthesis	Overexpression	Tobacco, carrot	Li et al. (2003), Kumar et al. (2004)

ROS scavenging

Gene	Species	Protein	Function	Method	Organism	Reference
Nt107	N. tabacum	Gluthation-S-transferase/gluthation peroxidase	GSSG synthesis	Overexpression	Tobacco	Roxas et al. (1997)
GS2	O. sativa	Glutamine synthetase	Glutamine synthesis	Overexpression	Rice	Hoshida et al. (2000)
GlyI	B. juncea	Glyoxylase	Methylglyoxal detoxification	Overexpression	Tobacco	Veena and Sopory (1999)
GlyII	O. sativa	Glyoxylase	Methylglyoxal detoxification	Overexpression	Tobacco	Singla-Pareek et al. (2003)
AtNRTC	A. thaliana	NADPH thioredoxin reductase	ROS detoxification	Mutation	Arabidopsis	Serrato et al. (2004)
RCI3	A. thaliana	Cationic peroxidase	Antioxidant	Overexpression, antisense	Arabidopsis	Llorente et al. (2002)

Protection of cell integrity

Gene	Species	Protein	Function	Method	Organism	Reference
HVA1	H. vulgare	LEA protein	Protein protection	Overexpression	Rice	Xu et al. (1996)
TPX2	N. tabacum	Peroxidase	Cell wall biosynthesis	Overexpression	Tobacco	Amaya et al. (1999)
SOS5	A. thaliana	Putative cell surface adhesion protein	Cell expansion	Mutation	Arabidopsis	Shi et al. (2003a, 2003b)

(Continued)

Table 3.1 (Continued)

Name	Source species	Gene product	Function	Identification method	Harbourg species	Reference
Regulatory genes						
NPK1	N. tabacum	MAPKK kinase	Signaling	Overexpression	Tobacco	Kovtun et al. (2000)
MKK2	A. thaliana	MAPK kinase	Signaling	Mutation, overexpression	Arabidopsis	Tiege et al. (2004)
AtGSK1	A. thaliana	Shaggy-like kinase	Signaling	Overexpression	Arabidopsis	Piao et al. (2001)
SOS2	A. thaliana	Serine/threonine protein kinase	SOS1 regulator	Mutation	Arabidopsis	Liu et al. (2000)
SOS3	A. thaliana	Ca^{2+} binding protein	Ca^{2+} sensor/SOS2 activator	Mutation	Arabidopsis	Liu and Zhu (1997)
SOS4	A. thaliana	Piridoxal kinase	Vitamin B6 biosynthesis	Mutation	Arabidopsis	Shi et al. (2002)
OsCDPK7	O. sativa	Calcium-dependent protein kinase	Signaling	Overexpression	Rice	Saijo et al. (2000)
CBL1	A. thaliana	Calcineurin B	Ca^{+2} sensor	Mutation, overexpression	Arabidopsis	Cheong et al. (2003), Albrecht et al. (2003)
OsISAP1	O. sativa	Zinc finger protein	Gene regulation	Overexpression	Tobacco	Mukhopadhyay et al. (2004)
STO1	A. thaliana	Zinc finger protein	Gene regulation	Overexpression	Arabidopsis	Nagaoka et al. (2003)
DREB1A	A. thaliana, O. sativa	Transcription factor	Gene regulation	Overexpression	Arabidopsis, rice	Kasuga et al. (1999), Dubouzet et al. (2003)
Alfin1	M. sativa	Transcription factor	Gene regulation	Overexpression	Alfalfa	Winicov (2000)
CBF2	A. thaliana	Transcription factor	CBF1 and CBF3 regulation	Mutation	Arabidopsis	Novillo et al. (2004)
Tsi1	N. tabacum	EREBP/AP2-type transcription factor	Induction of pathogenesis related genes	Overexpression	Tobacco	Park et al. (2001)

Gene	Species	Protein	Function	Type	Organism	Reference
CpMyb10	C. plantaginerum	Transcription factor	Gene regulation	Overexpression	Arabidopsis	Villalobos et al. (2004)
AtMRP5	A. thaliana	ATP-binding cassette transporter	Putative sulfonylurea receptor	Mutation	Arabidopsis	Lee et al. (2004)
Other salt-tolerance determinants						
FRY1	A. thaliana	Inositol polyphosphate-1-phosphatase	Inositol-1,4,5-trisphosphate catabolism	Mutation	Arabidopsis	Xiong et al. (2001)
TUR1	S. polyrrhiza	D-myo-inositol-3-phosphate synthase	Inositol synthesis	Overexpression	Arabidopsis	Smart and Flores (1997)
PINO1	P. coarctata	L-myo-Inositol-1-phosphate synthase	Inositol synthesis	Overexpression	Tobacco	Majee et al. (2004)
IMT1	M. crystallinum	Myo-inositol-O-metyltransferase	D-ononitol synthesis	Overexpression	Tobacco	Sheveleva et al. (1997)
AtHAL3	A. thaliana	FMN-binding protein	Coenzyme A biosynthesis	Overexpression	Arabidopsis	Espinosa-Ruiz et al. (1999)
AtRAB7	A. thaliana	Rab GTPase	Vesicle trafficking regulation	Overexpression	Arabidopsis	Mazel et al. (2004)
STT3	A. thaliana	Subunit isoform of oligosaccharyl transferase	Protein N-glycosylation	Mutation	Arabidopsis	Koiwa et al. (2003)

3.4.1   Effector genes

3.4.1.1 Na^+ homeostasis

Plant physiologists have predicted for some time the different transport determinants that are necessary for the control of Na^+ and Cl^- homeostasis in saline environments (Flowers *et al.*, 1977, 1986; Niu *et al.*, 1995). Although all the transport systems necessary for ion homeostasis have not been verified by molecular genetic criteria, substantial insight has been gained about how Na^+ homeostasis is achieved, including the functions of the plasma membrane and tonoplast H^+ pumps and Na^+/H^+ antiporters, and a Na^+ influx system (Blumwald, 2003; Zhu, 2003). Emerging also is an understanding about how Na^+ homeostasis occurs *in planta* (Zhu *et al.*, 2003). Much less is known about how intracellular or organismal Cl^- homeostasis is achieved (Niu *et al.*, 1995).

Genetic and physiological evidence support the likelihood that HKT1 mediates Na^+ influx across the plasma membrane (Rus *et al.*, 2001; Laurie *et al.*, 2002). Involvement of *AtHKT1* in Na^+ homeostasis was established when dysfunctional alleles were determined to suppress Na^+-hypersensitivity of *sos3* plants, implicating AtHKT1 as a Na^+ influx system in plant roots (Rus *et al.*, 2001). Antisense wheat lines with reduced TaHKT1 activity exhibit improved salt tolerance and reduced Na^+ uptake into roots (Laurie *et al.*, 2002). The *hkt1* mutations result in higher Na^+ shoot content and lower Na^+ root content (Maser *et al.*, 2002; Berthomieu *et al.*, 2003; Rus *et al.*, 2004). Although it is clear that HKT proteins control Na^+ homeostasis *in planta*, all of its functions are yet to be determined. In fact, genetic evidence indicates that AtHKT1 is also a regulator of K^+ homeostasis (Rus *et al.*, 2001, 2004).

SOS1 was the first locus identified in screen for Arabidopsis salt hypersensitive mutants (Wu *et al.*, 1996) and determined to encode a Na^+/H^+ antiporter SOS1 that mediates energy-dependent Na^+ transport across the plasma membrane (Shi *et al.*, 2000; Qiu *et al.*, 2002, 2004; Quintero *et al.*, 2002). *SOS1* expression is upregulated by NaCl stress through a process that includes mRNA stability (Shi *et al.*, 2003a, 2003b). Transgenic Arabidopsis plants overexpressing *SOS1* gene displayed enhanced tolerance, confirming the importance of Na^+ extrusion for salt adaptation (Shi *et al.*, 2003a, 2003b).

SOS1 uses the H^+ electrochemical gradient across the plasma membrane produced by the P-type H^+-ATPases; AHA family in Arabidopsis (Palmgren, 2002; Baxter *et al.*, 2003). Confirmation for the role of the plasma membrane H^+ pump in salt tolerance was established using an Arabidopsis line with a T-DNA insertion in *AHA4*. Although *aha4-1* plants exhibited a slight reduction in root and shoot growth compared to wild-type plants, indicating the importance of the H^+ pump for normal plant growth, an Na^+ sensitive phenotype was clearly evident (Vitart *et al.*, 2001).

Vacuolar compartmentalization of Na^+ is mediated by *NHX* antiporters (Blumwald, 2003). The molecular identity of the AtNHX1 antiporter was deduced from database sequence information and function confirmed by genetic and physiological evidence (Apse *et al.*, 1999; Gaxiola *et al.*, 1999; Quintero *et al.*, 2000; Apse & Blumwald, 2002; Ohta *et al.*, 2002; Shi & Zhu, 2002; Blumwald, 2003). NHX antiporters in Arabidopsis are encoded by a gene family comprising of six members (Yokoi *et al.*, 2002; Aharon *et al.*, 2003). Gain- and loss-of-function evidence indicates that AtNHX1 mediates salt adaptation but the physiological roles of the other *NHX* antiporters remain to be determined (Apse *et al.*, 1999, 2003; Apse & Blumwald, 2002; Yokoi *et al.*, 2002; Blumwald, 2003). In addition to *AtNHX1*, overexpression of *NHX1* gene orthologs from other species also resulted in salt tolerance (Ohta *et al.*, 2002; Wu *et al.*, 2004).

Tonoplast-localized NHX antiporters utilize the H^+ electrochemical gradient generated by H^+ pumps, the V-type ATPase and pyrophosphatase. Gain-of-function evidence established that the Arabidopsis pyrophosphatase facilitates Na^+/H^+ necessary for vacuolar compartmentalization and salt tolerance (Gaxiola *et al.*, 2001).

Recent evidence implicates the plasma membrane Na^+/H^+ antiporter *SOS1* and HKT1 in the regulation of Na^+ distribution between the shoot and root (Figure 3.1). Under moderate salinity (25 mM NaCl), *SOS1* is likely functioning in the loading of Na^+ into the xylem for transport to the shoot where the ion is loaded into leaf cells (Shi *et al.*, 2002). However, under severe salt stress (100 mM NaCl), it is proposed that *SOS1* functions to restrict Na^+ accumulation in the shoot by limiting net Na^+ uptake at the root tip and loading into the xylem sap (Shi *et al.*, 2002). A model for *AtHKT1* function indicates the involvement in the redistribution of Na^+ from the shoot to the root, apparently by facilitating Na^+ loading into the phloem in the shoot and unloading in the root (Berthomieu *et al.*, 2003).

The Ca^{2+}-dependent *Salt Overly Sensitive* (SOS) signaling pathway in Arabidopsis is one of the most well-defined ion homeostasis cascades in any organism (Qiu *et al.*, 2002; Zhu, 2002; Shi *et al.*, 2003a, 2003b). SOS3 is an EF hand Ca^{2+} binding protein that is activated by an intracellular Ca^{2+} transient. The Ca^{2+} sensor SCaBP5 (SOS3 family) and its interacting kinase PKS3 (SOS2 family) are components of a regulatory circuit that negatively controls ABA-induced Ca^{2+} channel gating, which is presumed necessary for Ca^{2+} oscillations that activate the *SOS3* and other signaling pathways required for salt adaptation (Guo *et al.*, 2002; Zhu, 2002). *SOS3* activates the *SOS2* serine/threonine kinase and recruits it to the plasma membrane where *SOS2* phosphorylates *SOS1* activating the Na^+/H^+ antiporter (Qiu *et al.*, 2002; Quintero *et al.*, 2002; Zhu, 2003). The SOS pathway also induces *SOS1* expression, apparently by enhancing mRNA stability (Shi *et al.*, 2000, 2003a, 2003b; Zhu, 2003).

3.4.1.2   Genes involved in osmotic homeostasis: Synthesis of compatible solutes

As indicated previously, salt tolerance requires that compatible solutes accumulate in the cytosol and organelles where these function in osmotic adjustment and osmoprotection (Rhodes & Hanson, 1993). With exceptions like K^+, most compatible osmolytes are organic solutes. Genes that encode enzymes that catalyze the biosynthesis of compatible solutes enhance salt and/or drought tolerance in gain-of-function strategies (Borsani et al., 2003; Flowers, 2004). In particular, genes from microorganisms have been widely used (Hasegawa et al., 2000), but here the focus is plant genes.

Accumulation of proline is a common plant response to environmental stresses, and proline is implicated to play a role in the adaptive response to salt (Delauney & Verma, 1993). Higher plants synthesize proline via two different pathways. In one pathway, glutamate is converted to Δ^1-pyrroline-5-carboxylate (P5C) by Δ^1-pyrroline-5-carboxylate synthetase (P5CS); the enzyme that catalyzes the rate-limiting step in proline biosynthesis in response to hyperosmotic stress conditions (Szoke et al., 1992; Kishor et al., 1995). P5C is then reduced to proline by Δ^1-pyrroline-5-carboxylate reductase (P5CR). In the other pathway ornithine is converted to P5C through δ-transamination, which is subsequently reduced to proline. Strong positive correlation between P5CS transcript abundance and proline levels have been confirmed numerous times (Hare et al., 1999). Upon relief from hyperosmotic stress, proline is rapidly degraded by the sequential action of the mitochondrial enzymes proline dehydrogenase (ProDH) and P5C dehydrogenase. In these conditions, the expression of ProDH is induced and the expression of P5C inhibited (Verbruggen et al., 1996; Peng et al., 1996).

Increased proline content has been achieved by either increasing its synthesis or reducing its degradation using molecular genetic manipulation of P5CS and PDH (proline dehydrogenase) genes, respectively. Overexpression of a P5CS from mungbean in transgenic tobacco plants caused the accumulation of proline up to 18-fold over control plants resulting in an enhanced biomass production under salt stress (Kishor et al., 1995). Alternatively, antisense suppression of proline degradation improved tolerance to salinity (Nanjo et al., 1999). Salinity tolerance was measured as the capacity to maintain the leaf turgor in the presence of high NaCl (600 mM) (Nanjo et al., 1999).

3.4.1.3   Genes involved in ROS scavenging

Salinity generates an increase in ROS that induces deleterious effects on cell metabolism (Polle, 1997; Borsani et al., 2001a, 2001b). The role of SOD in salt stress protection was examined using transgenic rice plants

(Tanaka *et al.*, 1999). Total SOD activity was maintained at an elevated level and ascorbate peroxidase activity increased upon salt stress. It was determined that salt tolerance was correlated with higher PSII activity in chloroplasts of transgenic plants compared to the wild-type plants. These results support the likelihood that increased levels of ascorbate peroxidase and chloroplastic SOD are important factors for salt tolerance of rice (Tanaka *et al.*, 1999).

Transgenic tobacco seedlings overexpressing *Nt107* encoding a glutathione-*S*-transferase (GST)/glutathione peroxidase (GPX) had GST- and GPX-specific activities approximately twofold higher than the wild-type seedlings. The GSH/GSH + GSSG ratio was lower in the transgenic plants but despite the increased oxidized glutathione pool transgenic seedlings exhibited increased salt tolerance (Roxas *et al.*, 1997). Overexpression of glutathione reductase in transgenic plants led to the elevated levels of GSH increased tolerance to salt and oxidative stress of leaves (Foyer *et al.*, 1991). Therefore, the influence of the glutathione oxidation state on salt tolerance of plants is presently unclear.

An alternative strategy to cope with oxidative damage under salt stress is to suppress the production of ROS. Although a role for photorespiration is still controversial, it may function as a possible route for the dissipation of the excess of light energy or reducing power (Osmond & Grace, 1995). Several studies suggest that the rate-limiting step in photorespiration is the reassimilation of ammonia catalyzed by the chloroplastic glutamine synthetase (Wallsgrove *et al.*, 1987). Rice plants transformed with a chloroplastic glutamine synthetase (*GS2*) gene from rice (Hoshida *et al.*, 2000) accumulated about 1.5-fold more *GS2* than control plants, had increased photorespiratory capacity and enhanced salt tolerance (Hoshida *et al.*, 2000).

The glyoxalase system is proposed to be involved in processes such as protection against α-oxaldehyde cytotoxicity, regulation of cell division and proliferation, etc. (reviewed in Thornalley, 1993). Induction of *glyoxalase I* gene expression in response to salt and osmotic stress occurs in tomato (Espartero *et al.*, 1995). Transgenic plants overexpressing *glyoxalase I* exhibit increased tolerance to high salt, as measured in a detached leaf disc senescence assay (Veena & Sopory, 1999).

Thioredoxins are small 12–13 kDa proteins with a redox active disulfide bridge, and are widely distributed in all organisms (Schurmann & Jacquot, 2000). Due to their disulfide/dithiol interchange activity, thioredoxins interact with target proteins and are involved in the regulation of a large number of cellular processes (Balmer & Buchanan, 2002). Using the rice genome sequence, a novel chloroplastic NADPH thioredoxin reductase (*OsNTRC*) has been recently identified (Serrato *et al.*, 2004). Knock out mutations of the Arabidopsis ortholog *AtNTRC* resulted in plant hypersensitivity to methyl viologen, drought and salt stress (Serrato *et al.*, 2004).

3.4.1.4   Genes involved in protection of cell integrity

The *HVA1* gene encoding a LEA protein is induced by ABA and several stresses including salt (Hong *et al.*, 1992). Transgenic rice plants expressing *HVA1*, driven by the constitutive promoter from the rice *ACTIN1* gene, have increased tolerance to salt stress (Xu *et al.*, 1996). The mechanism involved in the action is not clear but it is proposed that the improved tolerance could be due to a stabilization of cell structure.

Cell wall composition is changed substantially in response to environmental changes. Because cell wall properties that affect water permeability and extensibility affect cell growth during salt stress (Iraki *et al.*, 1989), cell wall alterations may be crucial to stress tolerance. Overexpression in tobacco of a cell wall peroxidase from tomato (TPX2) significantly increased the germination rate under salt stress (Amaya *et al.*, 1999). The greater water-holding capacity of transgenic seeds may have resulted in a higher germination rate under conditions where the availability of water is limited (Amaya *et al.*, 1999).

Under salt stress, the root tips of *SOS5* mutant plants swell and root growth is arrested (Shi *et al.*, 2003a, 2003b). The root-swelling phenotype is caused by the abnormal expansion of epidermal, cortical and endodermal cells. The walls of cells are thinner in the *SOS5* plants, and those between neighboring epidermal and cortical cells in roots of *SOS5* plants are less organized. The *SOS5* gene encodes an arabinogalactan protein, which is a putative cell surface adhesion protein that is required for normal cell expansion (Shi *et al.*, 2003a, 2003b).

3.4.2   Regulatory genes

3.4.2.1   Kinases

Plant proteins homologous to those of the HOG pathway in yeast have been identified. A putative MAPK from *Pisum sativum* (PsMAPK), which is 47% identical to Hog1p, functionally complements the salt growth defect of yeast *hog1* (Popping *et al.*, 1996). Combinations of Arabidopsis proteins MEKK1 and MEK1, or MAPKKK and MAPKK, can functionally complement the salt growth defect of yeast *pbs2* (Ichimura *et al.*, 1998). ANP1, a MAPKKK from Arabidopsis is activated by H_2O_2 and in turn activates two stress-regulated MAPKs, AtMPK3 and AtMPK6 (Kotvun *et al.*, 2000). Overexpression of NPK1, the tobacco ANP1 ortholog, enhanced salt tolerance (Kotvun *et al.*, 2000). The Arabidopsis MAPK kinase 2 (MKK2) and the downstream MPK4 and MPK6 were isolated by functional complementation of osmosensitive yeast mutants. MKK2 is activated by cold, salt and by MEKK1 (Teige *et al.*, 2004). Overexpression of wild type or constitutively active MKK2 resulted in enhanced freezing and salt tolerance (Teige *et al.*, 2004). In contrast, Arabidopsis mutations in the *MKK2* gene resulted in hypersensitivity to salt and cold stresses (Teige *et al.*, 2004).

Functional complementation of the salt-sensitive calcineurin mutant of yeast with Arabidopsis genes identified AtGSK1, which encodes a GSK3/ shaggy-like protein kinase (Piao et al., 1999). NaCl and ABA but not KCl induces the expression of AtGSK1 and transgenic Arabidopsis plants over-expressing AtGSK1 exhibit enhanced NaCl tolerance (Piao et al., 2001).

The SOS2 gene encodes a serine/threonine protein kinase with an N-terminal catalytic domain similar to that of the yeast SNF1 kinase (Liu et al., 2000). A Ca^{2+} transient activates SOS3, which in turn interacts with SOS2 suppressing auto inhibition of the kinase activity. The SOS2/SOS3 kinase complex is targeted to the plasma membrane where SOS2 activates SOS1 by phosphorylation (Quintero et al., 2002).

The SOS4 gene encodes a pyridoxal (PL) kinase that is involved in the biosynthesis of PL-5-phosphate (PLP), an active form of vitamin B6 (Shi & Zhu, 2002b). Besides being a cofactor for many cellular enzymes, PLP is also a ligand that regulates the activity of certain ion transporters in animal cells. The authors propose that this property might be related to the salt tolerance function of SOS4 (Shi & Zhu, 2002b).

3.4.2.2   Transcription factors

The transcription factor DREB1A (Dehydration Response Element Binding) specifically interacts with the DRE (Dehydration Response Element) promoter sequences and induces expression of stress tolerance genes with DRE elements in the promoter. Overexpression of DREB1A in Arabidopsis under the control of the CaMV 35S promoter activated the expression of many stress tolerance genes but these plants exhibited morphological abnormalities in the absence of stress (Liu et al., 1998). In contrast, plants expressing DREB1A under the control of the hyperosmotically inducible rd29A promoter were normal and exhibited a high degree of tolerance to salt stress (Kasuga et al., 1999). Overexpression of the rice ortholog OsDREB1A in Arabidopsis-induced over-expression of target stress-inducible genes similar to Arabidopsis DREB1A and resulted in plants with greater tolerance to drought, high-salt and freezing stresses illustrating functional conservation between dicots and monocots (Dubouzet et al., 2003).

In contrast to DREB1A, normal growth was not adversely affected in transgenic alfalfa plants that were overexpressing the Alfin1 transcription factor under the control of the 35S promoter (Winicov, 2000). These plants had both enhanced expression of the salt-inducible MsPRP2 gene in roots and an increase in root growth under salt stress.

Tsi1 transcription factor of tobacco was identified as a salt-induced transcript using mRNA differential display analysis (Park et al., 2001). Tsi1 is sequence homologous to EREBP/AP2 transcription factors and binds specifically to the GCC and the DRE/CRT promoter sequences. Interestingly, overexpression of the *Tsi1*-gene-induced expression of several pathogenesis-related genes under normal conditions, resulting in improved tolerance to salt and pathogens. These results suggest that *Tsi1* might be involved in the regulation of both abiotic and biotic stress signal transduction pathways (Park *et al.*, 2001). Additional transcription factors related with salt tolerance have been recently reported (Lee *et al.*, 2004; Villalobos *et al.*, 2004).

3.4.2.3   Other salt tolerance determinants

The *HAL2* gene encodes a (2′), 5′-bisphosphate nucleotidase and inositol-1-polyphosphatase enzyme, which functions in the catabolism of inositol-1,4, 5-trisphosphate (IP$_3$) and is implicated to be a target for Li$^+$ and Na$^+$ toxicity in yeast (Murguia *et al.*, 1995). Evidence for the importance of this enzyme in salt tolerance came from the identification of the Arabidopsis *SAL1* by functional complementation of the yeast *ena1-4* strain, a yeast strain that lacks the major Na$^+$- and Li$^+$-extrusion system. *SAL1* is the plant ortholog of *HAL2* (Quintero *et al.*, 1996). The *SAL1* mutation is allelic to the *fry1* mutation, which results in super-induction of ABA- and stress-responsive genes (Xiong *et al.*, 2001). Seed germination and postembryonic development of *fry1* are more sensitive to ABA or hyperosmotic stress. The mutant plants are also compromised in tolerance to freezing, drought and salt stresses (Xiong *et al.*, 2001).

OSM1 (*osmotic stress-sensitive mutant*)/SYP61 is a member of the syntaxin family of Arabidopsis and loss-of-function experimentation has identified the protein as a salt and hyperosmotic stress tolerance determinant (Zhu *et al.*, 2002). This implicates substantial necessity for membrane trafficking docking in salt and hyperosmotic stress adaptation. It is not clear yet what transport vesicles or target membranes are involved.

Mutations in the Arabidopsis *STT3a* gene cause NaCl/osmotic sensitivity that is characterized by reduced cell division in the root meristem (Koiwa *et al.*, 2003). *STT3a* and *STT3b* encode an essential subunit of the oligosaccharyl transferase complex that is involved in *N*-glycosylation of proteins (Koiwa *et al.*, 2003). However, unlike *STT3a* alleles, dysfunctional *STT3b* is not NaCl/osmotic sensitive indicating that the members of the family have both overlapping and essential functions in plant growth and developmental but protein glycosylation by *STT3a* is required for proper protein folding and cell cycle progression during salt and osmotic stress recovery.

3.5   Global analysis of transcriptional activation of salt-responsive genes

Upon salt-stress imposition, the capacity of plants for rapid development of osmotic and ionic homeostasis, as well as damage repair, is necessary for survival. During this period of adaptation, new physiological and biochemical steady states are established in order for completion of the life cycle (Bohnert *et al.*, 1995; Hasegawa *et al.*, 2000). A premise for some time has been that many of the genes that are important for salt tolerance can be identified because their transcription regulation is a component of the salt-stress response.

ABA regulates many aspects of plant growth and development, some of which may contribute to stress adaptation. For example, synthesis of LEA proteins, that is necessary to preserve viability of embryos, is regulated by ABA (Phillips *et al.*, 1997). Abiotic stresses, including high-salinity-induce ABA accumulation (Xiong & Zhu, 2003). Turgor reduction induced by the hyperosmotic component of salt stress is probably most responsible for the increase in endogenous ABA. However, ABA may be induced, to some degree, by ionic stress at least in roots, where accumulation of the phytohormone is reported to be ten times higher than in shoots (Jia *et al.*, 2002). This accumulation of ABA occurs mainly through the induction of ABA biosynthetic genes although ABA degradation also seems to be suppressed (Xiong & Zhu, 2003). An increase in ABA content induces rapid changes in plants, including the promotion of stomatal closure, which minimizes transpirational water loss. At the cellular level ABA also regulates water balance by inducing genes involved in osmolyte biosynthesis (Hasegawa *et al.*, 2000; Shinozaki & Yamaguchi-Shinozaki, 2000; Zhu, 2002). Cellular changes also include the dehydration tolerance through the induction of LEA-like protein genes (Hasegawa *et al.*, 2000).

Insight about the complexity and the linkage of plant responses, perhaps adaptive processes, to salt, drought and cold stresses and ABA comes from the use of microarrays that allows the study of genome-wide expression profiling. An analysis of 7000 full-length Arabidopsis cDNAs determined that the abundance of 245 transcripts was ABA-induced by fivefold or more (Seki *et al.*, 2002a, 2002b). Among the ABA-inducible genes identified, 22 encoded transcription factors, confirming that important regulatory processes involving this phytohormone are based on transcriptional control. In the same report, gene expression profiling of plants was determined after treatment with salt, cold or drought stress (Seki *et al.*, 2002a, 2002b). Many ABA-inducible genes were upregulated after drought (155) or salinity (133) treatment, while fewer of these genes were induced by cold stress (25). These results indicate that there is substantial overlap of genes whose expression is induced by drought and salinity stresses, and ABA (Shinozaki & Yamaguchi-Shinozaki, 2000; Seki *et al.*, 2002a, 2002b, 2004). Similar results were recently obtained with rice (Rabbani *et al.*, 2003). An analysis using a rice cDNA microarray of 1700 independent cDNAs determined that 36, 62, 57 and 43 genes were upregulated by cold, drought, high salinity and ABA, respectively, and this much commonality in genes that are induced by drought and salinity stresses, and ABA (Rabbani *et al.*, 2003).

In rice, transcript changes in response to high salinity were investigated using microarray analysis (Kawasaki *et al.*, 2001). The microarray contained 1728 cDNAs from libraries of salt-stressed roots. In this study, a time course of gene expression was obtained between 15 min and 1 wk after salt stress in a salt-tolerant rice variety (Pokkali) and a salt-sensitive rice variety (IR29). Approximately, 10% of the transcripts in Pokkali were significantly upregulated or downregulated within 1 h of salt stress. As plants adapted to salt stress, the initial differences between controls and stressed plants became less pronounced as the plants adapted over time. This adaptation to salt stress based on gene expression was supported by the observation that IR29 exhibited delayed gene expression (Kawasaki *et al.*, 2001). Responses to drought and salinity in barley (*Hordeum vulgare* L. cv. Tokak) of 1463 cDNAs were monitored by microarray hybridization (Ozturk *et al.*, 2002). In this report, around 15% of genes were either up- or down-regulated under drought stress, while NaCl led to a change in 5% of the transcripts indicating less overlap that previously shown for rice and Arabidopsis.

Microarray analysis was also used to identify genes that are induced during rehydration after a dehydration stress episode (Oono *et al.*, 2003). These genes are likely to be implicated in the recovery process after drought or high salt stress, or are necessary for osmotic downshock recovery. A total of 152 rehydration-inducible genes were identified and classified into three major groups: genes encoding regulatory proteins; genes encoding functional proteins involved in the recovery process from dehydration-induced damage and genes encoding functional proteins involved in plant growth (Oono *et al.*, 2003). Kawasaki *et al.* (2001) proposed that genes implicated in the recovery process are those that continue to have elevated expression after 7 days of salt stress because plants at this stage have resumed faster growth and can be considered to be adapted.

Because of the high DNA sequence identity between Arabidopsis and *T. halophila* (Inan *et al.*, 2004), cDNA microarrays of Arabidopsis can be used for the expression profiling of salt cress genes (Taji *et al.*, 2004). Only six genes were strongly induced in response to high salt in salt cress, whereas in similar conditions 40 genes were induced in Arabidopsis (Taji *et al.*, 2004). Many genes that are induced in Arabidopsis by salt stress were constitutively expressed in salt cress, including *SOS1* and *AtP5CS*. Constitutive expression of *AtP5CS*, a key enzyme in proline biosynthesis, resulted in higher levels of this amino acid in salt cress, even in the absence of salt stress (Taji *et al.*, 2004). Stress tolerance of salt cress may be due, in part, to constitutive expression of salt tolerance determinant genes that must be induced by the stress induced in Arabidopsis.

References

Adams, P., Nelson, D. E., Yamada, S., Chmara, W., Jensen, R. G., Bohnert, H. J. & Griffiths, H. (1998) Tansley Review No. 97. Growth and development of *Mesembryanthemum crystallinum* (Aizoaceae). *New Phytologist*, **138**, 171–190.

Aharon, G. S., Apse, M. P., Duan, S. L., Hua, X. J. & Blumwald, E. (2003) Characterization of a family of vacuolar Na^+/H^+ antiporters in *Arabidopsis thaliana*. *Plant and Soil*, **253**, 245–256.

Albrecht, V., Weinl, S., Blazevic, D., D'Angelo, C., Batistic, O., Kolukisaoglu, U., Bock, R., Schulz, B., Harter, K. & Kudla, J. (2003) The calcium sensor CBL1 integrates plant responses to abiotic stresses. *Plant Journal*, **36**, 457–470.

Amaya, I., Botella, M. A., de la Calle, M., Medina, M. I., Heredia, A., Bressan, R. A., Hasegawa, P. M., Quesada, M. A. & Valpuesta, V. (1999) Improved germination under osmotic stress of tobacco plants overexpressing a cell wall peroxidase. *FEBS Letters*, **457**, 80–84.

Amtmann, A. & Sanders, D. (1999) Mechanisms of Na^+ uptake by plant cells. *Advances in Botanical Research*, **29**, 75–112.

Amtmann, A., Fischer, M., Marsh, E. L., Stefanovic, A., Sanders, D. & Schachtman, D. P. (2001) The wheat cDNA LCT1 generates hypersensitivity to sodium in a salt-sensitive yeast strain. *Plant Physiology*, **126**, 1061–1071.

Amzallag, G. N., Lerner, H. R. & Poljakoff-Mayber, A. (1990) Induction of increased salt tolerance in *Sorghum bicolor* by NaCl treatment. *Journal of Experimental Botany*, **41**, 29–34.

Apse, M. P. & Blumwald, E. (2002) Engineering salt tolerance in plants. *Current Opinion in Biotechnology*, **13**, 146–150.

Apse, M. P., Aharon, G. S., Snedden, W. A. & Blumwald, E. (1999) Salt tolerance conferred by over-expression of a vacuolar Na^+/H^+ antiport in Arabidopsis. *Science*, **285**, 1256–1258.

Apse, M. P., Sottosanto, J. B. & Blumwald, E. (2003) Vacuolar cation/H^+ exchange, ion homeostasis, and leaf development are altered in a T-DNA insertional mutant of AtNHX1, the Arabidopsis vacuolar Na^+/H^+ antiporter. *Plant Journal*, **36**, 229–239.

Azaizeh, H. & Steudle, E. (1991) Effects of salinity on water transport of excised maize (*Zea mays* L.) roots. *Plant Physiology*, **97**, 1136–1145.

Balmer, Y. & Buchanan, B. B. (2002) Yet another plant thioredoxin. *Trends in Plant Science*, **7**, 191–193.

Baxter, I., Tchiu, J., Sussman, M. R., Boutry, M., Palmgren, M. G., Gribskov, M., Harper, J. F. & Axelsen, K. B. (2003) Genomic comparison of P-type ATPase ion pumps in Arabidopsis and rice. *Plant Physiology*, **132**, 618–628.

Berthomieu, P., Conejero, G., Nublat, A., Brackenbury, W. J., Lambert, C., Savio. C., Uozumi, N., Oiki, S., Yamada, K., Cellier, F., Gosti, F., Simonneau, T., Essah, P. A., Tester, M., Very, A.-A., Sentenac, H. & Casse, F. (2003) Functional analysis of *AtHKT1* in *Arabidopsis* shows that Na^+ recirculation by the phloem is crucial for salt tolerance. *EMBO Journal*, **22**, 2004–2014.

Binzel, M. L., Hasegawa, P. M., Handa, A. K., & Bressan, R. A. (1985) Adaptation of tobacco cells to NaCl. *Plant Physiology*, **79**, 118–125.

Binzel, M. L., Hess, F. D., Bressan, R. A. & Hasegawa, P. M. (1988) Intracellular compartmentation of ions in salt-adapted tobacco cells. *Plant Physiology*, **86**, 607–614.

Blumwald, E. (2003) Engineering salt tolerance in plants. *Biotechnology and Genetic Engineering Reviews*, **20**, 261–275.

Blumwald, E., Aharon, G. S. & Apse, M. P. (2000) Sodium transport in plant cells. *Biochimica et Biophysica Acta*, **1465**, 140–151.

Bohnert, H. J., Nelson, D. E. & Jensen, R. G. (1995) Adaptations to environmental stresses. *Plant Cell*, **7**, 1099–1111.

Borsani, O., Cuartero, J., Fernandez, J. A., Valpuesta V. & Botella, M. A. (2001a) Identification of two loci in tomato reveals distinct mechanisms for salt tolerance. *Plant Cell*, **13**, 873–887.

Borsani, O., Cuartero, J., Valpuesta, V. & Botella, M. A. (2002) Tomato *tos1* mutation identifies a gene essential for osmotic tolerance and abscisic acid sensitivity. *Plant Journal*, **32**, 905–914.

Borsani O., Valpuesta, V. & Botella, M. A. (2001b) Evidence for a role of salicylic acid in the oxidative damage generated by NaCl and osmotic stress in *Arabidopsis* seedlings. *Plant Physiology*, **126**, 1024–1030.

Borsani, O., Valpuesta, V. & Botella, M. A. (2003) Developing salt tolerant plants in a new century: a molecular biology approach. *Plant Cell Tissue and Organ Culture*, **73**, 101–115.

Bressan, R. A., Nelson, D. E., Iraki, N. M., LaRosa, P. C., Singh, N. K., Hasegawa, P. M. & Carpita, N. C. (1990) Reduced cell expansion and changes in cell walls of plant cells adapted to NaCl. In F. Kattermann (ed.) *Environmental Injury to Plants*. Academic Press, New York, pp. 137–171.

Bressan, R. A., Zhang, C., Zhang, H., Hasegawa, P. M., Bohnert, H. J. & Zhu, J. K. (2001) Learning from the Arabidopsis experience. The next gene search paradigm. *Plant Physiology*, **127**, 1354–1360.

Chen, T. H. & Murata, N. (2002) Enhancement of tolerance of abiotic stress by metabolic engineering of betaines and other compatible solutes. *Current Opinion in Plant Biology*, **5**, 250–257.

Cheong, Y. H., Kim, K. N., Pandey, G. K., Gupta, R., Grant, J. J. & Luan, S. (2003) CBL1, a calcium sensor that differentially regulates salt, drought, and cold responses in Arabidopsis. *Plant Cell*, **15**, 1833–1845.

Chinnusamy, V., Schumaker, K. & Zhu, J. K. (2004) Molecular genetic perspectives on cross-talk and specificity in abiotic stress signalling in plants. *Journal of Experimental Botany*, **55**, 225–236.

Cruz, R. T., Jordan, W. R. & Drew, M. C. (1992) Structural changes and associated reduction of hydraulic conductance in roots of *Sorghum bicolor* L. following exposure to water deficit. *Plant Physiology*, **99**, 203–212.

Cuartero, J. & Fernandez-Muñoz, R. (1999) Tomato and salinity. *Scientia Horticulturae*, **78**, 83–125.

Cuartero, J., Yeo, A. R. & Flowers, T. J. (1992) Selection of donors for salt-tolerance in tomato using physiological traits. *New Phytology*, **121**, 63–69.

Davenport, R. J. & Tester, M. (2000) A weakly voltage-dependent, nonselective cation channel mediates toxic sodium influx in wheat. *Plant Physiology*, **122**, 823–834.

Delaury, A. J. & Verma, D. P. S. (1993) Proline biosynthesis and osmoregulation in plants. *Plant Journal*, **4**, 215–223.

Dubcovsky, J., Galvez, A. F., & Dvorak, J. (1994) Comparison of the genetic organization of the early salt stress-responsive gene system in salt-tolerant *Lophopyrum elongatum* and salt-sensitive wheat. *Theoretical and Applied Genetics*, **87**, 957–964.

Dubouzet, J. G., Sakuma, Y., Ito, Y., Kasuga, M., Dubouzet, E. G., Miura, S., Seki, M., Shinozaki, K. & Yamaguchi-Shinozaki, K. (2003) OsDREB genes in rice, *Oryza sativa* L., encode transcription activators that function in drought-, high-salt- and cold-responsive gene expression. *Plant Journal*, **33**, 751–763.

Dvorak, J. & Zhang, H. B. (1992) Application of molecular tools for study of the phylogeny of diploid and polyploidy taxa in Triticeae. *Hereditas*, **116**, 37–42.

Dvorak, J., Noaman, M. M., Goyal, S. & Gorham, J. (1994) Enhancement of the salt tolerance of *Triticum turgidum* by the kna1 locus transferred from the *Triticum aestivum* chromosome 44 by homoeologous recombination. *Theoretical and Applied Genetics*, **87**, 872–877.

Epstein, E. (1961) The essential role of calcium in selective cation transport by plant cells. *Plant Physiology*, **36**, 437–444.

Epstein, E. (1998) How calcium enhances plant salt tolerance. *Science*, **280**, 1906–1907.

Epstein, E., Norlyn, J. D., Rush, D. W., Kingsbury, R., Kelly, D. B. & Wrana, A. F. (1980) Saline culture of crops: a genetic approach. *Science*, **210**, 399–404.

Espartero, J., Sanchez-Aguayo, I. & Pardo, J. M. (1995) Molecular characterization of glyoxalase-I from a higher plant; upregulation by stress. *Plant Molecular Biology*, **29**, 1223–1233.

Espinosa-Ruiz, A., Belles, J. M., Serrano, R. & Culianez-MacIa, F. A. (1999) *Arabidopsis thaliana* AtHAL3: a flavoprotein related to salt and osmotic tolerance and plant growth. *Plant Journal*, **20**, 529–539.

Flowers, T. J. (2004) Improving crop salt tolerance. *Journal of Experimental Botany*, **55**, 307–319.

Flowers, T. J. & Yeo, A. R. (1995) Breeding for salinity resistance in crop plants: where next? *Australian Journal of Plant Physiology*, **22**, 875–884.

Flowers, T. J., Hajibagheri, M. A. & Clipson, N. J. W. (1986) Halophytes. *Quarterly Review of Biology*, **61**, 313–337.

Flowers, T. J., Koyama, M. L., Flowers, S. A., Sudhakar, C., Singh, K. P. & Yeo, A. R. (2000) QTL: their place in engineering tolerance of rice to salinity. *Journal of Experimental Botany*, **51**, 99–106.

Flowers, T. J., Troke, P. F., & Yeo, A. R. (1977) The mechanism of salt tolerance in halophytes. *Annual Review of Plant Physiology*, **28**, 89–121.

Foolad, M. R. & Lin, G. Y. (1997) Genetic potential for salt tolerance during germination in Lycopersicon species. *Hortscience*, **32**, 296–300.

Foolad, M. R., Zhang, L. P. & Lin, G. Y. (2001) Identification and validation of QTLs for salt tolerance during vegetative growth in tomato by selective genotyping. *Genome*, **44**, 444–454.

Foolad, M. R., Zhang, L. P. & Subbiah, P. (2003) Genetics of drought tolerance during seed germination in tomato: inheritance and QTL mapping. *Genome*, **46**, 536–545.

Foyer, C., Lelandais, M., Galap, C. & Kunert, K. (1991) Effects of elevated cytosolic glutathione reductase activity on cellular glutathione pool and photosynthesis in leaves under normal and stress conditions. *Plant Physiology*, **97**, 863–872.

Gaxiola, R. A., Fink, G. R. & Hirschi, K. D. (2002) Genetic manipulation of vacuolar proton pumps and transporters. *Plant Physiology*, **129**, 967–973.

Gaxiola, R. A., Li, J., Undurraga, S., Dang, L. M., Allen, G. J., Alper, S. L. & Fink, G. R. (2001) Drought- and salt-tolerant plants result from overexpression of the AVP1 H^+-pump. *Proceedings of the National Academy of Sciences of the United States of America*, **98**, 11444–11449.

Gaxiola, R. A., Rao, R., Sherman, A., Grisafi, P., Alper, S. L. & Fink, G. R. (1999) The *Arabidopsis thaliana* proton transporters, AtNHX1 and Avp1, can function in cation detoxification in yeast. *Proceedings of the National Academy of Sciences of the United States of America*, **96**, 1480–1485.

Ghassemi, F., Jakeman, A. J. & Nix, H. A. (1995) *Salinisation of Land and Water Resources Human Causes Extent Management and Case Studies*. CABI Publishing, Wallingford, Oxon, p. 526.

Greenway, H. & Munns, R. (1980) Mechanisms of salt tolerance in nonhalophytes. *Annual Review of Plant Physiology*, **31**, 149–190.

Guo, Y., Xiong, L., Song, C. P., Gong, D., Halfter, U. & Zhu, J. K. (2002) A calcium sensor and its interacting protein kinase are global regulators of abscisic acid signaling in Arabidopsis. *Developmental Cell*, **3**, 233–244.

Hare, P. D., Cress, W. A., & van Staden, J. (1999) Proline synthesis and degradation: a model system for elucidating stress-related signal transduction. *Journal of Experimental Botany*, **50**, 413–434.

Hasegawa, P. M., Bressan, R. A., Nelsen, D. E., Samaras, Y. & Rhodes, D. (1994) Tissue culture in the improvement of salt tolerance in plants. In A. R. Yeo & T. J. Flowers (eds) *Soil Mineral Stresses. Approaches to Crop Improvement. Monographs on Theoretical and Applied Genetics*, Vol. 21. Springer-Verlag, Berlin, pp. 83–125.

Hasegawa, P. M., Bressan, R. A., Zhu, J.-K. & Bohnert, H. J. (2000) Plant cellular and molecular responses to high salinity. *Annual Review of Plant Physiology and Plant Molecular Biology*, **51**, 463–499.

Hirschi, K. (2004) The calcium conundrum. Both versatile nutrient and specific signal. *Plant Physiology*, **136**, 2438–2442.

Hirsch, R. E., Lewis, B. D., Spalding, E. P. & Sussman, M. R. (1998) A role for the AKT1 potassium channel in plant nutrition. *Science*, **280**, 918–921.

Hong, B., Barg, R. & Ho, T. H. (1992) Developmental and organ-specific expression of an ABA- and stress-induced protein in barley. *Plant Molecular Biology*, **18**, 663–674.

Hoshida, H., Tanaka, Y., Hibino, T., Hayashi, Y., Tanaka, A., Takabe, T. & Takabe, T. (2000) Enhanced tolerance to salt stress in transgenic rice that overexpresses chloroplast glutamine synthetase. *Plant Molecular Biology*, **43**, 103–111.

Ichimura, K., Mizoguchi, T., Irie, K., Morris, P., Giraudat, J., Matsumoto, K. & Shinozaki, K. (1998) Isolation of ATMEKK1 (a MAP kinase kinase kinase)-interacting proteins and analysis of a MAP kinase cascade in *Arabidopsis*. *Biochemical and Biophysical Research Communications*, **253**, 532–543.

Inan, G., Zhang, Q., Li, P., Wang, Z., Cao, Z., Zhang, H., Zhang, C., Quist, T. M., Goodwin, S. M., Zhu, J., Shi, H., Damsz, B., Charbaji, T., Gong, Q., Ma, S., Fredricksen, M., Galbraith, D. W., Jenks, M. A., Rhodes, D., Hasegawa, P. M., Bohnert, H. J., Joly, R. J., Bressan, R. A. & Zhu, J.-K. (2004) Salt cress. A halophyte and cryophyte Arabidopsis relative model system and its applicability to molecular genetic analyses of growth and development of extremophiles. *Plant Physiology*, **135**, 1718–1737.

Ingram, J. & Bartels, D. (1996) The molecular basis of dehydration tolerance in plants. *Annual Review of Plant Physiology and Plant Molecular Biology*, **47**, 377–403.

Iraki, N., Bressan, R. Hasegawa, P. & Carpita, N. (1989) Alteration of physical and chemical structure of the primary cell wall of growth-limited plant cells adapted to osmotic stress. *Plant Physiology*, **91**, 29–47.

Jeschke, W. D. (1984) K^+–Na^+ exchange at cellular membranes, intracellular compartmentation of cations and salt tolerance. In R. C. Staples & G. H. Toenniessen (eds) *Salinity Tolerance in Plants: Strategies for Crop Improvement*. Wiley & Sons, New York, pp. 37–66.

Jia, W., Wang, Y., Zhang, S. & Zhang, J. (2002) Salt-stress-induced ABA accumulation is more sensitively triggered in roots than in shoots. *Journal of Experimental Botany*, **53**, 2201–2206.

Johnson, D., Smith, S. & Dobrenz, A. (1992) Genetic and phenotypic relationships in response to NaCl at different developmental stages in alfalfa. *Theoretical and Applied Genetics*, **83**, 833–838.

Jones, R. A. (1986) High salt tolerance potential in *Lycopersicon* species during germination. *Euphytica*, **35**, 575–582.

Karley, A. J., Leigh, R. A. & Sanders, D. (2000) Differential ion accumulation and ion fluxes in the mesophyll and epidermis of barley. *Plant Physiology*, **122**, 835–844.

Kasuga, M., Liu, Q., Miura, S., Yamaguchi-Shinozaki, K. & Shinozaki, K. (1999) Improving plant drought, salt, and freezing tolerance by gene transfer of a single stress-inducible transcription factor. *Nature Biotechnology*, **17**, 287–291.

Kawasaki, S., Borchert, C., Deyholos, M., Wang, H., Brazille, S., Kawai, K., Galbraith, D. & Bohnert, H. J. (2001) Gene expression profiles during the initial phase of salt stress in rice. *Plant Cell*, **13**, 889–905.

Kinrade, T. B. (1998) Three mechanisms for the calcium alleviation of mineral toxicities. *Plant Physiology*, **118**, 513–520.

Kinrade, T. B. (1999) Interactions among Ca^{2+}, Na^+ and K^+ in salinity toxicity: quantitative resolution of multiple toxic and ameliorate effects. *Journal of Experimental Botany*, **50**, 1495–1505.

Kishor, P., Hong, Z., Miao, G. H., Hu, C. & Verma, D. (1995) Overexpression of [delta]-pyrroline-5-carboxylate synthetase increases proline production and confers osmotolerance in transgenic plants. *Plant Physiology*, **108**, 1387–1394.

Knight, H., Trewavas, A. J. & Knight, M. R. (1997) Calcium signaling in *Arabidopsis thaliana* responding to drought and salinity. *Plant Journal*, **12**, 1067–1078.

Koiwa, H., Li, F., McCully, M. G., Mendoza, I., Koizumi, N., Manabe, Y., Nakagawa, Y., Zhu, J., Rus, A., Pardo, J. M., Bressan, R. A. & Hasegawa, P. M. (2003) The STT3a subunit isoform of the Arabidopsis oligosaccharyl transferase controls adaptive responses to salt/osmotic stress. *Plant Cell*, **15**, 2273–2284.

Kovtun, Y., Chiu, W. L., Tena, G. & Sheen, J. (2000) Functional analysis of oxidative stress-activated mitogen-activated protein kinase cascade in plants. *Proceedings of the National Academy of Sciences of the United States America*, **97**, 2940–2945.

Koyama, M. L., Levesley, A., Koebner, R. M., Flowers, T. J. & Yeo, A. R. (2001) Quantitative trait loci for component physiological traits determining salt tolerance in rice. *Plant Physiology*, **125**, 406–422.

Kumar, S., Dhingra, A. & Daniell, H. (2004) Plastid-expressed betaine aldehyde dehydrogenase gene in carrot cultured cells, roots, and leaves confers enhanced salt tolerance. *Plant Physiology*, **136,** 2843–2854.

Läuchli, A. (1990) Calcium, salinity and plasma membrane. In R. T. Leonard & P. K. Kepler (eds) *Calcium in Plant Growth and Development. The American Society of Plant Physiologists Symposium Series*, Vol. 4. American Society of Plant Physiologists, Rockville, MD, pp. 26–35.

Läuchli, A. (1999) Salinity – potassium interactions in crop plants. In D. M. Oosterhuis & G. A. Berkowitz (eds) *Frontiers in Potassium Nutrition: New Perspectives on the Effects of Potassium on Physiology of Plants*. Potash & Phosphate Institute, Norcross, GA, pp. 71–76.

Läuchli, A. (1990) Calcium, salinity and plasma membrane. In R. T. Leonald & P. K. Hepler (eds) *Calcium in Plant Growth and Development. The American Society of Plant Physiologists Symposium Series*, Vol. 4. American Society of Plant Physiologists, Rockville, MD, pp. 26–35.

Laurie, S., Feeney, K. A., Maathuis, F. J., Heard, P. J., Brown, S. J. & Leigh, R. A. (2002) A role for HKT1 in sodium uptake by wheat roots. *Plant Journal*, **32**, 139–149.

Lee, E. K., Kwon, M., Ko, J. H., Yi, H., Hwang, M. G., Chang, S. & Cho, M. H. (2004) Binding of sulfonylurea by AtMRP5, an Arabidopsis multidrug resistance-related protein that functions in salt tolerance. *Plant Physiology*, **134**, 528–538.

Li, Q. L., Gao, X. R., Yu, X. H., Wang, X. Z. & An, L. J. (2003) Molecular cloning and characterization of betaine aldehyde dehydrogenase gene from *Suaeda liaotungensis* and its use in improved tolerance to salinity in transgenic tobacco. *Biotechnology Letters*, **25**, 1431–1436.

Liu, J. & Zhu, J. K. (1997) An arabidopsis mutant that requires increased calcium for potassium nutrition and salt tolerance *Proceedings of the National Academy of Sciences of the United States of America*, **94**, 14960–14964.

Liu, J. & Zhu, J. K. (1998) A calcium sensor homolog required for plant salt tolerance. *Science*, **280**, 1943–1945.

Liu, J., Ishitani, M., Halfter, U., Kim, C. S. & Zhu, J. K. (2000) The *Arabidopsis thaliana* SOS2 gene encodes a protein kinase that is required for salt tolerance. *Proceedings of the National Academy of Sciences of the United States of America*, **97**, 3730–3734.

Liu, Q., Kasuga, M., Sakuma, Y., Abe, H., Miura, S., Yamaguchi-Shinozaki, K. & Shinozaki, K. (1998) Two transcription factors, DREB1 and DREB2, with an EREBP/AP2 DNA binding domain separate two cellular signal transduction pathways in drought- and low temperature-responsive gene expression, respectively, in Arabidopsis. *Plant Cell*, **10**, 1391–1406.

Llorente, F., Lopez-Cobollo, R. M., Catala, R., Martinez-Zapater, J. M. & Salinas, J. (2002) A novel cold-inducible gene from Arabidopsis, RCI3, encodes a peroxidase that constitutes a component for stress tolerance. *Plant Journal*, **32**, 13–24.

Lovelock, C. E. & Ball, M. C. (2002) Influence of salinity on photosynthesis of halophytes. In A. Läuchli & U. Lüttge (eds) *Salinity: Environment–Plants–Molecules*. Kluwer Academic Publishers, Dordrecht, The Netherlands, pp. 315–339.

Maathuis, F. J. M. & Amtmann, A. (1999) K^+ nutrition and Na^+ toxicity: the basis of cellular K^+/Na^+ ratios. *Annals of Botany*, **84**, 123–133.

Majee, M., Maitra, S., Dastidar, K. G., Pattnaik, S., Chatterjee, A., Hait, N. C., Das, K. P. & Majumder, A. L. (2004) A novel salt-tolerant L-myo-inositol-1-phosphate synthase from *Porteresia coarctata* (Roxb.) Tateoka, a halophytic wild rice: molecular cloning, bacterial overexpression, characterization, and functional introgression into tobacco-conferring salt tolerance phenotype. *Journal of Biological Chemistry*, **279**, 28539–28552.

Maser, P., Eckelman, B., Vaidyanathan, R., Horie, T., Fairbairn, D. J., Kubo, M., Yamagami, M., Yamaguchi, K., Nishimura, M., Uozumi, N., Robertson, W., Sussman, M. R. & Schroeder, J. I. (2002) Altered shoot/root Na^+ distribution and bifurcating salt sensitivity in Arabidopsis by genetic disruption of the Na^+ transporter AtHKT1. *FEBS Letters*, **531**(2), 157–161.

Matsumoto, T. K., Ellsmore, A. J., Cessna, S. G., Low, P. S., Pardo, J. M., Bressan, R. A. & Hasegawa, P. M. (2002) An osmotically induced cytosolic Ca^{2+} transient activates calcineurin signaling to mediate ion homeostasis and salt tolerance of *Saccharomyces cerevisiae*. *Journal of Biological Chemistry*, **277**, 33075–33308.

Mazel, A., Leshem, Y., Tiwari, B. S. & Levine, A. (2004) Induction of salt and osmotic stress tolerance by overexpression of an intracellular vesicle trafficking protein AtRab7 (AtRabG3e). *Plant Physiology*, **134**, 118–128.

Moran, J. F., Becana, M., Iturbe-Ormatexe, I., Fechilla, S., Klucas, R. V. & Aparicio-Tejo, P. (1994) Drought induces oxidative stress in pea plants. *Planta*, **194**, 346–352.

Mukhopadhyay, A., Vij, S. & Tyagi, A. K. (2004) Overexpression of a zinc-finger protein gene from rice confers tolerance to cold, dehydration, and salt stress in transgenic tobacco. *Proceedings of the National Academy of Sciences of the United States of America*, **101**, 6309–6314.

Munns, R. (2002) Comparative physiology of salt and water stress. *Plant, Cell and Environment*, **25**, 239–250.

Munns, R. & Termaat, A. (1986) Whole-plant responses to salinity. *Australian Journal of Plant Physiology*, **13**, 143–160.

Munns, R., Husain, S., Rivelli, A. R., James, R., Condon, A. G., Lindsay, M., Lagudah, E., Schachtman, D. & Hare, R. (2002) Avenues for increasing salt tolerance of crops, and the role of physiologically-based selection traits. *Plant Soil*, **247**, 93–105.

Murguia, J. R., Belles J. M. & Serrano, R. (1995) A salt-sensitive 3'(2'),5'-bisphosphate nucleotidase involved in sulfate activation. *Science*, **267**, 232–234.

Nagaoka, S. & Takano, T. (2003) Salt tolerance-related protein STO binds to a Myb transcription factor homologue and confers salt tolerance in Arabidopsis. *Journal of Experimental Botany*, **54**, 2231–2237.

Nanjo, T., Kobayashi, M., Yoshiba, Y., Kakubari, Y., Yamaguchi-Shinozaki, K. & Shinozaki, K. (1999) Antisense suppression of proline degradation improves tolerance to freezing and salinity in *Arabidopsis thaliana*. *FEBS Letters*, **461**, 205–210.

Navarre, C. & Goffeau, A. (2000) Membrane hyperpolarization induced by deletion of PMP3, a highly conserved small protein of yeast plasma membrane. *EMBO Journal*, **19**, 2515–2524.

Niu, X., Bressan, R. A., Hasegawa, P. M. & Pardo, J. M. (1995) Ion homeostasis in NaCl stress environments. *Plant Physiology*, **109**, 735–742.

Novillo, F., Alonso, J. M., Ecker, J. R. & Salinas, J. (2004) CBF2/DREB1C is a negative regulator of CBF1/DREB1B and CBF3/DREB1A expression and plays a central role in stress tolerance in Arabidopsis. *Proceedings of the National Academy of Sciences of the United States of America*, **101**, 3985–3990.

Ohta, M., Hayashi, Y., Nakashima, A., Hamada, A., Tanaka, A., Nakamura, T. & Hayakawa, T. (2002) Introduction of a Na^+/H^+ antiporter gene from Atriplex gemelini confers salt tolerance to rice. *FEBS Letters*, **532**, 279–282.

Oono, Y., Seki, M., Nanjo, T., Narusaka, M., Fujita, M., Satoh, R., Satou, M., Sakurai, T., Ishida, J., Akiyama, K., Iida, K., Maruyama, K., Satoh, S., Yamaguchi-Shinozaki, K. & Shinozaki, K. (2003) Monitoring expression profiles of Arabidopsis gene expression during rehydration process after dehydration using ca. 7000 full-length cDNA microarray. *Plant Journal*, **34**, 868–887.

Osmond, C. B. & Grace, S. C. (1995) Perspectives on photoinhibition and photorespiration in the field: quintessential inefficiencies of the light and dark reactions of photosynthesis? *Journal of Experimental Botany*, **46**, 1351–1362.

Ozturk, Z. N., Talame, V., Deyholos, M., Michalowski, C. B., Galbraith, D. W., Gozukirmizi, N., Tuberosa, R. & Bohnert, H. J. (2002) Monitoring large-scale changes in transcript abundance in drought- and salt-stressed barley. *Plant Molecular Biology*, **48**, 551–573.

Palmgren, M. G. (2002) Plant plasma membrane H^+-ATPases: powerhouses for nutrient uptake. *Annual Review of Plant Physiology and Plant Molecular Biology*, **52**, 817–845.

Park, J. M., Park, C. J., Lee, S. B., Ham, B. K., Shin, R. & Paek, K. H. (2001) Overexpression of the tobacco Tsi1 gene encoding an EREBP/AP2-type transcription factor enhances resistance against pathogen attack and osmotic stress in tobacco. *Plant Cell*, **13**, 1035–1046.

Peng, Z., Lu, Q. & Verma, D. P. (1996) Reciprocal regulation of delta 1-pyrroline-5-carboxylate synthetase and praline dehydrogenase genes controls proline levels during and after osmotic stress in plants. *Molecular and General Genetics*, **253**, 334–341.

Phillips, J., Artsaenko, O., Fiedler, U., Horstmann, C., Mock, H.-P., Muntz, K. & Conrad, U. (1997) Seed-specific immunomodulation of abscisic acid activity induces a developmental switch. *EMBO Journal*, **16**, 4489–4496.

Piao, H. L., Lim, J. H., Kim, S. J., Cheong, G.-W. & Hwang, I. (2001) Constitutive overexpression of AtGSK1 induces NaCl stress responses in the absence of NaCl stress and results in enhanced NaCl tolerance in Arabidopsis. *Plant Journal*, **27**, 305–314.

Piao, H. L., Pih, K. T., Lim, J. H., Kang, S. G., Jin, J. B., Kim, S. H. & Hwang, I. (1999) An *Arabidopsis GSK3/shaggy*-like gene that complements yeast salt stress-sensitive mutants is induced by NaCl and abscisic acid. *Plant Physiology*, **119**, 1527–1534.

Polle, A. (1997) Defense against photo-oxidative damage in plants. In J. G. Scandalios (ed.) *Oxidative Stress and the Molecular Biology of Antioxidant Defenses*. Cold Spring Harbor Laboratory Press, Cold Spring Harbor, New York, pp. 623–666.

Popping, B., Gibbons, T. & Watson, M. D. (1996) The *Pisum sativum* MAP kinase homologue (*PsMAPK*) rescues the *Saccharomyces cerevisiae* hog1 deletion mutant under conditions of high osmotic stress. *Plant Molecular Biology*, **31**, 355–363.

Qiu, Q. S., Guo, Y., Dietrich, M. A., Schumaker, K. S. & Zhu, J. K. (2002) Regulation of SOS1, a plasma membrane Na^+/H^+ exchanger in *Arabidopsis thaliana*, by SOS2 and SOS3. *Proceedings of the National Academy of Sciences of the United States of America*, **99**, 8436–8441.

Qiu, Q. S., Guo, Y., Quintero, F. J., Pardo, J. M., Schumaker, K. S. & Zhu, J. K. (2004) Regulation of vacuolar Na^+/H^+ exchange in *Arabidopsis thaliana* by the salt-overly-sensitive (SOS) pathway. *Journal of Biological Chemistry*, **279**, 207–215.

Quintero, F. J., Blatt, M. R. & Pardo, J. M. (2000) Functional conservation between yeast and plant endosomal Na^+/H^+ antiporters. *FEBS Letters*, **471**, 224–228.

Quintero, F. J., Garciadeblas, B. & Rodriguez-Navarro, A. (1996) The SAL1 gene of Arabidopsis, encoding an enzyme with 3′(2′),5′-bisphosphate nucleotidase and inositol polyphosphate 1-phosphatase activities, increases salt tolerance in yeast. *Plant Cell*, **8**, 529–537.

Quintero, F. J., Ohta, M., Shi, H., Zhu, J. K. & Pardo, J. M. (2002) Reconstitution in yeast of the Arabidopsis SOS signaling pathway for Na^+ homeostasis. *Proceedings of the National Academy of Sciences of the United States of America*, **99**, 9061–9066.

Rabbani, M. A., Maruyama, K., Abe, H., Khan, M. A., Katsura, K., Ito, Y., Yoshiwara, K., Seki, M., Shinozaki, K. & Yamaguchi-Shinozaki, K. (2003) Monitoring expression profiles of rice genes under cold, drought, and high-salinity stresses and abscisic acid application using cDNA microarray and RNA gel-blot analyses. *Plant Physiology*, **133**, 1755–1767.

Rhoades, J. D. & Loveday (1990) Salinity in irrigated agriculture. In B. A. Stewart & D. R. Nielsen (eds) *Irrigation of Agricultural Crops. ASA Monograph.* pp. 1089–1142.

Rhodes, D. & Hanson, A. D. (1993) Quaternary ammonium and tertiary sulfonium compounds in higher plants. *Annual Review of Plant Physiology and Plant Molecular Biology*, **44**, 357–384.

Rodriguez-Navarro, A. (2000) Potassium transport in fungi and plants. *Biochimica et Biophysica Acta*, **1469**, 1–30.

Roxas, V. P., Lodhi, S. A., Garrett, D. K., Mahan, J. R. and Allen, R. D. (2000) Stress tolerance in transgenic tobacco seedlings that overexpress glutathione S-transferase/glutathione peroxidase. *Plant Cell Physiology*, **41**, 1229–1234.

Roxas, V. P., Smith, R. K., Jr, Allen, E. R. and Allen, R. D. (1997) Overexpression of glutathione S-transferase/glutathione peroxidase enhances the growth of transgenic tobacco seedlings during stress. *Nature Biotechnology*, **15**, 988–991.

Rubio, L., Rosado, A., Linares-Rueda, A., Borsani, O., Garcia-Sanchez, M. J., Valpuesta, V., Fernandez, J. A. & Botella, M. A. (2004) Regulation of K^+ transport in tomato roots by the TSS1 locus. Implications in salt tolerance. *Plant Physiology*, **134**, 452–459.

Rus, A., Lee, B.-H., Munoz-Mayor, A., Sharkhuu, A., Miura, K., Zhu, J.-K., Bressan, R. A. & Hasegawa, P. M. (2004) AtHKT1 facilitates Na^+ homeostasis and K^+ nutrition *in planta*. *Plant Physiology*, **136**, 2500–2511.

Rus, A., Yokoi, S., Sharkhuu, A., Reddy, M., Lee, B. H., Matsumoto, T. K., Koiwa, H., Zhu, J.-K., Bressan, R. A. & Hasegawa, P. M. (2001) AtHKT1 is a salt tolerance determinant that controls $Na^{(+)}$ entry into plant roots. *Proceedings of the National Academy of Sciences of the United States of America*, **98**, 14150–14155.

Saijo, Y., Hata, S., Kyozuka, J., Shimamoto, K. & Izui, K. (2000) Over-expression of a single Ca^{2+}-dependent protein kinase confers both cold and salt/drought tolerance on rice plants. *Plant Journal*, **23**, 319–327.

Sanders, D., Pelloux, J., Brownlee, C. & Harper, J. F. (2002) Calcium at the crossroads of signaling. *Plant Cell*, **14**(Suppl), S401–S417.

Schachtman, D. P. (2000) Molecular insights into the structure and function of plant $K^{(+)}$ transport mechanisms. *Biochimica et Biophysica Acta*, **1465**, 127–139.

Schachtman, D. P. & Schroeder, J. I. (1994) Structure and transport mechanism of a high-affinity potassium uptake transporter from higher plants. *Nature*, **370**, 655–658.

Schachtman, D. P., Kumar, R., Schroeder, J. I. & Marsh, E. L. (1997) Molecular and functional characterization of a novel low-affinity cation transporter (LCT1) in higher plants. *Proceedings of the National Academy of Sciences of the United States of America*, **94**, 11079–11084.

Schurmann, P. & Jacquot, J. P. (2000) Plant thioredoxin systems revisited. *Annual Review of Plant Physiology and Plant Molecular Biology*, **51**, 371–400.

Seki, M., Ishida, J., Narusaka, M., Fujita, M., Nanjo, T., Umezawa, T., Kamiya, A., Nakajima, M., Enju, A., Sakurai, T., Satou, M., Akiyama, K., Yamaguchi-Shinozaki, K., Carninci, P., Kawai, J., Hayashizaki, Y. & Shinozaki, K. (2002a) Monitoring the expression pattern of around 7,000 Arabidopsis genes under ABA treatments using a full-length cDNA microarray. *Functional and Integrative Genomics*, **2**, 282–291.

Seki, M., Narusaka, M., Ishida, J., Nanjo, T., Fujita, M., Oono, Y., Kamiya, A., Nakajima, M., Enju, A., Sakurai, T., Satou, M., Akiyama, K., Taji, T., Yamaguchi-Shinozaki, K., Carninci, P., Kawai, J., Hayashizaki, Y. & Shinozaki, K. (2002b) Monitoring the expression profiles of 7000 Arabidopsis genes under drought, cold and high-salinity stresses using a full-length cDNA microarray. *Plant Journal*, **31**, 279–292.

Seki, M., Satou, M., Sakurai, T., Akiyama, K., Iida, K., Ishida, J., Nakajima, M., Enju, A., Narusaka, M., Fujita, M., Oono, Y., Kamei, A., Yamaguchi-Shinozaki, K. & Shinozaki, K. (2004) RIKEN Arabidopsis full-length (RAFL) cDNA and its applications for expression profiling under abiotic stress conditions. *Journal of Experimental Botany*, **55**, 213–223.

Serrano, R. (1996) Salt tolerance in plants and microorganisms: toxicity targets and defense responses. *International Review of Cytology*, **165**, 1–52.

Serrano, R. & Rodriguez-Navarro, A. (2001) Ion homeostasis during salt stress in plants. *Current Opinion in Cell Biology*, **13**, 399–404.

Serrano, R., Mulet, J. M., Rios, G. Marquez, J. A., de Larriona, I. F., Leube, M. P., Mendizabal, I. Pascual-Ahuir, A., Proft, M. R. R. & Montesinos, C. (1999) A glimpse of the mechanism of ion homeostasis during salt stress. *Journal of Experimental Botany*, **50**, 1023–1036.

Serrato, A. J., Perez-Ruiz, J. M., Spinola, M. C. & Cejudo, F. J. (2004) A novel NADPH thioredoxin reductase, localized in the chloroplast, which deficiency causes hypersensitivity to abiotic stress in *Arabidopsis thaliana*. *Journal of Biological Chemistry*, July 28 (Epub ahead of print).

Sheveleva, E., Chmara, W., Bohnert, H. J. & Jensen, R. G. (1997) Increased salt and drought tolerance by D-ononitol production in transgenic *Nicotiana tabacum* L. *Plant Physiology*, **115**, 1211–1219.

Shi, H. & Zhu, J. K. (2002a) Regulation of expression of the vacuolar Na^+/H^+ antiporter gene AtNHX1 by salt stress and abscisic acid. *Plant Molecular Biology*, **50**, 543–550.

Shi, H. & Zhu, J. K. (2002b) SOS4, a pyridoxal kinase gene, is required for root hair development in Arabidopsis. *Plant Physiology*, **129**, 585–593.

Shi, H., Ishitani, M., Kim, C. & Zhu, J. K. (2000) The *Arabidopsis thaliana* salt tolerance gene SOS1 encodes a putative Na^+/H^+ antiporter. *Proceedings of the National Academy of Sciences of the United States of America*, **97**, 6896–6901.

Shi, H., Kim, Y., Guo, Y., Stevenson, B. & Zhu, J. K. (2003a) The Arabidopsis SOS5 locus encodes a putative cell surface adhesion protein and is required for normal cell expansion. *Plant Cell*, **15**, 19–32.

Shi, H., Lee, B. H., Wu, S. J. and Zhu, J. K. (2003b) Overexpression of a plasma membrane Na^+/H^+ antiporter gene improves salt tolerance in *Arabidopsis thaliana*. *Nature Biotechnology*, **21**, 81–85.

Shi, H., Quintero, F. J., Pardo, J. M. & Zhu, J. K. (2002) The putative plasma membrane $Na^{(+)}/H^{(+)}$ antiporter SOS1 controls long-distance $Na^{(+)}$ transport in plants. *Plant Cell*, **14**, 465–477.

Shinozaki, K. & Yamaguchi-Shinozaki, K. (2000) Molecular responses to dehydration and low temperature: differences and cross-talk between two stress signaling pathways. *Current Opinion in Plant Biology*, **3**, 217–223.

Shinozaki, K., Yamaguchi-Shinozaki, K. & Seki, M. (2003) Regulatory network of gene expression in the drought and cold stress responses. *Current Opinion in Plant Biology*, **6**, 410–417.

Singla-Pareek, S. L., Reddy, M. K. & Sopory, S. K. (2003) Genetic engineering of the glyoxalase pathway in tobacco leads to enhanced salinity tolerance. *Proceedings of the National Academy of Sciences of the United States of America*, **100**, 14672–14677.

Smart, C. C. & Flores, S. (1997) Overexpression of D-myo-inositol-3-phosphate synthase leads to elevated levels of inositol in Arabidopsis. *Plant Molecular Biology*, **33**, 811–820.

Szabolcs, I. (1994) Soil salinization. In M. Pressarkli (ed.) *Handbook of Plant Crop Stress*. Marcel Dekker, New York, pp. 3–11.

Szoke, A., Miao, G. H., Hong, Z. & Verma, D. P. S. (1992) Subcellular location of delta^1-pyrroline-5-carboxylate reductase in root/nodule and leaf of soybean. *Plant Physiology*, **99**, 1642–1649.

Taji, T., Seki, M., Satou, M., Sakurai, T., Kobayashi, M., Ishiyama, K., Narusaka, Y., Narusaka, M., Zhu, J.-K, Shinozaki, K. (2004) Comparative genomics in salt tolerance between Arabidopsis and Arabidopsis-relative halophyte salt cress using Arabidopsis microarray. *Plant Physiology*, **135**, 1697–1709.

Tanaka, Y., Hibino, T., Hayashi, Y., Tanaka, A., Kishitani, S., Takabe, T., Yokota, S. & Takabe, T. (1999) Salt tolerance of transgenic rice overexpressing yeast mitochondrial Mn-SOD in chloroplasts. *Plant Science*, **148**, 131–138.

Tarczynski, M. C., Jensen, R. G. & Bohnert, H. J. (1993) Stress protection of transgenic tobacco by production of the osmolyte mannitol. *Science*, **259**, 508–510.

Teige, M., Scheikl, E., Eulgem, T., Doczi, R., Ichimura, K., Shinozaki, K., Dangl, J. L. & Hirt, H. (2004) The MKK2 pathway mediates cold and salt stress signaling in Arabidopsis. *Molecular Cell*, **15**, 141–152.

Tester, M. & Davenport, R. (2003) Na^+ tolerance and Na^+ transport in higher plants. *Annals of Botany (London)*, **91**, 503–527.

Thornalley, P. J. (1993) The glyoxalase system in health and disease. *Molecular Aspects of Medicine*, **14**, 287–371.

Tsugane, K., Kobayashi, K., Niwa, Y., Ohba, Y., Wada, K. & Kobayashi, H. (1999) A recessive Arabidopsis mutant that grows photoautotrophically under salt stress shows enhanced active oxygen detoxification. *Plant Cell*, **11**, 1195–1206.

van Schilfgaarde, J. (1994) Irrigation – a blessing or a curse. *Agricultural Water Management*, **25**, 203–219.

Veena, Reddy, V. S. & Sopory, S. K. (1999) Glyoxalase I from *Brassica juncea*: molecular cloning, regulation and its over-expression confer tolerance in transgenic tobacco under stress. *Plant Journal*, **17**, 385–395.

Verbruggen, N., Hua, X. J., May, M. & Van Montagu, M. (1996) Environmental and developmental signals modulate proline homeostasis: evidence for a negative transcriptional regulator. *Proceedings of the National Academy of Sciences of the United States of America*, **93**, 8787–8791.

Villalobos, M. A., Bartels, D. & Iturriaga, G. (2004) Stress tolerance and glucose insensitive phenotypes in Arabidopsis overexpressing the CpMYB10 transcription factor gene. *Plant Physiology*, **135**, 309–324.

Vitart, V., Baxter, I., Doerner, P. & Harper, J. F. (2001) Evidence for a role in growth and salt resistance of a plasma membrane H^+-ATPase in the root endodermis. *Plant Journal*, **27**, 191–201.

Wallsgrove, R., Turner, J., Hall, N., Kendall, A. & Bright, S. (1987) Barley mutants lacking chloroplast glutamine synthetase. Biochemical and genetic analysis. *Plant Physiology*, **83**, 155–158.

Winicov, I. (2000) Alfin1 transcription factor overexpression enhances plant root growth under normal and saline conditions and improves salt tolerance in alfalfa. *Planta*, **210**, 416–422.

Wu, C. A., Yang, G. D., Meng, Q. W. & Zheng, C. C. (2004) The cotton GhNHX1 gene encoding a novel putative tonoplast $Na^{(+)}/H^{(+)}$ antiporter plays an important role in salt stress. *Plant Cell Physiology*, **45**, 600–607.

Wu, S.-J., Ding, L. & Zhu, J. K. (1996) *SOS1*, a genetic locus essential for salt tolerance and potassium acquisition. *Plant Cell*, **8**, 617–627.

Wyn Jones, R. G. & Pollard, A. (1983) Proteins, enzymes and inorganic ions. In A. Lauchli & R. L. Bieleskio (eds) *Encyclopedia of Plant Physiology. New Series Vol. 15B, Inorganic Plant Nutrition*. Springer-Verlag, Berlin, pp. 528–562.

Xiong, L. & Zhu, J. K. (2003) Regulation of abscisic acid biosynthesis. *Plant Physiology*, **133**, 29–36.

Xiong, L., Lee, B., Ishitani, M., Lee, H., Zhang, C. & Zhu, J. K. (2001) *FIERY1* encoding an inositol polyphosphate 1-phosphatase is a negative regulator of abscisic acid and stress signaling in Arabidopsis. *Genes and Development*, **15**, 1971–1984.

Xiong, L., Schumaker, K. S. & Zhu, J. K. (2002) Cell signaling during cold, drought, and salt stress. *Plant Cell*, **14**(Suppl.), S165–S183.

Xu, D., Duan, X., Wang, B., Hong, B., Ho, T. & Wu, R. (1996) Expression of a late embryogenesis abundant protein gene, HVA1, from barley confers tolerance to water deficit and salt stress in transgenic rice. *Plant Physiology*, **110**, 249–257.

Yadav, R., Flowers, T. J., & Yeo, A. R. (1996) The involvement of the transpirational bypass flow in sodium uptake by high- and low-sodium-transporting lines of rice developed through intravarietal selection. *Plant, Cell and Environment*, **22**, 329–336.

Yeo, A. R. & Flowers, T. J. (1986) The physiology of salinity resistance in rice (*Oryza sativa* L.) and a pyramiding approach to breeding varieties for saline soils. *Australian Journal of Plant Physiology*, **13**, 161–173.

Yeo, A. R., Flowers, S. A., Rao, G., Welfare, K., Senanayake, N. & Flowers, T. J. (1999) Silicon reduces sodium uptake in rice (*Oryza sativa* L.) in saline conditions and this is accounted for by a reduction in the transpirational bypass flow. *Plant, Cell and Environment*, **22**, 559–565.

Yokoi, S., Quintero, F. J., Cubero, B., Ruiz, M. T., Bressan, R. A., Hasegawa, P. M. & Pardo, J. M. (2002) Differential expression and function of *Arabidopsis thaliana* NHX Na^+/H^+ antiporters in the salt stress response. *Plant Journal*, **30**, 529–539.

Zhang, H. X. & Blumwald, E. (2001) Transgenic salt-tolerant tomato plants accumulate salt in foliage but not in fruit. *Nature Biotechnology*, **19**, 765–768.

Zhang, H. X., Hodson, J. N., Williams, J. P. & Blumwald, E. (2001) Engineering salt-tolerant Brassica plants: characterization of yield and seed oil quality in transgenic plants with increased vacuolar sodium accumulation. *Proceedings of the National Academy of Sciences of the United States of America*, **98**, 12832–12836.

Zhu, J. K. (2000) Genetic analysis of plant salt tolerance using Arabidopsis. *Plant Physiology*, **124**, 941–948.

Zhu, J. K. (2001) Plant salt tolerance. *Trends in Plant Science*, **6**, 66–71.

Zhu, J. K. (2002) Salt and drought stress signal transduction in plants. *Annual Review of Plant Physiology and Plant Molecular Biology*, **53**, 247–273.

Zhu, J. K. (2003) Regulation of ion homeostasis under salt stress. *Current Opinion in Plant Biology*, **6**, 441–445.

Zhu, J., Gong, Z., Zhang, C., Song, C.-P., Damsz, B., Inan, G., Koiwa, H., Zhu, J.-K., Hasegawa, P. M. & Bressan, R. A. (2002) OSM1/SYP61: a syntaxin protein in Arabidopsis controls abscisic-mediated and non-abscisic acid-mediate responses to abiotic stress. *Plant Cell*, **14**, 3009–3028.

Zhu, J. K., Hasegawa, P. M. & Bressan, R. A. (1997) Molecular aspects of osmotic stress. *Critical Reviews in Plant Sciences*, **16**, 253–277.

Zhu, J. K., Liu, J. & Xiong, L. (1998) Genetic analysis of salt tolerance in arabidopsis. Evidence for a critical role of potassium nutrition. *Plant Cell*, **10**, 1181–1191.

4 The CBF cold-response pathway

Sarah Fowler, Daniel Cook and Michael F. Thomashow

4.1 Introduction

Plants vary greatly in their abilities to cope with freezing temperatures (Levitt, 1980; Sakai & Larcher, 1987). Plants from tropical and subtropical regions – including agriculturally important plants such as cotton, soybean, maize and rice – have little or no capacity to withstand freezing temperatures. Indeed, many of these plants are chilling-sensitive, suffering injury when temperatures fall below 10°C. In contrast, herbaceous plants from temperate regions can survive freezing temperatures ranging from −5°C to −30°C, depending on the species, and woody plants can withstand even lower temperatures. Significantly, the maximum freezing tolerance of these plants is not constitutive, but is induced in response to low temperatures (below ∼ 10°C), a phenomenon known as 'cold acclimation'. For instance, rye plants grown at nonacclimating temperatures (∼ 22°C) are killed by freezing at −5°C, but after cold acclimation (2°C for 1–2 wk) are able to survive freezing temperatures down to about −30°C (Fowler *et al.*, 1977).

For many years, a fundamental goal of cold acclimation research has been to determine what occurs in response to low temperature that results in the enhancement of freezing tolerance. Earlier, comparative biochemical analyses indicated that extensive changes in cellular composition occurred with cold acclimation including alterations in lipid composition, increases in soluble protein content and accumulation of simple sugars, proline and other organic acids (Levitt, 1980; Sakai & Larcher, 1987). Results were consistent with some of these changes having roles in freezing tolerance. For instance, there was strong evidence that changes in lipid composition that occurred in response to low temperature stabilized membranes against certain forms of freeze-induced damage (Steponkus, 1984). Also, results indicated that compatible solutes such as proline and sucrose (Rudolph & Crowe, 1985; Carpenter & Crowe, 1988) and certain hydrophilic polypeptides (Hincha *et al.*, 1988, 1990) had cryoprotective effects. Consistent with this biochemical complexity were genetic analyses indicating that the ability of plants to cold acclimate involved the action of many genes (Thomashow, 1990). In hexaploid wheat, for instance, almost all of the chromosomes were implicated as contributing to freezing tolerance under one set of experimental conditions or another.

By the mid-1980s, the picture that had emerged was that cold acclimation was a complex biochemical, physiological and genetic process. However, a general consensus had not been reached as to which changes were integral to cold acclimation and which were important in other aspects of life at low temperature. Moreover, no individual gene had been identified that had a role in freezing tolerance. This made it difficult to critically test the hypotheses regarding the importance of a given change to the process of cold acclimation. Into this research vacuum rushed investigators employing a new approach: the isolation and characterization of genes that were induced in response to low temperature. The underlying hypothesis was that the increase in freezing tolerance that occurred with cold acclimation was likely to include the action of genes that were induced in response to low temperature. Recent studies with Arabidopsis have proved this hypothesis is correct. In particular, a low temperature regulatory pathway, the CBF cold-response pathway, was discovered and shown to participate in cold acclimation. Additional studies have shown that the major steps of the pathway are highly conserved in plants, but there are differences that may contribute to the variation in freezing tolerance observed in plants.

In this chapter, we focus on the function and regulation of the CBF cold-response pathway. Additional aspects of cold acclimation and freezing tolerance are covered in other recent reviews (Smallwood & Bowles, 2002; Stitt & Hurry, 2002; Xiong et al., 2002a; Shinozaki et al., 2003; Sung et al., 2003).

4.2 Arabidopsis CBF cold-response pathway

4.2.1 Discovery and overview

The Arabidopsis CBF cold-response pathway was discovered through investigations of cold-regulated gene expression. A number of earlier studies demonstrated that changes in gene expression occur in Arabidopsis in response to low temperature (Thomashow, 1999). These genes were given various designations including *COR* (cold-regulated), *LTI* (low-temperature-induced), *KIN* (cold-induced) and *RD* (responsive to dehydration). Within 2–4 hours of transferring plants to low temperature, the transcript levels for these genes were found to increase dramatically and remain elevated for weeks if plants were kept at low temperature. When plants were returned to warm temperatures, the transcript levels for these genes were shown to quickly (within hours) decrease to the levels found in nonacclimated plants.

Gene fusion studies with *COR15a* (Baker et al., 1994), *COR78/RD29a* (Horvath et al., 1993; Kiyosue et al., 1994) and *COR6.6/KIN2* (Wang & Cutler, 1995) established that the promoters of these cold-induced genes are activated in response to low temperature and drought. Additional studies indicated that the promoter regions of these genes included a DNA regulatory element, referred to as the CRT (C-repeat)/DRE (dehydration responsive element) (Baker et al., 1994; Yamaguchi-Shinozaki & Shinozaki, 1994), that imparted responsiveness to both low temperature and dehydration stress (Yamaguchi-Shinozaki & Shinozaki, 1994). The yeast one-hybrid method (Li & Herskowitz, 1993) was then used to identify a family of DNA-binding proteins that recognized the CRT/DRE element. The first three of these DNA-binding proteins were designated CBF1, CBF2 and CBF3 (Stockinger et al., 1997; Gilmour et al., 1998; Medina et al., 1999) or DREB1b, DREB1c and DREB1a (Liu et al., 1998), respectively. These proteins, which are encoded by three genes arranged in tandem on chromosome 4, are transcriptional activators that have masses of about 24 kDa, are about 90% identical in amino acid sequence and contain a conserved DNA-binding motif of about 60 amino acids designated the AP2/ERF domain (Riechmann & Meyerowitz, 1998).

When plants are grown at normal, warm temperatures ($\sim 22^\circ$C), the *CBF1-3* genes are not expressed. However, transcripts for all three genes accumulate rapidly (within 15 min) upon transferring plants to low temperature (e.g. 4°C). Thus, the CBF cold-response pathway is composed of two fundamental regulatory steps: first, *CBF1-3* genes are induced rapidly in response to low temperature; and second, the CBF transcriptional activators induce expression of genes containing CRT/DRE regulatory elements (Figure 4.1). Expression of the 'CBF regulon' of genes, i.e. those genes that are responsive to CBF expression, then leads to an increase in freezing tolerance (Jaglo-Ottosen et al., 1998; Liu et al., 1998; Kasuga et al., 1999).

Cold-induced expression of *CBF1-3* explains how the CRT/DRE imparts cold responsive gene expression. But how does the element impart dehydration-induced gene expression? It is accomplished through the action of additional AP2/ERF domain proteins: CBF4 (Haake et al., 2002), which is closely related to CBF1-3 (67% identity, 77% similarity), and DREB2a (Liu et al., 1998), which has little sequence similarity with the CBF1-3 proteins outside the DNA-binding domain. The *CBF4* and *DREB2a* genes are induced in response to drought and high salinity, but not low temperature. The question thus raised is whether expression of the CBF regulon also results in an increase in plant dehydration tolerance, and indeed, it does. Plants expressing the CBF regulon are not only more freezing tolerant, but are also more tolerant to drought, high salinity and other conditions that result in cellular dehydration (Liu et al., 1998; Kasuga et al., 1999; Haake et al., 2002). The connection between freezing tolerance and dehydration tolerance is that the injury to plant cells caused by freezing is largely due to the dehydration stress associated with freezing. As temperatures drop below 0°C, ice formation is generally initiated in the intercellular spaces due, in part, to the extracellular fluid having a higher freezing point (lower solute concentration) than the intracellular fluid. Because

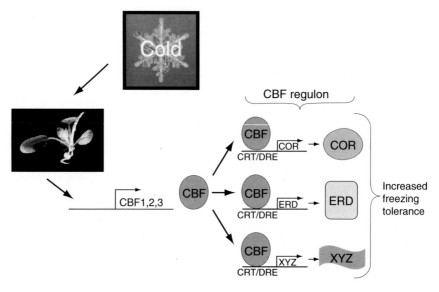

Figure 4.1 The Arabidopsis CBF cold-response pathway. Low temperature leads to rapid induction of the *CBF1, 2* and *3* genes, which in turn, results in expression of the CBF regulon of CRT/DRE-regulated genes (see text). Action of the CBF regulon, which includes *COR, ERD* and other cold-regulated genes (*XYZ;* see text), results in an increase in plant freezing tolerance.

the chemical potential of ice is less than that of liquid water at a given temperature, the formation of extracellular ice results in a drop in water potential outside the cell. Consequently, there is movement of unfrozen water from inside the cell to the intercellular spaces where it freezes. This process continues until chemical equilibrium is reached. The osmolarity of the remaining unfrozen intracellular and intercellular fluid is directly determined by the freezing temperature. At $-10°C$ it is approximately 5 osm, equivalent to about that of a 2.5 M solution of NaCl. Generally, freezing to $-10°C$ results in cells losing greater than 90% of their intracellular water (depending on the initial osmolarity of the cellular fluids). In short, freezing tolerance must include the activation of mechanisms that include tolerance to dehydration stress. Thus, it is not surprising to find that the CBF regulon offers cross-protection against drought and high salinity.

In sum, expression of the CBF regulon of genes helps protect cells against the potential damaging effects of dehydration associated with freezing, drought and high salinity. The regulon can be activated in response to low temperature through action of the CBF1-3/DREB1a-c transcription factors, or by drought and high salinity through the action of CBF4 and DREB2a transcription factors. The emphasis throughout the remainder of this chapter will be on the role of the CBF cold-response pathway in cold acclimation.

4.2.2 CBF proteins

4.2.2.1 General properties

The Arabidopsis genome encodes more than 140 proteins that contain an AP2/ERF DNA-binding domain (Riechmann *et al.*, 2000). The AP2/ERF domain consists of an α-helix and a three-stranded anti-parallel β-sheet, which contacts base pairs in the major groove and the sugar-phosphate backbones of the DNA (Allen *et al.*, 1998). Proteins containing the AP2/ERF domain, which has only been found in plant proteins, were initially classified into two groups based upon the number of AP2/ERF domains present in the protein (Riechmann & Meyerowitz, 1998). More recently, Sakuma *et al.* (2002) grouped 144 Arabidopsis AP2/ERF proteins into five subfamilies based on the amino acid sequence similarities of the AP2/ERF domains: the DREB subfamily (56 members), which includes the CBF/DREB1 proteins; the ERF subfamily (65 members); the AP2 subfamily (14 members); the RAV subfamily (six members) and 'others' (four members).

Sequences within the AP2/ERF domain have been implicated in determining the specificity of CBF/DREB1 binding to the CRT/DRE regulatory element. Two conserved amino acids in the AP2/ERF domain differ between the CBF1-3/DREB1a-c (DREB subfamily) and ERF (ERF subfamily) proteins. CBF1-3/DREB1a-c proteins have valine in the 14th position (V14) and glutamic acid in the 19th position (E19) while ERF proteins have alanine and aspartic acid, respectively (Sakuma *et al.*, 2002). Gel mobility shift assays with mutant CBF1-3/DREB1a-c proteins have been performed to test the importance of these amino acids. Mutant CBF3/DREB1a proteins in which V14 had been substituted for alanine (as in ERF), did not bind to the CRT/DRE indicating that V14 is a critical amino acid for determining the DNA-binding specificity of CBF3/DREB1a. However, mutation of the E19 for aspartic acid did not affect the binding of CBF3/DREB1a to the CRT/DRE (Sakuma *et al.*, 2002). Solution of the ERF-1 protein structure in complex with its DNA target has indicated that while the amino acids in the 14th and 19th position of the AP2/ERF domain are located in the second strand of the anti-parallel β-sheet, which contacts base pairs in the major groove of the DNA, these particular amino acids do not appear to actually contact the DNA (Allen *et al.*, 1998). However, Hao *et al.* (2002) report that the 14th amino acid of the AP2/ERF domain of AtERF1 shows significant van der Waals contact to amino acids that do contact DNA bases. Thus, different amino acids in the 14th position could impose a shift in the three-dimensional positions of the DNA-contacting residues and affect the DNA sequence recognition.

As noted, the CBF1-3/DREB1a-c proteins belong to the DREB subfamily of AP2/ERF proteins. However, they have a distinguishing feature that sets them apart from most of the other AP2/ERF proteins: unique amino acid sequences that bracket the AP2/ERF domain designated the CBF 'signature sequences' (Jaglo *et al.*, 2001). Immediately upstream of the AP2/ERF domain of these proteins is the consensus sequence PKK/RPAGRxKFxETRHP and immediately downstream is the sequence DSAWR. Only three other Arabidopsis proteins have close versions of this sequence, CBF4/DREB1d (Haake *et al.*, 2002; Sakuma *et al.*, 2002) and DREB1e and DREB1f (Sakuma *et al.*, 2002), indicating a close relationship between these proteins. The signature sequences presumably have an important functional role as they are conserved in CBF proteins from distantly related species, but the nature of this role remains to be determined.

Another AP2/ERF domain protein, DREB2a, has been shown to bind to the CRT/DRE elements (Liu *et al.*, 1998; Sakuma *et al.*, 2002). This protein does not have the CBF signature sequences and shares little sequence identity with the CBF/DREB1 proteins outside of the AP2/DRE-binding domain. The results of *in vitro* binding assays and transient gene expression tests using Arabidopsis protoplasts have shown that DREB2a binds to the CRT/DRE and activates expression of reporter genes (Liu & Zhu, 1998). However, constitutive overexpression of DREB2a only weakly induces expression of CRT/DRE-containing target genes in transgenic Arabidopsis plants. Thus, it has been proposed that stress-activated posttranslational modification such as phosphorylation of the DREB2a protein, might be necessary for DREB2a to function (Liu *et al.*, 1998). Seven additional DREB2 family members have been described and designated DREB2b through DREB2h (Liu *et al.*, 1998; Sakuma *et al.*, 2002).

4.2.2.2 Mechanism of action

How do the CBF proteins stimulate transcription of their target genes? Results suggest that they accomplish this, in part, through recruiting to gene promoters' transcriptional adaptor complexes that modify chromatin structure. The CBF1-3 proteins function as transcriptional activators in yeast indicating that they are able to interact with yeast proteins to stimulate transcription (Stockinger *et al.*, 1997; Gilmour *et al.*, 1998; Liu *et al.*, 1998). This observation led Stockinger *et al.* (2001) to test whether CBF activity in yeast required components of the Ada (mass $\approx$ 0.8 MDa) and/or SAGA (mass $\approx$ 1.8 MDa) transcriptional adaptor complexes (Sterner & Berger, 2000). These complexes include the histone acetyltransferase (HAT) protein Gcn5 and the transcriptional adaptor proteins Ada2 and Ada3. HAT proteins, which catalyze the addition of acetyl groups to specific lysine residues present in the amino-terminal tails of core histones, are thought to help in stimulating transcription by inducing changes in chromatin structure that result in promoters becoming more accessible to RNA polymerase II and other components of the transcriptional apparatus (Brown *et al.*, 2000). Stockinger *et al.* (2001) found that the Ada2, Ada3 and Gcn5 proteins are required for CBF1 to activate transcription in yeast, raising the question of whether Arabidopsis has transcriptional complexes related to Ada and/or SAGA. Indeed, Arabidopsis encodes two ADA2 proteins, designated ADA2a and ADA2b, and one GCN5 protein (it does not appear to encode a homologue of Ada3) (Stockinger *et al.*, 2001). The Arabidopsis GCN5 protein has intrinsic HAT activity, acetylating histone H3, and can physically interact *in vitro* with both the Arabidopsis ADA2a and ADA2b proteins (Stockinger *et al.*, 1997). Moreover, CBF1 interacts with the Arabidopsis GCN5 and ADA2 proteins suggesting that the CBF proteins function through the action of one or more protein complexes that include the GCN5 and/or ADA2 proteins.

To test this latter possibility, Vlachonasios *et al.* (2003) examined the expression of *CBF* and *COR* genes in Arabidopsis mutants carrying T-DNA insertions in either *ADA2b* or *GCN5*. The results indicated that expression of CBF-regulated *COR* genes was delayed and reduced in both *ada2b* and *gcn5* mutant plants. These observations, along with the finding that mutation of *GCN5* results in lower levels of histone H3 acetylation (Bertrand *et al.*, 2003), support the notion that the CBF proteins function, in part, through the recruitment of an Ada/SAGA-like complex to promoters that modifies chromatin structure. In addition, mutations that inactivate the *FVE* gene of Arabidopsis result in constitutive expression of *COR15a* and other CBF-targeted genes (Kim *et al.*, 2004). The significance here is that the FVE protein appears to function in a protein complex that deacetylates certain target genes including *FLC*, a key regulator of flowering time (Ausin *et al.*, 2004). A hypothesis put forward is that an FVE complex containing a histone deacetylase may keep *COR15a* and other cold-regulated genes in a repressed state due to deacetylation of histones and that this repression can be reversed through histone acetylation, which, in the case of *COR15a*, involves action of the CBF transcription factors recruiting a GCN5-containing complex to the gene promoter (Amasino, 2004; Kim *et al.*, 2004).

The *ada2b* and *gcn5* mutations have pleiotropic effects (Bertrand *et al.*, 2003; Vlachonasios *et al.*, 2003). Plants carrying mutations in these genes are dwarf in size, display aberrant root development and produce flowers with short petals and stamens. In addition, late-appearing flowers have been reported to display homeotic transformation of petals into stamens and sepals into filamentous structures and to produce ectopic carpels (Bertrand *et al.*, 2003). These phenotypes were traced to an expansion in the range of cells in which *WUSCHEL* and *AGAMOUS* were expressed, two regulatory genes with fundamental roles in flower formation (Bertrand *et al.*, 2003). In addition, approximately 5% of the 8200 genes assayed by DNA microarrays showed changes in expression in the *ada2b* and *gcn5* mutants, three-fourths of which were upregulated (Vlachonasios *et al.*, 2003). Significantly, only half of the affected genes were altered similarly in the two mutant lines indicating that the Arabidopsis ADA2b and GCN5 proteins have both similar and distinct functions and thus, may be components of both a common coactivator complex and separate complexes with distinct biological activities (Vlachonasios *et al.*, 2003). It is also noteworthy that nonacclimated *ada2* mutant plants were more freezing tolerant than nonacclimated wild-type plants, even though *COR* genes were not being expressed (Vlachonasios *et al.*, 2003). Thus, at warm, nonacclimating temperatures, ADA2b may be involved in repressing a freezing tolerance pathway that is independent of the CBF cold-response pathway.

A final point regards the role of the Arabidopsis *SFR6* gene in *CBF* action. Garry Warren and colleagues isolated a number of mutants of Arabidopsis that do not cold acclimate as effectively as wild-type plants and designated them *sfr* (sensitive to freezing) mutants (McKown *et al.*, 1996; Warren *et al.*, 1996). One of the mutations, *sfr6*, results in impaired CBF function. In plants carrying the recessive *sfr6* mutation, transcripts for the *CBF* genes accumulate normally in response to low temperature, but expression of CBF-targeted genes, including COR15a and COR78, is greatly impaired (Knight *et al.*, 1999; Boyce *et al.*, 2003). Thus, it appears that the SFR6 protein is required for the *CBF* genes to function. One interesting possibility is that SFR6 might be a member of an adaptor complex required for CBF action. Other possibilities include roles for SFR6 in translation of *CBF* transcripts and CBF protein stability.

4.2.3 *Function of the CBF cold-response pathway*

As noted earlier, expression of the CBF regulon results in an increase in plant freezing tolerance (Jaglo-Ottosen *et al.*, 1998; Liu *et al.*, 1998; Kasuga *et al.*, 1999; Gilmour *et al.*, 2000). Additionally, the CBF regulon appears to have a role in chilling tolerance. Whereas Arabidopsis *los4* mutant plants are sensitive to chilling temperatures, *los4* mutant plants overexpressing CBF3 are chilling tolerant (Gong *et al.*, 2002). How does expression of the CBF regulon bring about these effects? To address this question, efforts have been directed at identifying the genes that comprise the CBF regulon. Specifically, expression profiling experiments have been performed using a variety of different Affymetrix GeneChips and cDNA-based microarrays to identify genes that are upregulated in response to both CBF overexpression and low temperature; genes that are *bona fide* members of the CBF regulon would be expected to be responsive to both of these factors (Seki *et al.*, 2001; Fowler and Thomashow, 2002; Maruyama *et al.*, 2004; Vogel *et al.*, 2005). These studies have led to a total of 109 genes being assigned to the CBF regulon. Significantly, there are no obvious differences between the sets of genes controlled by CBF1, 2 and 3 since no qualitative differences in transcript accumulation were found using Affymetrix GeneChips containing probe sets for approximately 8000 Arabidopsis genes (Gilmour *et al.*, 2004).

The 109 genes that are currently assigned to the CBF regulon comprise a wide range of functional classes (Figure 4.2). The genes can be further

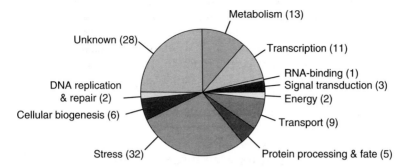

Figure 4.2 The Arabidopsis CBF regulon. The 109 genes designated members of the CBF regulon were assigned to a functional class based upon previously published classifications (Maruyama *et al*, 2004; Fowler & Thomashow, 2002; Vogel *et al.*, 2005) and information available from online databases (TAIR: http://www.arabidopsis.org/index.jsp, TIGR: http://www.tigr.org/tdb/e2k1/ath1/). The number of genes assigned to each category is given in parentheses.

grouped into four categories based on the nature of the proteins that they encode. The largest group, which includes more than 50% of the proteins, is 'unknown function'. The other three groups are cryoprotective proteins, regulatory proteins and biosynthetic proteins. Below, we summarize information on the latter three classes.

4.2.3.1 Cryoprotective proteins

COR/LEA/Dehydrin proteins: Among the genes that are the most highly induced during cold acclimation are the 'classical' *COR* (cold-regulated) genes, alternatively designated *KIN* (cold-induced), *RD* (responsive to dehydration), *LTI* (low-temperature-induced) or *ERD* (early, responsive to dehydration) (Thomashow, 1998). The proteins encoded by these genes are extremely hydrophilic and are either novel or are members of the dehydrin (Close, 1997) or LEA (Wise & Tunnacliffe, 2004) protein families. Expression profiling experiments have assigned 16 of these COR/LEA/dehydrin genes to the CBF regulon: *COR6.6/KIN2, KIN1, COR15a, COR15b, COR78/RD29A/ LTI79, RD29B/LTI65, COR47/RD17, ERD10/LTI45, Xero2, LEA14, At1g54410* (which encodes a dehydrin family protein) and five newly discovered *COR* genes, *COR8.5, COR17, COR35, COR33.5* and *COR42* (the numbers after the *COR* designation indicate the mass of the predicted polypeptide encoded by the gene) (Fowler & Thomashow, 2002; Maruyama *et al.*, 2004; Vogel *et al.*, 2005).

The functional activity of most of the COR/LEA/dehydrin proteins remains speculative. However, there is evidence that the polypeptide encoded by COR15a helps in stabilizing membranes against freezing injury (Artus *et al.*,

1996; Steponkus *et al.*, 1998). *COR15a* encodes a 15 kDa polypeptide, COR15a, that is targeted to the chloroplasts where it is processed to a 9.4 kDa mature polypeptide designated COR15am (Lin & Thomashow, 1992). Artus *et al.* (1996) found that constitutive expression of *COR15a* in transgenic Arabidopsis plants increased the freezing tolerance of chloroplasts about 2°C in nonacclimated plants. Additional analysis indicated that COR15am acts directly as a cryoprotective polypeptide (Steponkus *et al.*, 1998). *In vivo* and *in vitro* analyses indicated that COR15am decreases the propensity of membranes to form the hexagonal-II phase, a deleterious non-bilayer structure that occurs due to the cellular dehydration associated with freezing. It has been suggested that other COR, LEA and LEA-like polypeptides also act to stabilize membranes (Thomashow, 1999), though this hypothesis remains to be tested.

In addition to stabilizing membranes, there are reports that some COR and dehydrin proteins are able to protect other proteins against freeze–thaw inactivation *in vitro* (Bravo *et al.*, 2003; Hara *et al.*, 2003; Sanchez-Ballesta *et al.*, 2004). In addition, it has been hypothesized that COR/LEA/dehydrin proteins could act as a 'hydration buffer', sequestering ions and renaturing unfolded proteins (Bray, 1993; Wise & Tunnacliffe, 2004). The predicted structure of COR/LEA proteins provides few clues to their function as many of them appear 'natively unfolded' or 'intrinsically disordered' (Uversky *et al.*, 2000; Dunker *et al.*, 2001) though it is possible that these proteins take up a folded conformation upon dehydration or binding to a molecular target (Wise & Tunnacliffe, 2004). Novel computational tools to analyze nonglobular proteins have been utilized as a means to get around this limitation. Wise and Tunnacliffe (2004) used Protein or Oligonucleotide Probability Profiles (POPPs) (Wise, 2002), which allow a comparison of proteins on the basis of similarities in peptide composition rather than similarities in sequence, and found that the LEA proteins grouped into superfamilies. The SwissProt database was then used to search for proteins with similar POPPs enabling identification of functional themes associated with each superfamily. For example, one group of LEA proteins matched proteins involved with DNA recognition as well as binding to other polymers. This type of analysis, therefore, provides novel, experimentally testable hypotheses that may help to determine the molecular function of LEA/COR proteins.

Other stress-related proteins: The CBF regulon also includes genes belonging to the COR413 (Breton *et al.*, 2003) and early Arabidopsis aluminum-induced gene 1 (EARLI) (Wilkosz & Schlappi, 2000; Breton *et al.*, 2003; Bubier & Schlappi, 2004) families that have been proposed to encode proteins having roles in low temperature tolerance. The COR413 gene family comprises three genes (At1g29395, At2g15970 and At1g29390) that encode multi-spanning transmembrane proteins proposed to be targeted to the plasma or thylakoid membrane. While the exact function of these proteins is unclear, it is thought that since they are integral membrane proteins, they may play a structural role in stabilizing membrane lipid bilayers at low temperatures. An alternative hypothesis based on the structural similarity of the COR413 proteins to a family of mammalian G-protein coupled receptors proposes a role for these genes in signal transduction in response to low temperature (Breton *et al.*, 2003).

EARLI1 and two apparent homologues (At4g12470, At4g12490) are present in tandem array on chromosome 4 (Fowler & Thomashow, 2002; Maruyama *et al.*, 2004; Vogel *et al.*, 2005). These proteins are small, hydrophobic, rich in proline residues and contain a potential membrane-spanning domain. While overexpression of EARLI1 does not increase freezing tolerance of whole plants, it does protect cells from freezing damage as demonstrated by a decrease in electrolyte leakage at $-2°C$ to $-4°C$ compared to wild-type plants (Bubier & Schlappi, 2004). How this is accomplished is unknown.

4.2.3.2 Regulatory proteins

Transcription factors: Approximately 10% of the CBF regulon comprises genes encoding proteins with putative roles in transcriptional regulation (Seki *et al.*, 2001; Fowler & Thomashow, 2002; Maruyama *et al.*, 2004; Vogel *et al.*, 2005). Included are putative transcription factors from a wide variety of protein families: two AP2 domain proteins; six zinc finger proteins; a homologue of the Snapdragon floral meristem identity gene, Squamosa; a member of the Myb family of transcription factors. It has been suggested that these proteins might control subregulons of the CBF regulon (Fowler & Thomashow, 2002). In this regard, it is of note that 24% of CBF regulon genes do not have the core CCGAC sequence of the CRT/DRE-element within 1 kb of the start of transcription (Fowler & Thomashow, 2002). These genes are, therefore, candidates for being members of a CBF subregulon controlled by one of the transcription factors induced by the CBF activators. However, the existence of these subregulons remains to be demonstrated conclusively.

Signal transduction pathway components: The CBF regulon includes genes that encode phosphoinositide-specific phospholipase C-1 (AtPLC1; At5g58670), CBL (calcineurin B-like)-interacting protein kinase 25 (CIPK25; At5g25110) and a putative protein phosphatase 2C (At5g27930). As discussed later, low temperature signaling in Arabidopsis has been shown to involve transient cold-induced increases in calcium levels (Knight *et al.*, 1996; Tahtiharju *et al.*, 1997). In this regard, it is of interest that AtPLC1 has been proposed to function in calcium signaling pathways that transduce abiotic stress signals (Hirayama *et al.*, 1995; Xiong *et al.*, 2002a). In addition, members of the CBL-interacting protein kinase family, of which CIPK25 is a member, are thought to be the targets of calcium signals transduced via CBL calcium sensor proteins (Kolukisaoglu *et al.*, 2004). There are at least 25 members of this family in Arabidopsis, many of which have been reported to have roles in stress signal transduction (Kolukisaoglu *et al.*, 2004). For example, SOS2 (AtCIPK24) interacts with SOS3 (AtCBL4) to mediate salt-stress signaling (Liu & Zhu, 1998; Halfter *et al.*, 2000). More directly, AtCBL1 has been identified as a regulator of plant cold-stress signaling (Albrecht *et al.*, 2003; Cheong *et al.*, 2003). However, no role for AtCIPK25 in stress signaling has been reported to our knowledge and yeast two-hybrid assays suggest that AtCIPK25 does not interact with AtCBL1 (Kolukisaoglu *et al.*, 2004). Finally, the putative protein phosphatase 2C (encoded by At5g27930) contains a PP2C domain suggesting this protein is likely to have phosphatase activity. Phosphatases of this family have been implicated as negative regulators of low temperature signaling in Arabidopsis (Tahtiharju & Palva, 2001), but their precise role is unclear.

Five CBF regulon genes encode proteins that have been predicted to have protease or protease inhibitor activity (Seki *et al.*, 2001; Fowler & Thomashow, 2002; Maruyama *et al.*, 2004; Vogel *et al.*, 2005). A possible explanation for this is that these proteins could have a regulatory role during low temperature stress by modifying activity of a transcriptional regulator or component of a signal transduction pathway. It is also possible that cold-induced proteases degrade cellular proteins that have been denatured by low temperature. In support of this notion, Levy *et al.* (2004) identified a mutant in the chloroplast proteolytic machinery which exhibited impaired photoprotection and photosynthesis upon cold stress.

4.2.3.3 Biosynthetic proteins

Proline biosynthesis: A biochemical change that has been shown to occur with cold acclimation in Arabidopsis (Xin & Browse, 1998; Wanner & Junttila, 1999; Gilmour *et al.*, 2000) and other plants (Koster & Lynch, 1992) is an increase in free proline levels. Proline is a compatible solute that has cryoprotective properties, able to protect proteins and membranes against freezing damage *in vitro* (Rudolph & Crowe, 1985; Carpenter & Crowe, 1988). Moreover, Arabidopsis plants that produce elevated levels of proline are more freezing tolerant than control plants (Nanjo *et al.*, 1999). An enzyme with a key role in determining the levels of free proline is Δ^1-pyrroline-5-carboxylate synthetase (Yoshiba *et al.*, 1997). Arabidopsis has two genes, *P5CS1* and *2*, that encode Δ^1-pyrroline-5-carboxylate synthetase, the latter of which, *P5CS2*, is a member of the CBF regulon (Fowler & Thomashow, 2002). Low temperature and overexpression of CBF1, 2 or 3 results in increased levels of *P5CS2* transcripts and increased levels of free proline, which presumably contributes to the increase in freezing tolerance displayed by CBF-overexpressing plants (Xin & Browse, 1998; Gilmour *et al.*, 2000; Seki *et al.*, 2001; Fowler & Thomashow, 2002; Gilmour *et al.*, 2004).

Sugar metabolism: Another biochemical change that occurs during cold acclimation in Arabidopsis (Wanner & Junttila, 1999; Gilmour *et al.*, 2000) and a wide range of other plants (Levitt, 1980; Sakai & Larcher, 1987) is the accumulation of simple sugars including sucrose and raffinose. Like proline, sucrose and other sugars are compatible solutes that are able to stabilize membranes and proteins against freezing damage *in vitro* (Strauss & Hauser, 1986; Carpenter & Crowe, 1988). Overexpression of CBF1, 2 or 3 in non-acclimated Arabidopsis increases the levels of sugars including sucrose and raffinose, which also occurs with cold acclimation (Gilmour *et al.*, 2000, 2004). The increase in raffinose involves expression of galactinol synthase, which catalyzes the first committed step in the synthesis of raffinose (Fowler & Thomashow, 2002; Taji *et al.*, 2002). Transcripts for genes encoding two isoforms of galactinol synthase accumulate in response to both low temperature and CBF3 overexpression, along with several genes (At1g08890, At4g35300, At4g17550) that encode putative sugar transporters (Fowler & Thomashow, 2002; Seki *et al.*, 2001, 2002; Vogel *et al.*, 2005).

Fatty acid metabolism: The plasma membrane is a key site of freeze-induced cellular injury (Steponkus, 1984). In Arabidopsis, as in most plants, changes in membrane lipid composition occur during growth at low temperatures (Uemura *et al.*, 1995). Increased levels of unsaturated fatty acids, which lead to increased membrane fluidity, are associated with cold acclimation and increased freezing tolerance (Uemura & Steponkus, 1994; Uemura *et al.*, 1995; Smallwood & Bowles, 2002). The Arabidopsis *ADS2* gene encodes a $\Delta 9$-acyl-lipid desaturase (Fukuchi-Mizutani *et al.*, 1998) and is a member of the CBF regulon, as it is upregulated by both low temperature and CBF expression (Maruyama *et al.*, 2004). The gene, however, does not contain the CRT/DRE core sequence in the region of the promoter 1000 bp upstream of the start codon, and therefore, may not be a direct target of CBF activation, but may instead be upregulated by the action of a transcription factor that is regulated by CBF.

4.2.4   *Regulation of CBF gene expression in response to low temperature*

Accumulation of transcripts for *CBF1*, *CBF2* and *CBF3* can be detected within 15 min of transferring plants to low temperature (Gilmour *et al.*, 1998). Such rapid effects argue strongly that the 'thermometer' and 'signal transducer' involved in low temperature regulation of the *CBF* genes are present at warm, noninducing temperatures. Gilmour *et al.* (1998) thus suggested that a transcription factor, tentatively designated inducer of CBF expression (ICE), is present at warm temperature that recognizes DNA regulatory elements within the *CBF* promoter, the 'ICE boxes', and that either it, or another protein with which it interacts, is activated in response to low temperature by a signal transduction pathway that is also present at warm temperatures. Since then, it has been shown that the promoters of the *CBF* genes are responsive to low temperature (Shinwari *et al.*, 1998; Zarka *et al.*, 2003); *cis*-acting ICE DNA regulatory sequences have been determined (Zarka *et al.*, 2003); and proteins with roles in *CBF* expression, including the first ICE transcription factor (Chinnusamy *et al.*, 2003) have been identified. Progress in these areas is summarized below.

4.2.4.1 DNA regulatory elements controlling CBF expression

Gene fusion studies established that the promoters for *CBF1-3* are induced in response to low temperature (Shinwari *et al.*, 1998). Further mutational analysis of the *CBF2* promoter resulted in the identification of a 125 bp region that is sufficient to impart cold-responsive gene expression (Zarka *et al.*, 2003). Within this 125 bp segment are two regions, designated ICEr1 and ICEr2 (induction of CBF expression region 1 or 2), that are required for the segment to impart a robust cold response. The sequences are within 'boxes' that have been shown to be highly conserved among the CBF1, 2 and 3 promoters (Shinwari *et al.*, 1998). Presumably, there are transcription factors in nonacclimated plants that are able to bind to these elements. There are no obvious known transcription factor binding sites within the ICEr2 sequence, but ICEr1 contains the sequence CACATG, which includes a consensus recognition site for bHLH proteins, CANNTG (Massari & Murre, 2000). Thus, ICEr1 could be a site for binding of ICE1 (Chinnusamy *et al.*, 2003) or a related protein, described later.

4.2.4.2 Proteins with positive roles in CBF expression

ICE1: ICE1 is a nuclear gene that encodes a MYC-like bHLH protein that appears to have a fundamental role in regulating *CBF3* (Chinnusamy *et al.*, 2003). A dominant-negative allele of *ICE1*, *ice1*-1, results in almost complete elimination of *CBF3* transcript accumulation in response to low temperature. The *CBF3* promoter has five bHLH-binding sites, all of which can bind the ICE1 protein (Chinnusamy *et al.*, 2003). Significantly, however, accumulation of *CBF1* and *CBF2* transcripts in response to low temperature is little affected by the *ice1*-1 mutation (*CBF2* transcript levels may actually be somewhat higher in the *ice1*-1 mutant) (Chinnusamy *et al.*, 2003). Thus, ICE1 does not have a critical role in cold-induced expression of these two genes. However, Arabidopsis encodes an estimated 139 bHLH proteins (Riechmann *et al.*, 2000), some of which have DNA-binding domains that are quite similar to that present in ICE1 (Chinnusamy *et al.*, 2003). Moreover, the promoters of *CBF1* and *CBF2* contain potential bHLH-binding sites (the ICEr1 region of CBF2 includes a bHLH-binding site). Thus, bHLH proteins other than ICE1 are candidate regulators of *CBF1* and 2.

LOS4: *LOS4* is a nuclear gene that encodes an RNA helicase-like protein that positively affects *CBF* gene expression (Gong *et al.*, 2002). The recessive *los4*-1 mutation results in reduced accumulation of *CBF* transcripts in response to low temperature as well as reduced accumulation of transcripts for CBF-induced genes such as *RD29A* and *COR15A* (Gong *et al.*, 2002). Not unexpectedly, *los4*-1 mutant plants are less able to cold acclimate than wild-type plants, but in addition, they are chilling-sensitive. Overexpression of CBF3 suppresses the chilling-sensitive phenotype indicating a role for the CBF regulon in chilling tolerance (Gong *et al.*, 2002). The mechanism whereby the LOS4 protein affects accumulation of *CBF* transcripts in response to low temperature is not known.

4.2.4.3   *Proteins with negative roles in CBF expression*

HOS1: *HOS1* is a nuclear encoded gene that has a role in cold-induced expression of the *CBF* genes (Ishitani *et al.*, 1998). Recessive *hos1* mutations result in enhanced low-temperature-induced accumulation of *CBF* transcripts indicating that the HOS1 protein has a role in downregulating expression of the *CBF* genes. How this is accomplished is not known. However, Lee *et al.* (2002) have suggested that it involves the regulation of protein degradation. The HOS1 protein encodes a novel protein that has a RING finger motif. Since many proteins with RING fingers have E3 ubiquitin ligase activity, Lee *et al.* (2002) proposed that HOS1 may be involved in targeting a protein for degradation by the 26S proteosome. They showed, using HOS1-GFP fusions, that the HOS1 protein is located in the cytoplasm at warm temperatures, but at cold temperature, it accumulates in the nucleus. Thus, it was proposed that at low temperature HOS1 enters the nucleus and targets for degradation a factor involved in the induction of *CBF* transcription, thereby downregulating the expression of *CBF* genes and also, consequently, the *COR* genes (CBF-regulated genes). One obvious possibility is that ICE1 is a target for HOS1. Surprisingly, *hos1* mutants are less freezing tolerant than wild-type plants. The reason for this is not immediately obvious, since the expression of at least some of the CBF cold-response pathway genes are induced to higher levels in the mutant.

LOS1: *LOS1* is a nuclear gene that also affects the expression of *CBF* genes (Guo *et al.*, 2002). The recessive *los1*-1 mutation results in an increase in *CBF* transcript accumulation in response to low temperature. Thus, like *HOS1*, *LOS1* would appear to have a role in downregulating *CBF* expression. However, unlike the recessive *hos1* mutation, the recessive *los1*-1 mutation results in a decrease in cold-induced *COR* gene expression (Guo *et al.*, 2002). A possible explanation for these different effects is suggested by the fact that *LOS1* encodes a translation elongation factor 2-like protein. At warm temperatures, the recessive *los1*-1 mutation has little, if any, effect on protein synthesis. However, at low temperature, protein synthesis is nearly eliminated by the *los1*-1 mutation. Thus, it is not surprising that cold-induction of *COR* and other *CBF*-targeted genes is dramatically reduced in *los1*-1 plants; presumably the *CBF* proteins are not synthesized at low temperature in this mutant and thus cannot participate in the induction of *COR* and other cold-regulated genes. The reason why *CBF* transcripts accumulate to higher levels in cold-treated *los1*-1 plants, as compared to cold-treated wild-type plants, remains to be determined. However, as suggested by Guo *et al.* (2002), it would seem to indicate that the CBF proteins either feedback-suppress their own transcription or induce the expression of proteins that downregulate *CBF* expression. Recent results with CBF2 support the former notion. Finally, as would be anticipated, the *los1*-1 mutation results in a decrease in the ability of the plants to cold acclimate.

CBF2: A recent study has provided evidence that CBF2 is a negative regulator of *CBF1* and *CBF3* expression (Novillo *et al.*, 2004). In particular, it was reported that a T-DNA insertion that inactivated the expression of the *CBF2* gene resulted in constitutive expression of *CBF1* and *CBF3*, as well as *CBF*-targeted genes *LTI78*, *KIN1* and *COR47*. The *cbf2* mutant plants were more freezing tolerant than wild-type plants when compared after growth under either nonacclimating or cold-acclimating conditions. Complementation of the *cbf2* phenotype with the *CBF2* gene under the control of its native promoter indicated that the T-DNA insertion in the *cbf2* line was not directly inducing expression of *CBF1* and *CBF3*. Novillo *et al.* (2004) reported that maximal induction of *CBF2* in response to low temperature occurred approximately 1 h after that of *CBF1* and *CBF3*. Thus, the investigators proposed that upon exposure to low temperature, CBF regulators such as ICE1 rapidly induce the expression of *CBF1* and *3* followed a short while later by the induction of *CBF2*, which then suppresses transcription of *CBF1* and *CBF3*. This model supposes that the negative regulation of *CBF1* and *CBF3* by *CBF2* may be required for the tightly controlled and transient expression of *CBF* genes. Repression of *CBF1* and *3* expression by *CBF2* may explain how elimination of protein synthesis in the *los1* mutant leads to increased accumulation of *CBF* transcripts (Guo *et al.*, 2002).

CAX1: The *CAX1* (cation exchanger 1) nuclear gene also appears to be a negative regulator of *CBF1*, *2* and *3* (Catalá *et al.*, 2003). Transcript levels for *CBF1*, *2* and *3* are considerably higher in cold-treated plants carrying the recessive *cax1-3* and *cax1-4* mutations, as compared to cold-treated wild-type plants (Catalá *et al.*, 2003). Accordingly, the transcript levels for the CBF-target genes – *KIN1*, *LTI78* and *COR47* – are also considerably higher. Not surprisingly, the increase in *CBF* and CBF-target gene expression in the *cax1-3* and *cax1-4* mutant plants correlates with an increase in freezing tolerance after cold acclimation relative to wild-type plants. *CAX1* encodes a vacuolar Ca^{2+}/H^{+} antiporter localized to the vacuolar membrane (Cheng *et al.*, 2003). It has been proposed to be involved in the regulation of intracellular Ca^{2+} levels and in particular to have a role in returning cytosolic Ca^{2+} concentrations to basal levels following a transient increase in response to external stimuli such as low temperature (Hirschi, 1999). It has, therefore, been proposed that when plants are exposed to low temperature, restoration of $[Ca^{2+}]_{cyt}$ to basal levels via CAX1 is required for the correct generation of a Ca^{2+} signal, which ultimately affects the expression of *CBF* genes (Catalá *et al.*, 2003). Mutations in *CAX1* might prevent the restoration leading to aberrantly high $[Ca^{2+}]_{cyt}$ and excessive accumulation of *CBF* transcripts (Catalá *et al.*, 2003).

4.2.4.4 Other potential CBF regulatory proteins

FIERY1 (*FRY1*, Xiong *et al.*, 2001) and *FIERY2* (*FRY2*, Xiong *et al.*, 2002b) are nuclear genes that encode proteins that potentially have roles in regulating *CBF* expression. The recessive *fry*-1 mutation was reported to result in slightly sustained accumulation of *CBF2* transcripts; the levels in the mutant were 1.8 times higher than those found in wild-type plants after a 6 h cold treatment (Xiong *et al.*, 2001). Similarly, recessive *fry2* mutations were reported to result in higher levels of *CBF1, 2* and *3* transcripts at 6 h of cold-treatment, though again, the effects were small (Xiong *et al.*, 2002b). *FRY1* was shown to correspond to the *SAL1* gene (Quintero *et al.*, 1996) that had previously been shown to encode a bifunctional protein with inositol polyphosphate-1-phosphatase activity. The levels of inositol-1,4,5-trisphosphate (IP3) in unstressed *fry1*-1 plants are about tenfold higher than those in nonstressed wild-type plants (Xiong *et al.*, 2002a). The increase in IP3 levels caused by the *fry1*-1 mutation might cause upregulation of the *CBF* genes by causing an increase in calcium levels (see below for the role of calcium in cold-regulated gene expression), but this remains to be determined. *FRY2* encodes a novel protein with two double-stranded RNA-binding domains and a conserved domain with similarity to the RNPII CTD phoshatase, FCP1, found in yeast and animals. Thus, FRY2 is probably involved in transcription, but how it affects cold-regulated expression of the *CBF* genes is unknown.

4.2.4.5 Light and circadian rhythms

Kim *et al.* (2002) have reported a role for light in controlling expression of the *CBF* cold-response pathway. The investigators found that cold-induced expression of a CRT/DRE::GUS reporter fusion (the GUS reporter was placed under control of four copies of the CRT/DRE regulatory element) was much higher when plants were incubated in the light as opposed to the dark. Moreover, they found that in the light, Arabidopsis *phyB* mutants carrying the CRT/DRE::GUS reporter expressed much lower levels of the *GUS* reporter gene in response to low temperature than did wild-type plants carrying the reporter gene. These results suggest that PHYB has a role in mediating a light-signaling event involved in cold-induced gene expression mediated through the CRT/DRE. The question raised is whether low temperature induction of the *CBF* genes is also affected by light. The results of Kim *et al.* (2002) indicate that they are; accumulation of *CBF1, 2* and *3* transcripts after a 1 h cold treatment was considerably greater when plants were treated in the light as compared to the dark. Additional study will be required to understand the molecular basis for the observed effects of light on *CBF* and CRT/DRE-driven gene expression.

CBF expression is also influenced by the circadian rhythms (Harmer *et al.*, 2000). Transcript levels for *CBF3* exhibit circadian-regulated cycling at warm temperatures where it is normally considered to be uninduced. Transcript levels for CBF-targeted *COR* genes also cycle, but with the peak phase delayed by approximately 8 h. Such a delay would be expected if the clock-controlled peak of *CBF* induction was followed by a peak in CBF protein which then induced a peak of *COR* gene expression. These results are particularly intriguing since circadian rhythms in chilling and freezing tolerance have been described for several plant species including cotton (Rikin *et al.*, 1993) and soybean (Couderchet & Koukkari, 1987). In both these species, the clock regulates the development of a low-temperature resistant phase that peaks at the end of the light phase. While cycling of low-temperature tolerance has not been observed in Arabidopsis, the phase of low-temperature resistance in soybean and cotton coincides with the clock-regulated peak of CBF-target gene induction observed in Arabidopsis (Harmer *et al.*, 2000).

4.2.4.6 Role of calcium

The induction of at least some Arabidopsis cold-regulated genes, including the CBF-targeted gene *COR6.6*, has been linked to calcium signaling (Knight *et al.*, 1996; Tahtiharju *et al.*, 1997). In particular, transfer of Arabidopsis from warm to cold temperature results in a rapid, transient elevation of cytoplasmic calcium levels, which is required for induction of *COR6.6* (Knight *et al.*, 1996; Tahtiharju *et al.*, 1997). Similar observations have been made for cold-regulated genes in alfalfa and *Brassica napus* (Monroy & Dhindsa, 1995; Sangwan *et al.*, 2001). A question that now needs to be addressed is whether calcium-dependent low temperature induction of gene expression proceeds through the CBF cold-response pathway. That is, does the transient increase in calcium brought about by exposure to low temperature result initially in induction of the *CBF* genes followed by the action of CBF proteins at the promoters of *COR6.6* and other CBF-targeted genes? The requisite tools would seem to be available to answer this question in the near future.

4.2.4.7 Role of ABA

The CBF cold-response pathway is generally thought to be an ABA-independent pathway (Liu *et al.*, 1998; Medina *et al.*, 1999). This conclusion is supported by findings indicating that neither the *CBF1-3* genes nor the CRT/DRE element are responsive to ABA (Liu *et al.*, 1998; Medina *et al.*, 1999). Recent results, however, suggest that the situation may be more complicated. In particular, Knight *et al.* (2004) have been able to detect an increase in *CBF1-3* transcript levels in plants treated with exogenously applied ABA and have shown that the promoters for each of the three genes is ABA responsive. At present, it is unclear why these experiments yielded results different from those previously described by two other research groups (Liu *et al.*, 1998; Medina *et al.*, 1999). It is possible that differences in culture conditions or developmental stage of growth are involved. Whatever the precise reason, the possibility has been raised that the CBF cold-response pathway is not completely independent of ABA action, which may explain, in part, the genetic 'cross-talk' that has been observed between cold- and ABA-regulated expression of the CBF-target gene, *COR78/RD29a* (Ishitani *et al.*, 1997).

4.3 Conservation of the CBF cold-response pathway

As mentioned previously, plants vary greatly in their ability to survive cold temperatures. It is now clear that the CBF cold-response pathway is a component of cold tolerance in Arabidopsis. An important question that then arises is whether the pathway contributes to cold tolerance in other plants as well. In addition, can the differences in freezing, as well as chilling tolerance displayed by plants be traced, at least in part, to differences in the CBF cold-response pathways in these plants? Studies to address these questions are in their infancy. However, the picture that is beginning to emerge is that at least portions of the pathway are highly conserved in plants. There are, for instance, cold-regulated genes encoding CBF-like proteins (that include the CBF signature sequences) in *B. napus* (Jaglo *et al.*, 2001; Gao *et al.*, 2002), tomato (Jaglo *et al.*, 2001; Zhang *et al.*, 2004), wheat (Jaglo *et al.*, 2001) and rice (Dubouzet *et al.*, 2003). In addition, there are indications that differences in the CBF cold-response pathway may contribute to differences in cold tolerance between Arabidopsis and tomato (Zhang *et al.*, 2004). Below is a summary of results from selected systems (which are currently the best developed) to illustrate these points.

4.3.1 Brassica napus

Like Arabidopsis, *B. napus* is a member of the *Brassicaceae* family. In addition, it can cold acclimate. Thus, it might be expected that *B. napus* would have a CBF cold-response pathway. Indeed, *B. napus* encodes at least four CBF-like proteins that have the CBF signature sequences (Jaglo *et al.*, 2001; Gao *et al.*, 2002). The proteins are ~90% identical in amino acid sequence to each other and ~75% identical in sequence to Arabidopsis CBF1. An alignment of the *B. napus* proteins with Arabidopsis CBF1 shows that the sequence identity extends throughout the protein, though the greatest identity is within the AP2/ERF DNA-binding domain. As occurs with the *CBF1-3* genes in Arabidopsis, transcript levels for the *B. napus CBF*-like genes accumulate rapidly (within 30 min) upon exposing plants to low temperature, reaching maximum levels within a few hours of plants being transferred to low temperature, after which time they decrease, but at 24 h remain elevated over the level found in nonacclimated plants (Jaglo *et al.*, 2001; Gao *et al.*, 2002). Constitutive expression of Arabidopsis *CBF1*, *CBF2* or *CBF3* in transgenic *B. napus* activates expression of homologues of CBF-targeted Arabidopsis *COR* genes, namely *Bn115* and *Bn28*, which are homologues of Arabidopsis *COR15a* and COR6.6, respectively (Jaglo *et al.*, 2001). In addition, the expression of Arabidopsis *CBF* genes in transgenic *B. napus* results in an increase in freezing tolerance (Jaglo *et al.*, 2001). Electrolyte leakage experiments indicated that leaf tissue from nonacclimated control *B. napus* plants had an EL_{50} value of about $-2.1°C$, whereas leaf tissue from nonacclimated CBF-expressing plants had an EL_{50} value of about $-4.7°C$. *CBF* expression also caused an increase in the freezing tolerance of cold-acclimated plants. Leaf tissue from cold-acclimated control *B. napus* plants had an EL_{50} value of about $-8.1°C$, whereas leaf tissue from cold-acclimated *CBF*-expressing plants had an EL_{50} value of $-12.7°C$. Taken together, the results indicate that *B. napus* has a CBF cold-response pathway that is closely related to the Arabidopsis CBF cold-response pathway.

4.3.2   Tomato

In contrast to Arabidopsis and *B. napus*, tomato (*Lycopersicon esculentum*) is a chilling-sensitive plant that cannot cold acclimate. A question thus raised is whether tomato has a functional CBF response pathway. Studies indicate that it does, but suggest that the CBF regulon of tomato is smaller in gene number and less diverse than the CBF regulon of Arabidopsis, differences that may contribute to the differences in freezing tolerance exhibited by these plants (Zhang *et al.*, 2004). Specifically, like Arabidopsis, tomato has three genes encoding *CBF*-like proteins, designated *LeCBF1-3*, that are present in tandem array in the genome (Jaglo *et al.*, 2001; Zhang *et al.*, 2004). The amino acid sequences of the three tomato CBF proteins are 70–84% identical to each other and 51–59% identical to the sequence of the Arabidopsis CBF1-3 proteins and contain the PKK/RPAGRxKFxETRHP and DSAWR CBF signature sequences. Like the *CBF1-3* genes of Arabidopsis, the *LeCBF1* gene of tomato is induced in response to low temperature and mechanical agitation; *LeCBF2* and *LeCBF3* are not induced in response to low temperature, but are in response to mechanical agitation. Like the *CBF1-3* genes of Arabidopsis, the *LeCBF1-3* genes of tomato are not induced in response to drought or high salinity (Jaglo *et al.*, 2001; Zhang *et al.*, 2004). Finally, constitutive expression of *LeCBF1* in transgenic Arabidopsis plants activates expression of *COR* genes and increases plant freezing tolerance indicating that *LeCBF1* encodes a functional CBF protein (Zhang *et al.*, 2004).

Taken together, these results indicate that tomato has the ability to sense low temperature and activate expression of at least one functional *CBF* gene. Thus, tomato would appear to have a functional upstream portion of the CBF cold-response pathway. To assess downstream components, transgenic tomato plants that constitutively express *CBF* genes under the control of the CaMV 35S promoter have been created and tested for stress tolerance and changes in gene expression. The results indicate that constitutive expression of either *AtCBF1*, *AtCBF3* or *LeCBF1* in transgenic tomato has no effect on freezing tolerance (Hsieh *et al.*, 2002b; Zhang *et al.*, 2004). Further, gene expression studies on transgenic tomato plants overexpressing either *LeCBF1* or *AtCBF3* resulted in the identification of only three genes that were members of the tomato CBF regulon (i.e. were induced in response to both low temperature and CBF overexpression) (Zhang *et al.*, 2004); two of the CBF regulon genes were dehydrins, each of which had at least one potential CRT/DRE element in their promoter. In these experiments, a cDNA microarray representing approximately 8000 tomato genes was used to monitor gene expression; about 25% of the total genes were tested for expression. Thus, if the genes on the microarray are generally representative of the genome, then there might be some 10–15 genes that comprise the CBF regulon in tomato, which is still considerably less than the 109 genes that have been assigned to the Arabidopsis CBF regulon. Additional experiments indicated that of eight tomato genes that were likely orthologs of Arabidopsis CBF regulon genes, none were responsive to CBF overexpression in tomato. From these results, it was concluded that tomato has a complete CBF cold-response pathway, but that the tomato CBF regulon differs from that of Arabidopsis, appearing to be considerably smaller and less diverse in function (Zhang *et al.*, 2004).

What might account for the apparent differences in size and complexity of the tomato and Arabidopsis CBF regulons? One possibility is differences in the proteins required for CBF activity. As described earlier, in Arabidopsis, the *SFR6* gene is required for the CBF1-3 transcription factors to stimulate expression of the *COR* genes (Knight *et al.*, 1999; Boyce *et al.*, 2003). Perhaps tomato encodes a 'weak' allele of *SFR6* (or other gene required for high-level CBF function), which contributes to most CBF-targeted genes being expressed at low levels. Alternatively, the apparent differences in tomato and Arabidopsis CBF regulons might reflect differences in the distribution of functional

CRT/DRE elements within the genomes of these two plants. Additional study is required to critically distinguish between these two possibilities.

One last point is that overexpression of *AtCBF1* in transgenic tomato plants has been reported to result in an increase in both chilling and drought tolerance (Hsieh *et al.*, 2002a, 2002b). Thus, it would appear that the tomato CBF regulon has some functions in common with those of the Arabidopsis CBF regulon.

4.3.3   Rice

Rice (*Oryza sativa*) is not only one of the world's most important crop species but it has become the major model for monocotyledonous crop species due to its significant commercial value and relatively small ($\sim$430 Mb) diploid genome. Rice is freezing-sensitive (it cannot cold acclimate) and is considered to be chilling-sensitive, though cultivars vary greatly in their tolerance to low temperature; *indica* rice subspecies, associated with tropical environments, are more sensitive to low temperature than *japonica* subspecies (Andaya & Mackill, 2003a, 2003b). Despite the difference in cold tolerance between rice and Arabidopsis, screens of the rice genome sequence and cDNA libraries have led to the identification of a family of four genes encoding CBF-like proteins, OsDREB1a-d (Choi *et al.*, 2002; Dubouzet *et al.*, 2003). The OsDREB1 proteins share extensive homology with CBF proteins from Arabidopsis and include the CBF signature sequences PKK/RPAGRxKFxETRHP and DSAWR, though three of the four rice sequences have 'L' substituted for the last 'R' (Dubouzet *et al.*, 2003). Transcripts for *OsDREB1a* and *OsDREB1b* accumulated within 40 min of exposure to low temperature. *OsDREB1a* expression was also responsive to salt stress and wounding. In contrast to the Arabidopsis *CBF* genes, Arabidopsis *OsDREB1c* appeared to be constitutively expressed. Accumulation of *OsDREB1d* transcripts has not yet been detected, though a variety of abiotic stresses has been tested (Dubouzet *et al.*, 2003). The OsDREB1a protein was shown to bind the CRT/DRE sequence in a sequence specific manner and to increase freezing tolerance of whole plants when constitutively expressed in transgenic Arabidopsis plants. Expression of CBF-target genes was examined in these transgenic Arabidopsis plants using transcriptome profiling and northern analysis. Analysis of the promoter regions of the six genes identified as being upregulated by expression of *OsDREB1a*, and of several Arabidopsis CBF-target genes that were not upregulated, indicated that those genes with at least one GCCGAC sequence in their promoter region were OsDREB1a targets while those containing ACCGAC sequences were not. Gel mobility shift assays confirmed the preferential binding of OsDREB1a to GCCGAC compared to ACCGAC (Dubouzet *et al.*, 2003). These results indicate that as in chilling- and freezing-sensitive tomato, the upstream portion of the CBF cold-response pathway appears to be present in rice. It is now of significant importance to determine the nature of the CBF regulon in rice.

4.4 Concluding remarks

We have learned much about the Arabidopsis CBF cold-response pathway since the first report of a CBF transcription factor in 1997 (Stockinger *et al.*, 1997). As we hope is evident from the presentation above, much has been learned about the CBF transcription factors themselves, their regulation, and the genes that they control. However, much remains to be determined. We know little about the nature of the 'thermometer' used by Arabidopsis to sense temperature and regulate expression of the *CBF/DREB1* genes. We know little about how the CBF transcription factors stimulate gene expression. We know a considerable amount about the identity of the genes that comprise the CBF regulon in Arabidopsis, but for most of the genes, we lack an understanding of how they contribute to low temperature tolerance. We now know that at least portions of the CBF cold-response pathway are conserved among a wide range of plant species, but we do not know to what extent differences in these pathways contribute to the differences in cold tolerance observed in plants. Even in Arabidopsis, we are not certain to what extent low temperature tolerance is conditioned by the CBF cold-response pathway. Is the CBF cold-response pathway the major contributor to freezing and chilling tolerance or are there other genetic pathways that contribute as suggested by the *eskimo1* and *ada2* mutations? Indeed, expression profiling experiments indicate the existence of many cold-regulated genes that fall outside the CBF cold-response pathway (Fowler & Thomashow, 2002) and a recent study has identified a homeodomain transcription factor, HOS9, that appears to regulate a CBF-independent pathway that contributes to freezing tolerance (Zhu *et al.*, 2004). Finally, can we use the information that has been generated about the CBF cold-response pathway to improve the stress tolerance of agriculturally important plants? The next few years should bring significant new insights to these and related issues.

References

Albrecht, V., Weinl, S., Blazevic, D., D'Angelo, C., Batistic, O., Kolukisaoglu, U., Bock, R., Schulz, B., Harter, K. & Kudla, J. (2003) The calcium sensor CBL1 integrates plant responses to abiotic stresses. *The Plant Journal*, **36**, 457–470.

Allen, M. D., Yamasaki, K., Ohme-Takagi, M., Tateno, M. & Suzuki, M. (1998) A novel mode of DNA recognition by a beta-sheet revealed by the solution structure of the GCC-box binding domain in complex with DNA. *EMBO Journal*, **17**, 5484–5496.

Amasino, R. (2004) Take a cold flower. *Nature Genetics*, **36**, 111–112.

Andaya, V. C. & Mackill, D. J. (2003a) Mapping of QTLs associated with cold tolerance during the vegetative stage in rice. *Journal of Experimental Botany*, **54**, 2579–2585.

Andaya, V. C. & Mackill, D. J. (2003b) QTLs conferring cold tolerance at the booting stage of rice using recombinant inbred lines from a *japonica* × *indica* cross. *Theoretical and Applied Genetics*, **106**, 1084–1090.

Artus, N. N., Uemura, M., Steponkus, P. L., Gilmour, S. J., Lin, C. & Thomashow, M. F. (1996) Constitutive expression of the cold-regulated *Arabidopsis thaliana COR15a* gene affects both chloroplast and protoplast freezing tolerance. *Proceedings of the National Academy of Sciences of the United States of America*, **93**, 13404–13409.

Ausin, I., Alonso-Blanco, C., Jarillo, J. A., Ruiz-Garcia, L. & Martinez-Zapater, J. M. (2004) Regulation of flowering time by FVE, a retinoblastoma-associated protein. *Nature Genetics*, **36**, 162–166.

Baker, S. S., Wilhelm, K. S. & Thomashow, M. F. (1994) The 5′-region of *Arabidopsis thaliana cor15a* has *cis*-acting elements that confer cold-, drought- and ABA-regulated gene expression. *Plant Molecular Biology*, **24**, 701–713.

Bertrand, C., Bergounioux, C., Domenichini, S., Delarue, M. & Zhou, D. X. (2003) Arabidopsis histone acetyltransferase AtGCN5 regulates the floral meristem activity through the WUSCHEL/AGAMOUS pathway. *Journal of Biological Chemistry*, **278**, 28246–28251.

Boyce, J. M., Knight, H., Deyholos, M., Openshaw, M. R., Galbraith, D. W., Warren, G. & Knight, M. R. (2003) The *sfr6* mutant of Arabidopsis is defective in transcriptional activation via CBF/DREB1 and DREB2 and shows sensitivity to osmotic stress. *Plant Journal*, **34**, 395–406.

Bravo, L. A., Gallardo, J., Navarrete, A., Olave, N., Martinez, J., Alberdi, M., Close, T. J. & Corcuera, L. J. (2003) Cryoprotective activity of a cold-induced dehydrin purified from barley. *Physiologia Plantarum*, **118**, 262–269.

Bray, E. A. (1993) Molecular responses to water deficit. *Plant Physiology*, **103**, 1035–1040.

Breton, G., Danyluk, J., Charron, J. B. & Sarhan, F. (2003) Expression profiling and bioinformatic analyses of a novel stress-regulated multispanning transmembrane protein family from cereals and Arabidopsis. *Plant Physiology*, **132**, 64–74.

Brown, C. E., Lechner, T., Howe, L. & Workman, J. L. (2000) The many HATs of transcription coactivators. *Trends in Biochemical Sciences*, **25**, 15–19.

Bubier, J. & Schlappi, M. (2004) Cold induction of *EARLI*, a putative Arabidopsis lipid transfer protein, is light and calcium dependent. *Plant Cell and Environment*, **27**, 929–936.

Carpenter, J. F. & Crowe, J. H. (1988) The mechanism of cryoprotection of proteins by solutes. *Cryobiology*, **25**, 244–255.

Catalá, R., Santos, E., Alonso, J. M., Ecker, J. R., Martinez-Zapater, J. M. & Salinas, J. (2003) Mutations in the Ca^{2+}/H^{+} transporter CAX1 increase CBF/DREB1 expression and the cold-acclimation response in Arabidopsis. *Plant Cell*, **15**, 2940–2951.

Cheng, N. H., Pittman, J. K., Barkla, B. J., Shigaki, T. & Hirschi, K. D. (2003) The Arabidopsis *cax1* mutant exhibits impaired ion homeostasis, development, and hormonal responses and reveals interplay among vacuolar transporters. *Plant Cell*, **15**, 347–364.

Cheong, Y. H., Kim, K. N., Pandey, G. K., Gupta, R., Grant, J. J. & Luan, S. (2003) CBL1, a calcium sensor that differentially regulates salt, drought, and cold responses in Arabidopsis. *Plant Cell*, **15**, 1833–1845.

Chinnusamy, V., Ohta, M., Kanrar, S., Lee, B. H., Hong, X., Agarwal, M. & Zhu, J. K. (2003) ICE1: a regulator of cold-induced transcriptome and freezing tolerance in Arabidopsis. *Genes and Development*, **17**, 1043–1054.

Choi, D. W., Rodriguez, E. M. & Close, T. J. (2002) Barley *CBF3* gene identification, expression pattern, and map location. *Plant Physiology*, **129**, 1781–1787.

Close, T. J. (1997) Dehydrins: a commonality in the response of plants to dehydration and low temperature. *Physiologia Plantarum*, **100**, 291–296.

Couderchet, M. & Koukkari, W. L. (1987) Daily variations in the sensitivity of soybean seedlings to low temperature. *Chronobiology International*, **4**, 537–541.

Dubouzet, J. G., Sakuma, Y., Ito, Y., Kasuga, M., Dubouzet, E. G., Miura, S., Seki, M., Shinozaki, K. & Yamaguchi-Shinozaki, K. (2003) *OsDREB* genes in rice, *Oryza sativa* L., encode transcription activators that function in drought-, high-salt- and cold-responsive gene expression. *Plant Journal*, **33**, 751–763.

Dunker, A. K., Lawson, J. D., Brown, C. J., Williams, R. M., Romero, P., Oh, J. S., Oldfield, C. J., Campen, A. M., Ratliff, C. M., Hipps, K. W., Ausio, J., Nissen, M. S., Reeves, R., Kang, C., Kissinger, C. R., Bailey, R. W., Griswold, M. D., Chiu, W., Garner, E. C. & Obradovic, Z. (2001) Intrinsically disordered protein. *Journal of Molecular Graphics and Modelling*, **19**, 26–59.

Fowler, D. B., Dvorak, J. & Gusta, L. V. (1977) Comparative cold hardiness of several Triticum species and *Secale cereale* L. *Crop Science*, **17**, 941–943.

Fowler, S. & Thomashow, M. F. (2002) Arabidopsis transcriptome profiling indicates that multiple regulatory pathways are activated during cold acclimation in addition to the CBF cold response pathway. *Plant Cell*, **14**, 1675–1690.

Fukuchi-Mizutani, M., Tasaka, Y., Tanaka, Y., Ashikari, T., Kusumi, T. & Murata, N. (1998) Characterization of delta 9 acyl-lipid desaturase homologues from *Arabidopsis thaliana*. *Plant and Cell Physiology*, **39**, 247–253.

Gao, M. J., Allard, G., Byass, L., Flanagan, A. M. & Singh, J. (2002) Regulation and characterization of four CBF transcription factors from *Brassica napus*. *Plant Molecular Biology*, **49**, 459–471.

Gilmour, S. J., Fowler, S. G. & Thomashow, M. F. (2004) Arabidopsis transcriptional activators CBF1, CBF2 and CBF3 have matching functional activities. *Plant Molecular Biology*, **54**, 767–781.

Gilmour, S. J., Sebolt, A. M., Salazar, M. P., Everard, J. D. & Thomashow, M. F. (2000) Overexpression of the Arabidopsis CBF3 transcriptional activator mimics multiple biochemical changes associated with cold acclimation. *Plant Physiology*, **124**, 1854–1865.

Gilmour, S. J., Zarka, D. G., Stockinger, E. J., Salazar, M. P., Houghton, J. M. & Thomashow, M. F. (1998) Low temperature regulation of the *Arabidopsis* CBF family of AP2 transcriptional activators as an early step in cold-induced *COR* gene expression. *The Plant Journal*, **16**, 433–442.

Gong, Z. Z., Lee, H., Xiong, L. M., Jagendorf, A., Stevenson, B. & Zhu, J. K. (2002) RNA helicase-like protein as an early regulator of transcription factors for plant chilling and freezing tolerance. *Proceedings of the National Academy of Sciences of the United States of America*, **99**, 11507–11512.

Guo, Y., Xiong, L., Ishitani, M. & Zhu, J. K. (2002) An Arabidopsis mutation in translation elongation factor 2 causes superinduction of *CBF/DREB1* transcription factor genes but blocks the induction of their downstream targets under low temperatures. *Proceedings of the National Academy of Sciences of the United States of America*, **99**, 7786–7791.

Haake, V., Cook, D., Riechmann, J. L., Pineda, O., Thomashow, M. F. & Zhang, J. Z. (2002) Transcription factor CBF4 is a regulator of drought adaptation in Arabidopsis. *Plant Physiology*, **130**, 639–648.

Halfter, U., Ishitani, M. & Zhu, J. K. (2000) The Arabidopsis SOS2 protein kinase physically interacts with and is activated by the calcium-binding protein SOS3. *Proceedings of the National Academy of Sciences of the United States of America*, **97**, 3735–3740.

Hao, D. Y., Yamasaki, K., Sarai, A. & Ohme-Takagi, M. (2002) Determinants in the sequence specific binding of two plant transcription factors, CBF1 and NtERF2, to the DRE and GCC motifs. *Biochemistry*, **41**, 4202–4208.

Hara, M., Terashima, S., Fukaya, T. & Kuboi, T. (2003) Enhancement of cold tolerance and inhibition of lipid peroxidation by citrus dehydrin in transgenic tobacco. *Planta*, **217**, 290–298.

Harmer, S. L., Hogenesch, L. B., Straume, M., Chang, H. S., Han, B., Zhu, T., Wang, X., Kreps, J. A. & Kay, S. A. (2000) Orchestrated transcription of key pathways in Arabidopsis by the circadian clock. *Science*, **290**, 2110–2113.

Hincha, D. K., Heber, U. & Schmitt, J. M. (1990) Proteins from frost-hardy leaves protect thylakoids against mechanical freeze–thaw damage in vitro. *Planta*, **180**, 416–419.

Hincha, D. K., Schmitt, J. M. & Heber, U. (1988) Cryoprotective proteins in plants. *Cryobiology*, **25**, 557–558.

Hirayama, T., Ohto, C., Mizoguchi, T. & Shinozaki, K. (1995) A gene encoding a phosphatidylinositol-specific phospholipase C is induced by dehydration and salt stress in *Arabidopsis thaliana*. *Proceedings of the National Academy of Sciences of the United States of America*, **92**, 3903–3907.

Hirschi, K. D. (1999) Expression of Arabidopsis *CAX1* in tobacco: altered calcium homeostasis and increased stress sensitivity. *Plant Cell*, **11**, 2113–2122.

Horvath, D. P., McLarney, B. K. & Thomashow, M. F. (1993) Regulation of *Arabidopsis thaliana* L. (Heyn) *cor78* in response to low temperature. *Plant Physiology*, **103**, 1047–1053.

Hsieh, T. H., Lee, J. T., Charng, Y. Y. & Chan, M. T. (2002a) Tomato plants ectopically expressing Arabidopsis *CBF1* show enhanced resistance to water deficit stress. *Plant Physiology*, **130**, 618–626.

Hsieh, T. H., Lee, J. T., Yang, P. T., Chiu, L. H., Charng, Y. Y., Wang, Y. C. & Chan, M. T. (2002b) Heterology expression of the Arabidopsis *C-repeat/dehydration response element binding factor 1* gene confers elevated tolerance to chilling and oxidative stresses in transgenic tomato. *Plant Physiology*, **129**, 1086–1094.

Ishitani, M., Xiong, L. M., Stevenson, B. & Zhu, J. K. (1997) Genetic analysis of osmotic and cold stress signal transduction in Arabidopsis: interactions and convergence of abscisic acid-dependent and abscisic acid-independent pathways. *Plant Cell*, **9**, 1935–1949.

Ishitani, M., Xiong, L. M., Lee, H. J., Stevenson, B. & Zhu, J. K. (1998) *HOS1*, a genetic locus involved in cold-responsive gene expression in Arabidopsis. *Plant Cell*, **10**, 1151–1161.

Jaglo, K. R., Kleff, S., Amundsen, K. L., Zhang, X., Haake, V., Zhang, J. Z., Deits, T. & Thomashow, M. F. (2001) Components of the Arabidopsis C-repeat/dehydration-responsive element binding factor cold-response pathway are conserved in *Brassica napus* and other plant species. *Plant Physiology*, **127**, 910–917.

Jaglo-Ottosen, K. R., Gilmour, S. J., Zarka, D. G., Schabenberger, O. & Thomashow, M. F. (1998) *Arabidopsis CBF1* overexpression induces *COR* genes and enhances freezing tolerance. *Science*, **280**, 104–106.

Kasuga, M., Liu, Q., Miura, S., Yamaguchi-Shinozaki, K. & Shinozaki, K. (1999) Improving plant drought, salt, and freezing tolerance by gene transfer of a single stress-inducible transcription factor. *Nature Biotechnology*, **17**, 287–291.

Kim, H. J., Hyun, Y., Park, J. Y., Park, M. J., Park, M. K., Kim, M. D., Kim, H. J., Lee, M. H., Moon, J., Lee, I. & Kim, J. (2004) A genetic link between cold responses and flowering time through *FVE* in *Arabidopsis thaliana*. *Nature Genetics*, **36**, 167–171.

Kim, H. J., Kim, Y. K., Park, J. Y. & Kim, J. (2002) Light signalling mediated by phytochrome plays an important role in cold-induced gene expression through the C-repeat/dehydration responsive element (C/DRE) in *Arabidopsis thaliana*. *Plant Journal*, **29**, 693–704.

Kiyosue, T., Yamaguchi-Shinozaki, K. & Shinozaki, K. (1994) Characterization of two cDNAs (ERD10 and ERD14) corresponding to genes that respond rapidly to dehydration stress in *Arabidopsis thaliana*. *Plant and Cell Physiology*, **35**, 225–231.

Knight, H., Trewavas, A. J. & Knight, M. R. (1996) Cold calcium signaling in Arabidopsis involves two cellular pools and a change in calcium signature after acclimation. *Plant Cell*, **8**, 489–503.

Knight, H., Veale, E. L., Warren, G. J. & Knight, M. R. (1999) The *sfr6* mutation in Arabidopsis suppresses low-temperature induction of genes dependent on the CRT/DRE sequence motif. *Plant Cell*, **11**, 875–886.

Knight, H., Zarka, D. G., Okamoto, H., Thomashow, M. F. & Knight, M. R. (2004) Abscisic acid induces CBF gene transcription and subsequent induction of cold-regulated genes via the CRT promoter element. *Plant Physiology*, **135**, 1710–1717.

Kolukisaoglu, U., Weinl, S., Blazevic, D., Batistic, O. & Kudla, J. (2004) Calcium sensors and their interacting protein kinases: genomics of the Arabidopsis and rice CBL-CIPK signaling networks. *Plant Physiology*, **134**, 43–58.

Koster, K. L. & Lynch, D. V. (1992) Solute accumulation and compartmentation during the cold-acclimation of Puma rye. *Plant Physiology*, **98**, 108–113.

Lee, H., Guo, Y., Ohta, M., Xiong, L., Stevenson, B. & Zhu, J. K. (2002) *LOS2*, a genetic locus required for cold-responsive gene transcription encodes a bi-functional enolase. *EMBO Journal*, **21**, 2692–2702.

Levitt, J. (1980) *Responses of Plants to Environmental Stress. Chilling, Freezing and High Temperature Stresses*. Academic Press, New York.

Levy, M., Bachmair, A. & Adam, Z. (2004) A single recessive mutation in the proteolytic machinery of Arabidopsis chloroplasts impairs photoprotection and photosynthesis upon cold stress. *Planta*, **218**, 396–405.

Li, J. J. & Herskowitz, I. (1993) Isolation of Orc6, a component of the yeast origin recognition complex by a one-hybrid system. *Science*, **262**, 1870–1874.

Lin, C. & Thomashow, M. F. (1992) DNA sequence analysis of a complementary DNA for cold-regulated Arabidopsis gene *cor15* and characterization of the COR15 polypeptide. *Plant Physiology*, **99**, 519–525.

Liu, J. & Zhu, J. K. (1998) A calcium sensor homolog required for plant salt tolerance. *Science*, **280**, 1943–1945.

Liu, Q., Kasuga, M., Sakuma, Y., Abe, H., Miura, S., Yamaguchi-Shinozaki, K. & Shinozaki, K. (1998) Two transcription factors, DREB1 and DREB2, with an EREBP/AP2 DNA binding domain separate two cellular signal transduction pathways in drought- and low-temperature-responsive gene expression, respectively, in Arabidopsis. *Plant Cell*, **10**, 1391–1406.

Maruyama, K., Sakuma, Y., Kasuga, M., Ito, Y., Seki, M., Goda, H., Shimada, Y., Yoshida, S., Shinozaki, K. & Yamaguchi-Shinozaki, K. (2004) Identification of cold-inducible downstream genes of the Arabidopsis DREB1A/CBF3 transcriptional factor using two microarray systems. *Plant Journal*, **38**, 982–993.

Massari, M. E. & Murre, C. (2000) Helix-loop-helix proteins: regulators of transcription in eucaryotic organisms. *Molecular and Cellular Biology*, **20**, 429–440.

McKown, R., Kuroki, G. & Warren, G. (1996) Cold responses of *Arabidopsis* mutants impaired in freezing tolerance. *Journal of Experimental Botany*, **47**, 1919–1925.

Medina, J., Bargues, M., Terol, J., Perez-Alonso, M. & Salinas, J. (1999) The Arabidopsis CBF gene family is composed of three genes encoding AP2 domain-containing proteins whose expression is regulated by low temperature but not by abscisic acid or dehydration. *Plant Physiology*, **119**, 463–470.

Monroy, A. F. & Dhindsa, R. S. (1995) Low-temperature signal-transduction – Induction of cold acclimation-specific genes of alfalfa by calcium at 25-degrees-C. *Plant Cell*, **7**, 321–331.

Nanjo, T., Kobayashi, M., Yoshiba, Y., Kakubari, Y., Yamaguchi-Shinozaki, K. & Shinozaki, K. (1999) Antisense suppression of proline degradation improves tolerance to freezing and salinity in *Arabidopsis thaliana*. *FEBS Letters*, **461**, 205–210.

Novillo, F., Alonso, J. M., Ecker, J. R. & Salinas, J. (2004) CBF2/DREB1C is a negative regulator of *CBF1/DREB1B* and *CBF3/DREB1A* expression and plays a central role in, stress tolerance in *Arabidopsis*. *Proceedings of the National Academy of Sciences of the United States of America*, **101**, 3985–3990.

Quintero, F. J., Garciadeblas, B. & RodriguezNavarro, A. (1996) The SAL1 gene of Arabidopsis, encoding an enzyme with $3'(2'),5'$-bisphosphate nucleotidase and inositol polyphosphate 1-phosphatase activities, increases salt tolerance in yeast. *Plant Cell*, **8**, 529–537.

Riechmann, J. L. & Meyerowitz, E. M. (1998) The AP2/EREBP family of plant transcription factors. *Biological Chemistry*, **379**, 633–646.

Riechmann, J. L., Heard, J., Martin, G., Reuber, L., Jiang, C., Keddie, J., Adam, L., Pineda, O., Ratcliffe, O. J., Samaha, R. R., Creelman, R., Pilgrim, M., Broun, P., Zhang, J. Z., Ghandehari, D., Sherman, B. K. & Yu, G. (2000) Arabidopsis transcription factors: genome-wide comparative analysis among eukaryotes. *Science*, **290**, 2105–2110.

Rikin, A., Dillwith, J. W. & Bergman, D. K. (1993) Correlation between the circadian rhythm of resistance to extreme temperatures and changes in fatty acid composition in cotton seedlings. *Plant Physiology*, **101**, 31–36.

Rudolph, A. S. & Crowe, J. H. (1985) Membrane stabilization during freezing: the role of two natural cryoprotectants, trehalose and proline. *Cryobiology*, **22**, 367–377.

Sakai, A. & Larcher, W. (1987) *Frost Survival of Plants: Responses and Adaptation to Freezing Stress*. Springer-Verlag, Berlin.

Sakuma, Y., Liu, Q., Dubouzet, J. G., Abe, H., Shinozaki, K. & Yamaguchi-Shinozaki, K. (2002) DNA-binding specificity of the ERF/AP2 domain of Arabidopsis DREBs, transcription factors involved in dehydration- and cold-inducible gene expression. *Biochemical and Biophysical Research Communications*, **290**, 998–1009.

Sanchez-Ballesta, M. T., Rodrigo, M. J., LaFuente, M. T., Granell, A. & Zacarias, L. (2004) Dehydrin from citrus, which confers *in vitro* dehydration and freezing protection activity, is constitutive and highly expressed in the flavedo of fruit but responsive to cold and water stress in leaves. *Journal of Agricultural and Food Chemistry*, **52**, 1950–1957.

Sangwan, V., Foulds, I., Singh, J. & Dhindsa, R. S. (2001) Cold-activation of *Brassica napus* BN115 promoter is mediated by structural changes in membranes and cytoskeleton, and requires Ca^{2+} influx. *Plant Journal*, **27**, 1–12.

Seki, M., Narusaka, M., Abe, H., Kasuga, M., Yamaguchi-Shinozaki, K., Carninci, P., Hayashizaki, Y. & Shinozaki, K. (2001) Monitoring the expression pattern of 1300 Arabidopsis genes under drought and cold stresses by using a full-length cDNA microarray. *Plant Cell*, **13**, 61–72.

Seki, M., Narusaka, M., Ishida, J., Nanjo, T., Fujita, M., Oono, Y., Muramatsu, M., Hayashizaki, Y., Kawai, J., Carninci, P., Itoh, M., Ishii, Y., Arakawa, T., Shibata, K., Shinagawa, A. & Shinozaki, K. (2002) Monitoring the expression profiles of 7000 Arabidopsis genes under drought, cold and high-salinity stresses using a full-length cDNA microarray. *Plant Journal*, **31**, 279–292.

Shinozaki, K., Yamaguchi-Shinozaki, K. & Seki, M. (2003) Regulatory network of gene expression in the drought and cold stress responses. *Current Opinion in Plant Biology*, **6**, 410–417.

Shinwari, Z. K., Nakashima, K., Miura, S., Kasuga, M., Seki, M., Yamaguchi-Shinozaki, K. & Shinozaki, K. (1998) An Arabidopsis gene family encoding DRE/CRT binding proteins involved in low-temperature-responsive gene expression. *Biochemical and Biophysical Research Communications*, **250**, 161–170.

Smallwood, M. & Bowles, D. J. (2002) Plants in a cold climate. *Philosophical Transactions of the Royal Society of London Series B – Biological Sciences*, **357**, 831–847.

Steponkus, P. L. (1984) Role of the plasma-membrane in freezing-injury and cold-acclimation. *Annual Review of Plant Physiology and Plant Molecular Biology*, **35**, 543–584.

Steponkus, P. L., Uemura, M., Joseph, R. A., Gilmour, S. J. & Thomashow, M. F. (1998) Mode of action of the *COR15a* gene on the freezing tolerance of *Arabidopsis thaliana*. *Proceedings of the National Academy of Sciences of the United States of America*, **95**, 14570–14575.

Sterner, D. E. & Berger, S. L. (2000) Acetylation of histones and transcription-related factors. *Microbiology and Molecular Biology Reviews*, **64**, 435–459.

Stitt, M. & Hurry, V. (2002) A plant for all seasons: alterations in photosynthetic carbon metabolism during cold acclimation in Arabidopsis. *Current Opinion in Plant Biology*, **5**, 199–206.

Stockinger, E. J., Gilmour, S. J. & Thomashow, M. F. (1997) *Arabidopsis thaliana CBF1* encodes an AP2 domain-containing transcriptional activator that binds to the C-repeat/DRE, a *cis*-acting DNA regulatory element that stimulates transcription in response to low temperature and water deficit. *Proceedings of the National Academy of Sciences of the United States of America*, **94**, 1035–1040.

Stockinger, E. J., Mao, Y., Regier, M. K., Triezenberg, S. J. & Thomashow, M. F. (2001) Transcriptional adaptor and histone acetyltransferase proteins in Arabidopsis and their interactions with CBF1, a transcriptional activator involved in cold-regulated gene expression. *Nucleic Acids Research*, **29**, 1524–1533.

Strauss, G. & Hauser, H. (1986) Stabilization of lipid bilayer vesicles by sucrose during freezing. *Proceedings of the National Academy of Sciences of the United States of America*, **83**, 2422–2426.

Sung, D. Y., Kaplan, F., Lee, K. J. & Guy, C. L. (2003) Acquired tolerance to temperature extremes. *Trends in Plant Science*, **8**, 179–187.

Tahtiharju, S. & Palva, T. (2001) Antisense inhibition of protein phosphatase 2C accelerates cold acclimation in *Arabidopsis thaliana*. *Plant Journal*, **26**, 461–470.

Tahtiharju, S., Sangwan, V., Monroy, A. F., Dhindsa, R. S. & Borg, M. (1997) The induction of *kin* genes in cold-acclimating *Arabidopsis thaliana*. Evidence of a role for calcium. *Planta*, **203**, 442–447.

Taji, T., Ohsumi, C., Iuchi, S., Seki, M., Kasuga, M., Kobayashi, M., Yamaguchi-Shinozaki, K. & Shinozaki, K. (2002) Important roles of drought- and cold-inducible genes for galactinol synthase in stress tolerance in *Arabidopsis thaliana*. *Plant Journal*, **29**, 417–426.

Thomashow, M. F. (1990) Molecular genetics of cold acclimation in higher plants. In J. G. Scandalios (ed.) *Genomic Responses to Environmental Stress*. Academic Press, New York, pp. 99–131.

Thomashow, M. F. (1998) Role of cold-responsive genes in plant freezing tolerance. *Plant Physiology*, **118**, 1–8.

Thomashow, M. F. (1999) Plant cold acclimation: freezing tolerance genes and regulatory mechanisms. *Annual Review of Plant Physiology and Plant Molecular Biology*, **50**, 571–599.

Uemura, M. & Steponkus, P. L. (1994) A contrast of the plasma membrane lipid composition of oat and rye leaves in relation to freezing tolerance. *Plant Physiology*, **104**, 479–496.

Uemura, M., Joseph, R. A. & Steponkus, P. L. (1995) Cold acclimation of *Arabidopsis thaliana* (effect on plasma membrane lipid composition and freeze-induced lesions). *Plant Physiology*, **109**, 15–30.

Uversky, V. N., Gillespie, J. R. & Fink, A. L. (2000) Why are "natively unfolded" proteins unstructured under physiologic conditions? *Proteins*, **41**, 415–427.

Vlachonasios, K. E., Thomashow, M. F. & Triezenberg, S. J. (2003) Disruption mutations of *ADA2b* and *GCN5* transcriptional adaptor genes dramatically affect Arabidopsis growth, development, and gene expression. *Plant Cell*, **15**, 626–638.

Vogel, J. T., Zarka, D. G., Van Buskirk, H. A., Fowler, S. G. & Thomashow, M. F. (2005) Roles of the CBF2 and ZAT12 transcription factors in configuring the low temperature transcriptome of Arabidopsis. *Plant Journal*, **41**, 195–211.

Wang, H. & Cutler, A. J. (1995) Promoters from Kin1 and Cor6.6, 2 Arabidopsis-thaliana low-temperature-inducible and ABA-inducible genes, direct strong beta-glucuronidase expression in guard-cells, pollen and young developing seeds. *Plant Molecular Biology*, **28**, 619–634.

Wanner, L. A. & Junttila, O. (1999) Cold-induced freezing tolerance in Arabidopsis. *Plant Physiology*, **120**, 391–400.

Warren, G., McKown, R., Marin, A. L. & Teutonico, R. (1996) Isolation of mutations affecting the development of freezing tolerance in *Arabidopsis thaliana* (L.) Heynh. *Plant Physiology*, **111**, 1011–1019.

Wilkosz, R. & Schlappi, M. (2000) A gene expression screen identifies *EARLI1* as a novel vernalization-responsive gene in *Arabidopsis thaliana*. *Plant Molecular Biology*, **44**, 777–787.

Wise, M. J. (2002) The POPPs: clustering and searching using peptide probability profiles. *Bioinformatics*, **18**(Suppl. 1), S38–S45.

Wise, M. J. & Tunnacliffe, A. (2004) POPP the question: what do LEA proteins do? *Trends in Plant Science*, **9**, 13–17.

Xin, Z. & Browse, J. (1998) *Eskimo1* mutants of *Arabidopsis* are constitutively freezing-tolerant. *Proceedings of the National Academy of Sciences of the United States of America*, **95**, 7799–7804.

Xiong, L., Schumaker, K. S. & Zhu, J. K. (2002a) Cell signaling during cold, drought, and salt stress. *Plant Cell*, **14**(Suppl.), S165–S183.

Xiong, L. M., Lee, B. H., Ishitani, M., Lee, H., Zhang, C. Q. & Zhu, J. K. (2001) *FIERY1* encoding an inositol polyphosphate 1-phosphatase is a negative regulator of abscisic acid and stress signaling in Arabidopsis. *Genes and Development*, **15**, 1971–1984.

Xiong, L. M., Lee, H., Ishitani, M., Tanaka, Y., Stevenson, B., Koiwa, H., Bressan, R. A., Hasegawa, P. M. & Zhu, J. K. (2002b) Repression of stress-responsive genes by FIERY2, a novel transcriptional regulator in Arabidopsis. *Proceedings of the National Academy of Sciences of the United States of America*, **99**, 10899–10904.

Yamaguchi-Shinozaki, K. & Shinozaki, K. (1994) A novel *cis*-acting element in an Arabidopsis gene is involved in responsiveness to drought, low-temperature, or high-salt stress. *Plant Cell*, **6**, 251–264.

Yoshiba, Y., Kiyosue, T., Nakashima, K., Yamaguchi-Shinozaki, K. & Shinozaki, K. (1997) Regulation of levels of proline as an osmolyte in plants under water stress. *Plant and Cell Physiology*, **38**, 1095–1102.

Zarka, D. G., Vogel, J. T., Cook, D. & Thomashow, M. F. (2003) Cold induction of Arabidopsis CBF genes involves multiple ICE (inducer of *CBF* expression) promoter elements and a cold-regulatory circuit that is desensitized by low temperature. *Plant Physiology*, **133**, 910–918.

Zhang, X., Fowler, S. G., Cheng, H., Lou, Y., Rhee, S. Y., Stockinger, E. J. & Thomashow, M. F. (2004) Freezing sensitive tomato has a functional CBF cold response pathway, but a CBF regulon that differs from that of freezing tolerant Arabidopsis. *Plant Journal*.

Zhu, J. H., Shi, H. Z., Lee, B. H., Damsz, B., Cheng, S., Stirm, V., Zhu, J. K., Hasegawa, P. M. & Bressan, R. A. (2004) An *Arabidopsis* homeodomain transcription factor gene, HOS9, mediates cold tolerance through a CBF-independent pathway. *Proceedings of the National Academy of Sciences of the United States of America*, **101**, 9873–9878.

5 Plant responses to high temperature

Jane Larkindale, Michael Mishkind and Elizabeth Vierling

5.1 Introduction

Plants experience high temperature in many different ways and adaptation or acclimation to high temperature occurs over different time scales and levels of plant organization. Exposure to high temperature can be chronic or long term, as experienced in hotter habitats, or it can be more acute, as a result of seasonal or daily temperature extremes. It is also clear that different plant tissues and organs, and plants at different growth stages will be damaged in different ways depending on the heat susceptibility of the dominant cellular processes that are active at the time of the stress. Altogether, this means that 'heat stress' (and therefore, 'heat tolerance') is not a single phenomenon, but rather a varying set of complex perturbations of organismal homeostasis.

At the cellular level, heat affects a wide range of structures and functions. High temperatures alter lipid properties, causing membranes to become more fluid and thereby disrupting membrane processes. All proteins have an optimal temperature window for activity, so increased temperatures alter enzyme activity leading to imbalance in metabolic pathways, and eventually at high temperature proteins denature. Membrane and protein damage leads to the production of active oxygen species (AOS). This may not be effectively controlled through antioxidants at high temperatures, resulting in heat-induced oxidative damage in addition to the direct effects of heating. At the physiological level, this damage translates into reduced efficiency of photosynthesis, impaired translocation of assimilates and loss of carbon gain. These factors in turn combine to cause altered phenology, reproductive failure and accelerated senescence (Hall, 2001). Thus, it is to be expected that many different processes, involving many genes, are involved in plant responses to heat. It is also important to recognize that, in the natural environment, plants are often exposed to heat in conjunction with other stresses. This is especially true of heat and drought stress, which often co-occur. Heat stress may also be accompanied by high irradiance stress. Other environmental stresses also result in similar types of damage as those caused by heat. Therefore, it is expected that the responses of plants to different stresses might overlap, and that exposure to any one stress may affect subsequent reactions to another stress. Signaling and protective pathways from different stresses are therefore likely to intersect, sharing common components.

In this chapter, first we briefly consider how high temperatures limit plant growth and reproduction, and then address in detail some of the molecular and genetic studies of cellular responses to high temperature. Molecular studies are beginning to illuminate processes that are involved in sensing, responding and acclimating to heat over a short time scale. In addition to the summary presented here, readers are also referred to other recent reviews (Klueva et al., 2001; Iba, 2002; Sung & Guy, 2003; Wang et al., 2003, 2004).

One, now classic, response to acute heat stress is the production of heat shock proteins (HSPs), which function, at least in part, as molecular chaperones in cellular protein quality control (Boston et al., 1996). HSPs, however, are only one component of the response to high temperatures, and current data indicate that there are multiple pathways that contribute to the ability of plants to tolerate heat. There has also been a recent explosion in information about the signaling molecules that may be involved in responses to heat, and these components begin to offer insight into events occurring in a plant as temperatures begin to rise. Altogether, genomics, genetic and transgenic experiments are helping to define the roles of individual and suites of genes in the response of plants to high temperatures. As the numerous interacting pathways become clearer it will be possible to understand the interactions of heat and other stresses and to consider the best strategies for improving plant heat tolerance.

It is important to reemphasize that the cellular responses to be discussed here are happening in the context of a complex organism that has many functionally distinct cell and tissue types and a multistage life cycle. To gain a full appreciation of heat stress in the context of the whole plant and its physiological processes readers are encouraged to consult excellent older literature, including Turner and Kramer (1980), Raper and Kramer (1983), Patterson and Graham (1987), Blum (1988), Weis and Berry (1988), Hall (1990) and Nobel (1991). While understanding the cellular response is daunting enough, integrating our knowledge into an understanding of stress effects on the agricultural productivity and evolutionary fitness of plants is the ultimate challenge.

5.2   Physiological responses to high temperature

5.2.1   High temperature limits to optimal plant performance

Plants can experience wide fluctuations of temperature on a daily or seasonal basis. It is to be expected, therefore, that plants have evolved mechanisms through which cellular activity could be maintained at varying temperatures. These mechanisms have not been fully defined (Patterson & Graham, 1987). It has been estimated that the optimal thermal range for any plant is about 10°C wide (Mahan *et al.*, 1995). Exposure to temperatures outside this optimal range, though not necessarily lethal, can be considered stressful. The actual upper temperature limits for survival, which for temperate plant species is between 40°C and 55°C, depending on the length of exposure (Klueva *et al.*, 2001), act to limit plant distribution and the area available for agriculture.

Plants show differential sensitivity to high temperature depending on the severity, duration and developmental timing of the stress. This makes it difficult to estimate the damage done to crops by heat stress. The connection between plant water status and temperature also complicates analysis of the relative of contribution of heat and drought stress under field conditions (Carlson, 1990). However, for maize, the best single variable for estimating yield was reported to be the extent to which the daily maximum temperature exceeded 32°C during pollination and seed fill (Thompson, 1975; Dale, 1983). Similarly, the highest yields of soybean have been correlated with the coolest reproductive season temperatures (Martineau *et al.*, 1979), and optimal temperature for high yield in cereal crops has been determined to be 20–30°C (Keeling *et al.*, 1993). Not only moderate midday heat, but also high temperatures at night are damaging, leading to increases in carbon loss by respiration, decreased yield, alterations in timing of flowering, and even inhibition of flower bud development (Hall, 1990). These relatively mild high temperatures contrast with situations in which plants are challenged by severe stress temperatures, such as in the semiarid tropics where cereal grains can experience soil surface temperatures of 55°C (Peacock *et al.*, 1993). In closed canopy maize fields in the midwestern USA leaf temperatures can be over 40°C, while developing pea seeds in Idaho have been recorded to have temperatures up to 68°C (Hawthorn *et al.*, 1966). Altogether, it seems likely that many plants experience some form of heat stress during their life cycle.

5.2.2   *Heat sensitivity of photosynthesis*

The fact that heat stress limits photosynthetic output is well documented, and photosynthesis declines at temperatures well below those lethal to the plant. However, the mechanism by which this occurs remains controversial. Current data preclude distinguishing whether the decline in photosynthesis at elevated temperatures occurs by a common mechanism or a diversity of mechanisms. Numerous components of the photosynthetic apparatus display heat lability, but it has been difficult to determine which one is rate limiting under conditions of heat stress. It has long been proposed that PSII is the most heat-sensitive component of the photosynthetic apparatus. Other parameters implicated as weak links during heat stress include components that facilitate CO_2 transfer from the intercellular space to the chloroplast

(Bernacchi *et al.*, 2002), photosynthetic electron transport (Wise *et al.*, 2004) and loss of ribulose bisphosphate (RuBP) regeneration capacity (Wise *et al.*, 2004).

Recently, Salvucci and Crafts-Brandner (2004) have argued that the heat lability of various photosynthetic components is secondary to the failure of cells to maintain RuBP carboxylase/oxygenase (Rubisco) in an optimally activated state. Under conditions of heat stress Rubisco activity declines as a consequence of the association of catalytic 'misfire products' at the active site rather than temperature-induced enzyme inactivation. It is the well-characterized temperature sensitivity of Rubsico activase that precludes adequate reactivation of these dead-end complexes during periods of heat stress. Interestingly, in wheat and cotton leaves specific isoforms of activase accumulate during heat stress (Law & Crafts-Brandner, 2001; Law *et al.*, 2001), which may add stability to the multisubunit chaperone-like holoenzyme. Arguing against a role for Rubisco activase as a limiting factor in photosynthetic output during heat stress, Wise *et al.* (2004) report that Rubisco activity in field-grown cotton remains sufficient during heat stress to process levels of RuBP supplied to the enzyme, and hence maintain upstream photosynthetic processes. Clearly, more work will be required to sort out the factors that limit photosynthetic output during heat stress. Finally, it should be noted that an alternate function for Rubisco activase was suggested by Rokka *et al.* (2001) as a consequence of their finding that the enzyme was reported to relocate to thylakoid membranes under conditions of heat stress, where it appears to associate with thylakoid-bound polysomes. Whether it acts as a chaperone at this location has not yet been explored.

Interest in the consequences of global warming has led to analysis of the influence of CO_2 levels on the heat sensitivity of photosynthesis, with both negative and positive consequences documented. For example, the activation state of Rubisco declines with rising CO_2 concentrations, with the elevated CO_2 stimulating increases in ATP consumption that reduce ATP/ADP ratios, and thus possibly inhibit Rubisco activase (Crafts-Brandner & Salvucci, 2000). This effect would compound photosynthetic losses caused by the heat sensitivity of the enzyme. In contrast, a protective role for CO_2 was reported by Taub *et al.* (2000). Their results showed that leaves of *Cucumis sativus* and several other species grown in an atmosphere enriched to 750 ppm CO_2 retained significantly higher photosynthetic rates (measured as PSII activity or CO_2 uptake) after a heat shock at 40°C. The heat treatments in these experiments were performed in the dark, thus minimizing the possibly confounding effects of photoinhibition. Although mechanisms for this example of thermoprotection have not been evaluated experimentally, the authors point out that total osmolyte concentrations are known to increase in leaves grown at elevated CO_2 levels, thus suggesting a role for compatible solutes in the process.

5.2.3   Heat sensitivity of reproduction

In addition to the studies cited above concerning effects of chronic high temperature on crop yield, numerous studies document that successful fertilization and seed fill are more sensitive to high temperature than vegetative growth. Unfortunately, reduced fertility and seed fill at high temperature cannot be clearly pinpointed to a defect in a single function, but most likely varies with the type of heat stress and the plant species. Loss of fertility could result from problems in male meiosis, pollen germination, pollen tube growth, or megagametophyte defects among other factors. Likewise, effects on flower production, grain set, endosperm division, source photosynthesis, and assimilate transport and partitioning can all contribute to ultimate seed yield and weight. Even recent studies are still only trying to define the heat-sensitive processes in different plant species (Commuri & Jones, 2001; Kim *et al.*, 2001; Sato *et al.*, 2002; Cross *et al.*, 2003; Hurkman *et al.*, 2003; Zahedi *et al.*, 2003; Young *et al.*, 2004; Kobata & Uemuki, 2004).

Perhaps the most progress at the molecular level has been made in an examination of potential factors limiting starch production in wheat grain at high temperatures. Hurkman *et al.* (2003) observed transcript levels of starch biosynthetic enzymes comparing 24/17°C to 37/28°C and 37/17°C stress regimes. Although transcript levels of three isoforms of starch synthase were dramatically lowered by the heat treatment, this effect was not mirrored in the rate of starch accumulation. The overall time of grain-fill was severely shortened, however, and the type of starch granule was altered. Heat effects on wheat starch synthase were also reported by Zahedi *et al.* (2003) who concluded that temperature differences in starch synthase efficiency were correlated with differences in the temperature sensitivity of grain fill between two cultivars. However, the complexity of the reproductive process is obviously consistent with multiple gene effects.

5.3   Cellular acquired thermotolerance

An important and well-defined response to high temperature stress is the ability of plants and other organisms to rapidly acclimate to withstand normally lethal temperatures (Vierling, 1991). This phenomenon is termed acquired thermotolerance and should be clearly distinguished from the inherent differences between species in optimal growth temperatures. The acquisition of thermotolerance is a cell autonomous phenomenon and results from prior exposure to a conditioning pretreatment, which can be a short, sublethal high temperature or other moderate stress treatment. Thermotolerance can also be induced by a gradual increase in temperature prior to reaching the normally lethal temperature, as would be experienced in the natural environment. For example, many temperate plant species will die if directly exposed to 45°C for a period of a few hours, but will survive the same exposure if the temperature is gradually increased to 45°C, as would occur over the course of a day in nature. In the laboratory, the same result can be effected by short heat pretreatments (2 h or less) at temperatures below the lethal temperature (Hong & Vierling, 2000; Burke et al., 2000). Even in optimal environments plants may experience high temperatures that would be lethal or damaging in the absence of this rapid acclimation. Furthermore, because plants typically experience diurnal temperature fluctuations, the acquisition of thermotolerance may reflect a more general mechanism that contributes to homeostasis of metabolism on a daily basis. Acquired thermotolerance does indeed appear to have evolved in response to diurnal temperature changes, as the induced tolerance is ephemeral. It declines dramatically within hours, and virtually no tolerance remains 24 h after the inducing pretreatment. In considering cellular and molecular responses to temperature, potential differences between mechanisms involved in short-term acquired thermotolerance versus mechanisms involved in long-term adaptation to different temperature environments should be kept in mind.

5.4 Heat shock proteins/molecular chaperones

Perhaps the best-characterized response of plants and other organisms to high temperature stress is the production of HSPs (Vierling, 1991). Within seconds of reaching a critical temperature 5–10°C above the optimum for growth, transcription of HSPs is induced. Maximum transcript levels are observed within 1–2 h, after which transcript levels decline. The accumulation of proteins can be detected with sensitive antibodies less than an hour after initiation of the stress. The level of HSP transcripts and proteins are virtually a thermometer – levels are proportional to the degree of stress until the temperature at which lethality occurs (Chen et al., 1990; DeRocher et al., 1991).

There are five well-characterized classes of HSPs that have been defined in both plants and other organisms: Hsp100/ClpB, Hsp90, Hsp70/DnaK, Hsp60/ GroE and small HSPs (sHSPs) (alternate names in eukaryotes/prokaryotes are given). The Hsp70 and Hsp60 proteins are among the most highly conserved proteins in nature, consistent with a fundamental role in cellular response to stress (Külz, 2003). All of these HSPs have been shown to function as molecular chaperones, which are a diverse group of proteins that share the property of binding to other proteins that are in unstable structural states (Boston et al., 1996). Chaperones facilitate a range of processes including protein folding, transport of proteins across membranes, modulation of protein activity, regulation of protein degradation and prevention of irreversible protein aggregation. The latter activity is believed to be critical to survival of high temperature stress and to explain the induction of these proteins by high temperature. Genome analysis has provided a complete picture of the diversity of HSPs in plants, and Tables 5.1 and 5.2 summarize genetic and transgenic plant analyses that begin to address exactly how these proteins may contribute to heat tolerance.

5.4.1   Hsp100/ClpB

Interestingly, the Hsp100/ClpB class of chaperones is found in bacteria, yeast, certain parasitic protozoans and plants, but not in other higher eukaryotes (Schirmer *et al.*, 1996; Agarwal *et al.*, 2001, 2002). Perhaps plants have retained this class of chaperones through evolution because of their unique exposure to the environment compared to other higher eukaryotes. Plants have both cytosolic and chloroplast-targeted Hsp100/ClpB proteins, both of which have been shown to strongly induced by heat (Schirmer *et al.*, 1994; Keeler *et al.*, 2000). These proteins are hexameric members of the AAA+ proteins, which couple ATP binding and hydrolysis to a variety of protein-remodeling activities (Neuwald *et al.*, 1999; Lupas & Martin, 2002). The recent 3.0 Å structure of a *Thermus thermophilus* ClpB subunit and a cryo-EM reconstruction of the hexamer, has defined the structure of the different ClpB domains and their relative orientation (Lee *et al.*, 2003). Both *in vivo* and *in vitro* data in yeast and bacteria indicate that the protective function of these chaperones is a result of their ability to resolubilize protein aggregates in cooperation with the Hsp70/DnaK chaperone system (Maurizi & Xia, 2004). Recent data also support an interaction of the Hsp100/ClpB and sHSP chaperone systems in protein disaggregation (Mogk *et al.*, 2003a; 2003b). Although it is well accepted that Hsp100/ClpB proteins are involved in protein disaggregation, no critical cellular substrates have been identified in plants or any other organism. The cytosolic Hsp100/ClpB proteins are the only HSPs for which direct genetic evidence supports a significant role in acquired thermotolerance in plants (Table 5.1). In *Arabidopsis thaliana* analysis of both missense and insertional mutants of Hsp101 (*hot1* mutants), and of cosuppression and antisense transgenics (Table 5.2) have shown that the mutant plants fail to acquire thermotolerance at different growth stages (Hong & Vierling, 2000; Hong *et al.*, 2003; Queitsch *et al.*, 2000). Similarly, Hsp101 Mu insertional mutants in maize are defective in acquired thermotolerance (Nieto-Sotolo *et al.*, 2002). Constitutive expression of Hsp101 in Arabidopsis also enhances thermotolerance in the absence of an adaptive pretreatment (Queitsch *et al.*, 2000). The important role of plant Hsp100 in thermotolerance was predicted from previous experiments demonstrating that the homologous gene is required for thermotolerance in *Saccharomyces cerevisiae*, cyanobacteria and *Escherichia coli* and from the observation that plant Hsp100/ClpB could restore thermotolerance to a yeast *hsp104* mutant (Lee *et al.*, 1994; Schirmer *et al.*, 1994; Wells *et al.*, 1998). Surprisingly, the cytosolic Hsp100/ClpB null mutants in Arabidopsis and maize show no obvious developmental defects or

Table 5.1 Phenotypes of plants with mutations in chaperones and related proteins

Mutant (species)	Gene	Affect on heat tolerance	Conditions tested (measurement)	Comments	Reference
hot1 (*Arabidopsis*)	Hsp101	Reduced thermotolerance	2.5-day dark grown seedlings: 90 min, 38°C > 2 h, 22°C > 2 h, 45°C (hypocotyl elongation); imbibed seeds: 2 h, 45°C (germination); 10-day seedlings: 90 min, 38°C > 2 h, 22°C > 2 h, 45°C (photographed)	No phenotype at normal temperatures	Hong and Vierling (2000, 2001)
Hsp101 (maize)	Hsp101	Reduced thermotolerance	2.5-day seedlings: 1 h, 40°C > 1 h, 28°C > 1 h, 48 or 50°C; 2.5-day seedlings: 1 h, 40, 48 or 50°C (growth)	No phenotype at normal temperatures	Nieto-Sotelo *et al.* (2002)
Hsp90-1 (*Arabidopsis*)	Hsp90.1	Not tested	Not tested	No phenotype at normal temperatures; RPS2-mediated disease susceptible	Takahashi *et al.* (2003)
Hsp90.2 (*Arabidopsis*)	Hsp90.2	Not tested	Not tested	Modest developmental changes; reduced RPM1-mediated disease resistance	Hubert *et al.* (2003)
cr88 (*Arabidopsis*)	Plastid Hsp90	No change	Details not given	Chlorate resistant; plastids develop late; long hypocotyls in red light	Cao *et al.* (2003)
len1 (*Arabidopsis*)	Chaperonin 60 beta	Accelerated cell death in response to heat	5-week old plants: 2 h, 45°C (photographed)	Develops lesions on leaves and constitutively expresses systemic acquired resistance	Ishikawa *et al.* (2003)
slp (*Arabidopsis*)	Chaperonin 60 alpha	Not tested	Not tested	Defect in plastid development and embryo development	Apuya *et al.* (2001)
rtm2 (*Arabidopsis*)	sHsp-like	No effect	2.5-day dark grown seedlings: 1 h, 38°C > 1 h, 22°C > 1.5 h, 45°C (hypocotyl elongation)	Gene is not heat inducible but is required for virus resistance	Whitham *et al.* (2000)
emp2 (maize)	Heat shock binding protein 1	Not tested	Not tested	Embryos die, overexpression of HSPs	Fu *et al.* (2002)
squint (*Arabidopsis*)	Hsp90-interacting factor	No effect	Not given	None	Berardini *et al.* (2001)

Table 5.2 Transgenic plants with altered chaperone levels

Gene inserted (species)	Construct	Transformed species	Effect	Conditions tested (measurement)	Reference
Hsp101 (Arabidopsis)	35S:sense; 35S:antisense	Arabidopsis	Decreased thermotolerance (cosuppression or antisense)	2.5-day dark grown seedlings: 90 min, 38°C > 2 h, 22°C > 2 h, 45°C (hypocotyl elongation); 14-day plants: 90 min, 38°C > 2 h, 45°C (photographed); 30 h seedlings: 2 h, 47°C > 2 h (photographed)	Queitsch et al. (2000)
Hsp101 (Arabidopsis)	35S:sense	Arabidopsis	Increased thermotolerance	14-day plants: 30, 45 or 60 min, 45°C (photographed); 3-day seedlings:30 min, 47°C (photographed)	Queitsch et al. (2000)
Hsp101 (rice)	35S: sense	Rice	Increased thermotolerance	21-day plants: 1.5 h, 38°C (photographed)	Katiyar-Agarwal et al. (2003)
Hsp70 (Arabidopsis)	Heat-inducible promoter: antisense	Arabidopsis	Decreased thermotolerance	7-day seedlings: 1 h, 44, 46, 48, 50, 52, 54°C; 2 h, 35°C (photographed)	Lee and Schöffl (1996)
Hsc70-1 (Arabidopsis)	Cor78/Rd29A promoter:sense	Arabidopsis	Increased thermotolerance	28-day plants: 10 min, 42, 44, 46, 48, 50°C (ion leakage)	Sung and Guy (2003)
Hsc70-1 (Arabidopsis)	Cor78/Rd29A promoter: antisense	Arabidopsis	No surviving plants	Not applicable	Sung and Guy (2003)
BiP (ER) (tobacco)	35S:sense	Tobacco	Decrease in basal BiP transcripts; reduced unfolded protein response	Not tested for thermotolerance	Leborgne-Castel et al. (1999)
BiP (ER) (tobacco)	35S:antisense	Tobacco	Minor reductions in BiP were lethal under nonstress conditions	Not tested for thermotolerance	Leborgne-Castel et al. (1999)

Gene (source)	Construct	Species	Phenotype	Conditions	Reference
BiP (ER) (tobacco)	35S:sense	Tobacco	Tolerant to tunicamycin during germination and tolerant to water deficit	Not tested for thermotolerance	Alvim et al. (2001)
BiP (ER) (tobacco)	35S:antisense	Tobacco	Intolerant of drought; higher induction of antioxidative defenses	Not tested for thermotolerance	Alvim et al. (2001)
Hsp17.7-CI (carrot)	Endogenous promoter: antisense	Carrot	Decreased thermotolerance	Cell cultures: 2 h, 37°C > 1 h, 23°C > 30 min, 48°C (dry weight); culture explants: 1–3 h, 50°C (ion leakage)	Malik et al. (1999)
Hsp17.7-CI (carrot)	35S:sense	Carrot	Increased thermotolerance	Cell cultures: 2 h, 37°C > 1 h, 23°C > 30 min, 48°C (dry weight); culture explants: 1–3 h, 50°C (ion leakage)	Malik et al. (1999)
TLHS1:sHsp-CI (tobacco)	35S:sense	Tobacco	Increased heat tolerance of cotyledon opening	Seedlings: 1–4 h, 40 or 45°C (cotyledon opening)	Park and Hong (2002)
Hsp17.6A-CI (Arabidopsis)	35S:sense	Arabidopsis	Increased salt and drought, but not heat tolerance	7-day seedlings: 2 h, 37–43°C (viability)	Sun et al. (2001)
Plastid Hsp 21 (Arabidopsis)	35S:sense	Arabidopsis	Increased thermotolerance under high light conditions	21-day plants: 7 days, 4 h/day, 40°C 800 or 500 umol $m^2 s^{-1}$ (dry weight)	Härndahl et al. (1999)
Mitochondrial Hsp23.8 (tomato)	35S:sense	Tobacco	Increased thermotolerance	28-day plants: 2 h, 46 or 48°C (photographed)	Sanmiya et al. (2004)
Mitochondrial Hsp23.8 (tomato)	35S:antisense	Tobacco	Increased heat sensitivity	28-day plants: 2 h, 46 or 48°C (photographed)	Sanmiya et al. (2004)

growth rate disadvantage under optimal temperature conditions. Further, while essential for acquired thermotolerance to severe heat stress, there is no evidence that the protein confers any advantage to plants suffering from chronic, milder heat stress (Hong & Vierling, 2001). This observation emphasizes the point that different forms of heat stress require different protective responses. How Hsp100/ClpB functions in acquired thermotolerance is an important question that will require definition of the targets protected by this chaperone.

5.4.2   Hsp90

The Hsp90 chaperones are highly abundant proteins that are essential for viability in a number of eukaryotes, most likely including plants. Studies in animals and yeast have demonstrated that Hsp90 has a key role in modulating the activity of metastable client proteins involved in signal transduction, either by facilitating interactions with ligands, trafficking clients to membranes or altering client interactions with other cellular components. These activities depend on Hsp90 function within a dynamic multiprotein complex that includes Hsp70, Hsp40 (DnaJ) and other proteins (p60/sti/Hop; an immunophilin; p23; p48/Hip; p50/Cdc37). Sequence conservation of Hsp90 across eukaryotes and the demonstration that plant Hsp90 can function in the context of these dynamic complexes from mammalian cells, indicate the mechanism of action of Hsp90 in plants is homologous to that in other eukaryotes (Krishna & Gloor, 2001).

Arabidopsis has seven Hsp90 genes, encoding four cytosolic forms, and forms targeted to the chloroplast, mitochondrion and endoplasmic reticulum (Krishan & Gloor, 2001). Although some of these genes show significant heat induction, no direct connection between Hsp90s and heat tolerance has been defined. As listed in Table 5.1, mutations in genes for two of the cytosolic forms and the chloroplast form have been described. No data were reported concerning heat tolerance of the former mutants (Hubert et al., 2003; Takahashi et al., 2003), and the chloroplast Hsp90 mutant appeared wild type for heat tolerance (Cao et al., 2003).

As with Hsp100/ClpB, uncovering the client proteins with which Hsp90s interact in plants will be important for understanding any role for Hsp90s in heat stress tolerance. Although not related to heat tolerance, recent data have demonstrated that plant cytosolic Hsp90s are involved in chaperoning disease-resistance proteins in plants as demonstrated in both Arabidopsis and tobacco (Hubert et al., 2003; Lu et al., 2003; Takahashi et al., 2003; Liu et al., 2004). The client proteins are receptor/signaling molecules, consistent with what is known about client proteins in other systems. The same studies have also uncovered Hsp90-interacting proteins in plants that have domains similar to the interacting proteins described in mammals, again supporting conservation of mechanism.

Hsp90 in plants and animals has also been postulated to buffer genetic variation, and loss of Hsp90 function is proposed to release genetic variation for selection in evolution (Sangster *et al.*, 2004). This function is a consequence of Hsp90 modulating the activity of signaling proteins and the dependent pathways. Whether such a function could also control epigenetic variation in response to stress, resulting in stress-resistant phenotypes for some individuals remains to be explored.

5.4.3   *Hsp70/DnaK*

The Hsp70/DnaK chaperone family is probably the best understood of the molecular chaperones, and readers are referred to excellent reviews describing their structure and mechanism (Kelley, 1999; Mayer *et al.*, 2002). Genomic analysis of Arabidopsis defined 14 Hsp70 proteins, five cytosolic forms, three targeted to the endoplasmic reticulum (BiP), three targeted to the chloroplasts, two localized to mitochondria and one potential pseudogene (Lin *et al.*, 2001; Sung *et al.*, 2001). Phylogenetic analysis shows establishment of the different organellar forms occurred early in eukaryotic evolution. Detailed expression analysis in Arabidopsis revealed complex developmental and stress responsiveness of these different genes; all but one mitochondrial and one chloroplast HSP70 showed significant induction in response to heat (Sung *et al.*, 2001).

The number of Hsp70 genes in plants complicates genetic analysis of their role in plant heat tolerance. A complete set of gene knockout lines has not yet been assembled, and may indeed not be viable considering the established essential nature of these proteins in other organisms. Lee and Schöffl (1996) used heat-inducible Hsp70 antisense RNA to demonstrate a requirement for Hsp70 in thermotolerance. Another attempt to alter the levels of Hsp70-1 in transgenic Arabidopsis with a constitutive promoter led to negative effects on plant growth and viability (Sung *et al.*, 2003). Difficulty in achieving changes in expression suggests that Hsp70 levels are tightly regulated. Overexpression of Hsp70-1 did, however, result in enhanced heat tolerance under specific test conditions (Table 5.2). Overexpression of an endoplasmic reticulum Hsp70, BiP, has been demonstrated to abrogate ER stress (Leborgne-Castel *et al.*, 1999) and enhance drought, but not heat tolerance (Alvim *et al.*, 2001). In Chlamydomonas the chloroplast Hsp70 has been implicated in photoprotection of PSII during photoinhibition (Schroda *et al.*, 1999). Overall, it can be concluded that Hsp70s are important for plant heat tolerance, but manipulating expression of these essential proteins to enhance plant heat tolerance may be difficult.

5.4.4   *Hsp60/GroE*

The Hsp60, or chaperonin, family of chaperones have not been directly implicated in the acclimation of plants to heat stress. However, the drastic phenotypic consequences seen at normal temperatures when their expression is attenuated or eliminated (Table 5.1) suggest that their participation is essential in such fundamental processes as protein assembly and transport. These pleiotropic effects include defects in embryonic development (Apuya *et al.*, 2001) and slow growth (Zabaleta *et al.*, 1994a) and may complicate experimental approaches aimed at demonstrating a specific role in stress responses. Despite these difficulties, some evidence suggests that Group I chaperonins, which are found in plastids, mitochondria and eubacteria (Hill & Hemmingsen, 2001; Wang *et al.*, 2004) may function in the context of temperature stress. For example, heat induces expression of the mitochondrial chaperonin Cpn60(2) (Prasad & Stewart, 1992). The chloroplastic family member Cpn60-β3 is not heat-regulated (Zabaleta *et al.*, 1994b), but plants carrying a T-DNA insertion in this gene display heightened heat sensitivity as well as a tendency to form lesion mimics (Ishikawa *et al.*, 2003) (Table 5.1). Cyanobacteria may prove to be more tractable than vascular plants in determining a role for the chaperonins during heat stress as reports have found the chaperonin GroEL in these organisms to be heat inducible and to play a role in thermotolerance (Kovacs *et al.*, 2001; Asadulghani & Nakamoto, 2003).

The evolutionarily distinct, Group II chaperonins are found in the cytoplasm of eukaryotes and in the Archaea (Macario *et al.*, 1999; Lund *et al.*, 2003; Hill & Hemmingsen, 2001). The small literature concerning this group in plants (Hill & Hemmingsen, 2001; Wang *et al.*, 2004) does not report possible roles in heat stress. In the Archaea, however, these proteins are strongly induced during heat stress, and their synthesis correlates with thermotolerance (Macario *et al.*, 1999; Lund *et al.*, 2003).

Although generally considered to mediate protein folding, roles for chaperonins in various aspects of membrane assembly and function have recently emerged. Deaton *et al.* (2004) report that, *in vitro*, the Group I chaperonin GroEL coordinates the assembly of bacteriorhodopsin in cytoplasmic membranes, and thus, may play a similar role *in vivo*. In examining the cytoplasic localization of a Group II chaperonin from an archaeal hyperthermophile, Trent *et al.* (2003) found that it associates with membranes under normal and heat stress conditions and suggest a structural role in maintaining membrane stability and permeability.

5.4.5   The sHSP family of proteins

Compared to other eukaryotes, plants are characterized by an unusual complexity of proteins belonging to the sHSP family. The sHSPs and homologous α-crystallins are a ubiquitous group of stress proteins that range in size from ~16–42 kDa and contain an α-crystallin domain – a conserved C-terminal domain of approximately 90 amino acids. Another defining feature of the sHSPs is that, with few exceptions, the proteins form oligomeric complexes of

200–750 kDa (9 to > 24 subunits), with the size dependent on the specific protein (van Montfort *et al.*, 2002). In plants sHSPs are highly and rapidly heat-induced, becoming among the most abundant RNAs in the cell (Vierling, 1991). In addition, during heat stress, sHSP transcripts and proteins have also been detected under other stress conditions and at specific developmental stages, but they are not generally detectable in most vegetative tissues (Waters *et al.*, 1996).

Plant sHSPs appear to have evolved independently after the divergence of plants and animals (Waters *et al.*, 1996; Waters, 2003). While in other organisms sHSPs are found in the cytoplasm and at times in the nucleus, plants express both cytosolic sHSPs and specific isoforms targeted to intracellular organelles. There are at least two types of sHSPs in the cytosol, referred to as class I and class II proteins, which are typically present in multiple isoforms. Class I and II plant sHSPs share only ~50% identity in the α-crystallin domain and are estimated to have diverged over 400 million years ago (Waters & Vierling, 1999). How they may be specialized functionally is unknown. Three separate gene families have also been described that encode mitochondrion, plastid and endoplasmic reticulum-localized sHSPs, each with appropriate organelle targeting signals. Analysis of the Arabidopsis genome has revealed further complexity in the sHSP family, with additional classes of nuclear- and peroxisomal-targeted forms, along with less well-conserved classes of what appear to be additional cytosolic sHSPs (Scharf *et al.*, 2001). The evolutionary expansion of the plant sHSP family, which is already seen in primitive land plants, suggests that plant sHSP diversity may be the result of selection pressure for tolerance to many types of stress encountered by plants when they made the transition to growth on land (Waters, 2003).

Current models propose that sHSPs act as ATP-independent chaperones by binding to aggregating proteins and maintaining them accessible for refolding by Hsp70 (and cochaperones) and, under some circumstances the Hsp100/ClpB proteins. The X-ray crystal structure of a dodecameric, cytosolic sHSP from *Triticum aestivum* (wheat) provides the only high-resolution structure of an sHSP from eukaryotes (van Montfort *et al.*, 2001). This structural information, as well as considerable biochemical data from studies of the plant cytosolic proteins, has been pivotal for the development of the model of sHSP function (van Montfort *et al.*, 2002; Basha *et al.*, 2004). However, this model is derived primarily from *in vitro* reconstitution experiments, and as for Hsp100/ClpB, no specific, *in vivo* substrates protected by sHSPs have been defined. It has been suggested that protein components of the oxygen-evolving side of PSII are directly protected by the chloroplast-localized sHSP (Heckathorn *et al.*, 1998; Barua *et al.*, 2003). Others have questioned these results, however (Tanaka *et al.*, 2000; Härndahl & Sundby, 2001), and direct interactions seem unlikely given the apparent stromal location of the sHSP versus the luminal location of the oxygen-evolving apparatus. Altogether, considerable work remains to define sHSP target proteins.

In addition to their potential to interact with nonnative proteins in a chaperone capacity, a hypothesis has been put forward that sHSPs interact with membranes to reduce high-temperature-induced membrane fluidity (Vigh et al., 1998; Török et al., 2001). Membranes are clearly an important cellular component perturbed by heat, and a need for membrane stabilization is logical. Direct interaction of sHSPs other than α-crystallin with membranes has only been reported for *Synechocystis* Hsp16.6 (Török et al., 2001; Tsvetkova et al., 2002). Purified recombinant protein interacted with lipid vesicles as monitored by fluorescence anisotropy to reduce lipid fluidity, and also increased the surface pressure of lipid monolayers. There was evidence for both lipid specificity and for membrane penetration of the sHSP. The potential ability of sHSPs to stabilize directly photosynthetic membranes by lipid interaction could explain data supporting their role in photosystem II protection, but requires further investigation.

Direct evidence for a major role for sHSPs in stress tolerance has been difficult to obtain. There are no studies examining null mutations of genes from any of the major sHSP gene families, and due to gene redundancy, multiple mutations may be required before any phenotype is exhibited (Scharf et al., 2001). However, there is other evidence to support an important role for sHSPs in heat tolerance. Significantly, in the cyanobacterium *Synechocystis*, deletion of the only sHSP gene, Hsp16.6, results in a heat-sensitive phenotype (Giese & Vierling, 2002). Expression of cytosolic plant sHSPs also provides some protection to *E. coli* from heat stress (Yeh et al., 1997; Soto et al., 1999a, 199b). There have been relatively few studies examining plants with altered expression of specific sHSPs (Table 5.2). Class I cytosolic sHSPs (Malik et al., 1999; Park & Hong, 2002) and a mitochondrial sHSP (Sanmiya et al., 2004) have been linked to heat tolerance in transgenic experiments. Constitutive expression of the chloroplast sHSP provided some protection of plants from high light coupled with heat stress, but an increase in tolerance to heat alone could not be demonstrated (Härndahl et al., 1999). Similarly, constitutive expression of cytosolic class II sHSPs provided some protection from osmotic stress, but the same plants did not show enhanced heat tolerance (Sun et al., 2001).

5.5   Other components of the response to heat

While the induction of HSPs is clearly important for heat tolerance (Tables 5.1 and 5.2), there are many other processes involved in surviving heat stress. HSP deletion mutants in some species can still acquire thermotolerance (yeast, Smith & Yaffe, 1991; *Drosophila*, Zatsepina et al., 2001). In fact, in yeast, it has been shown that a knockout of HSF1, which regulates the expression of many of the major HSPs in this organism, does not prevent the acquisition of thermotolerance (Smith & Yaffe, 1991), and a *Drosophila* mutant with reduced HSP70 levels showed extremely high levels of thermotolerance (Zatsepina *et al.*, 2001). In addition, there are mutants defective in thermotolerance that show normal induction of HSPs (Lee & Park, 1998; Hong *et al.*, 2003) (Table 5.3), and chemical treatments that induce thermotolerance in a number of organisms do not necessarily induce HSPs (Chinese hamster, Borrelli *et al.*, 1996; yeast, Swan & Watson, 1999; human, Hershko *et al.*, 2003). These data and further experiments, as discussed below, indicate that processes other than the induction of HSPs are involved in the acquisition of thermotolerance.

5.5.1 Antioxidant production

In plants, there are a number of non-HSP transcripts that are upregulated by heat. In particular, the Arabidopsis cytosolic ascorbate peroxidase gene (APX1) has been shown not only to be heat upregulated, but also to contain a functional heat shock element (HSE) in its 5′ promoter region (Storozhenko *et al.*, 1998). This HSE is acted on by heat shock transcription factors (HSFs) (Panchuk *et al.*, 2002). Arabidopsis APX1 knockout plants have been created and found to be sensitive to high light stress, but their ability to survive high temperatures was not assessed (Pnueli *et al.*, 2003). These plants were, however, shown to accumulate HSPs normally under heat stress, but not under light stress. The barley APX1 gene, homologous to APX3 in Arabidopsis, which encodes a peroxisomal APX, is also heat inducible (Shi *et al.*, 2001). The specific function of the different APX genes under heat stress in plants has yet to be ascertained, but it is well established that heat causes secondary oxidative damage in plants (Dat *et al.*, 1998a, 1998b; Gong *et al.*, 1998; Larkindale & Knight, 2002), and APX is likely to be involved in limiting that oxidative damage.

In addition to APX, the activities of other antioxidant-regenerating enzymes are altered during heat shock in plants. In the grass *Agrostis palustris*, superoxide dismutase (SOD) increased during long-term heat stress, while APX, catalase and glutathione reductase activities decreased (Jiang & Huang, 2001). During the first week of long-term stress, however, APX and peroxidase increased, and catalase decreased in that species; SOD increased in activity more slowly (Larkindale & Huang, 2004a). In maize, the activities of all of these enzymes increased under heat stress (Gong *et al.*, 1998). Pretreatment with various chemicals that alter the antioxidant capacity of the cell also alter the ability of plants to tolerate subsequent heat stresses (Dat *et al.*, 1998a, 1998b; Gong *et al.*, 1998). Significantly, these treatments do not, in general, affect the ability of the plants to produce HSPs (Larkindale & Vierling, unpublished data), suggesting that the induction of thermotolerance through changes in the antioxidant capacity occurs through some pathway distinct from

Table 5.3 Other mutant plants with defects in thermotolerance

Mutant	Gene	Affect	Reference
NahG (Arabidopsis)	NahG transgene: salicylate hydroxylase	Decreased heat tolerance	Larkindale and Knight (2002), Clarke *et al.* (2004), Larkindale and Vierling (in preparation)
npr1 (Arabidopsis)	Involved in SA signaling in pathogenesis	Decreased heat tolerance, normal sHSP synthesis	Clarke *et al.* (2004) Larkindale and Vierling (in preparation)
abi1 (Arabidopsis)	Phosphatase involved in ABA signaling	Decreased heat tolerance	Larkindale and Knight (2001), Larkindale and Vierling (in preparation)
abi2 (Arabidopsis)	Phosphatase involved in ABA signaling	Decreased heat tolerance	Larkindale and Knight (2001), Larkindale and Vierling (in preparation)
abo-(Catharanthus roseus)	Constitutively produces ABA	Sensitive to heat, salinity and drought	Rai *et al.* (2001)
etr1 (Arabidopsis)	Mutant in ethylene signaling	Somewhat decreased heat tolerance, especially in nonacclimated plants	Larkindale and Knight (2001), Larkindale and Vierling (in preparation)
fad7 (Arabidopsis)	Omega-3 desaturase, decreased trienoic fatty acids in chloroplasts	Increased heat tolerance of photosynthesis	Murakami *et al.* (2000)
fad7/8 (Arabidopsis)	Omega-3 desaturase, decreased trienoic fatty acids in chloroplasts	Increased heat tolerance of photosynthesis	Murakami *et al.* (2000)
fad3 (Arabidopsis)	Omega-3 desaturase, decreased trienoic fatty acids in ER	No change	Murakami *et al.* (2000)

Gene	Description	Phenotype	Reference
STR7 (soybean)	Atrazine resistant mutant: increased saturation of chloroplast lipids	Increased heat tolerance of photosynthesis	Alfonso *et al.* (2001)
TU8 (Arabidopsis)	Deficient in glucosinolate synthesis and pathogen-induced auxin accumulation	Decreased heat tolerance, decreased expression of Hsp90	Ludwig-Müller *et al.* (2000)
ERECTA (Arabidopsis)	Gene involved in regulating adaxial–abaxial leaf polarity	Increased heat sensitivity	Qi *et al.* (2004)
ssadh (Arabidopsis)	Succinic-semialdehyde DHase – last enzyme in GABA pathway	Increased heat sensitivity, increased active oxygen	Bouche *et al.* (2003)
AtCHIP (Arabidopsis)	E3 ubiquitin ligase	Decreased heat tolerance	Yan *et al.* (2003)
AtTS02 (Arabidopsis)	Deficient in chlorophyll accumulation	Decreased heat tolerance and accumulation of Hsp27	Burke *et al.* (2000)
uvh6 (Arabidopsis)	XPD DNA helicase	Decreased heat tolerance, normal HSP synthesis	Jenkins *et al.* (1997), Larkindale and Vierling (in preparation)
hot2 (Arabidopsis)	Unknown	Decreased acquired thermotolerance, normal HSP synthesis	Hong *et al.* (2003)
hot3 (Arabidopsis)	Unknown	Decreased acquired thermotolerance	Hong *et al.* (2003)
hot4 (Arabidopsis)	Unknown	Decreased heat tolerance, normal HSP synthesis	Hong *et al.* (2003)

the induction of HSPs. Recent microarray studies of Arabidopsis plants deficient in APX, however, have found that certain HSPs are expressed atypically in these plants under other stress conditions, although expression of HSPs under heat stress occurs normally (Pnueli *et al.*, 2003). This suggests that while there appear to be distinct pathways for the induction of antioxidants and HSPs during heat stress, there may be some crosstalk between the pathways.

5.5.2   *Other heat-stress regulated genes*

The fact that heat causes protein denaturation suggests that components of the protein degradation machinery should also increase during heat stress. In fact, recent evidence from yeast points to the accumulation of denatured or aggregated proteins as the major cause of heat-induced toxicity in this organism, and enhanced cytoplasmic proteolysis was shown to compensate for the absence of HSP synthesis (Friant *et al.*, 2003; Riezman, 2004). These authors observed that overexpression of UB14 suppressed a sphingolipid-signaling mutant (*lcb1-100*), which has reduced HSP synthesis and enhanced thermosensitivity. Increasing the ubiquitin levels in this mutant yielded cells that were no longer thermosensitive, despite retaining low HSP levels (Friant *et al.*, 2003). This suggests that increased proteolysis in yeast can compensate for increased stress levels. A number of ubiquitin pathway genes are also upregulated during heat stress in some plants. Both ubiquitin and conjugated ubiquitin levels increased faster in a heat-tolerant soybean relative compared to commercial soybean lines (Ortiz & Cardemil, 2001). In Arabidopsis UBQ14 levels increased in parallel with the induction of HSP70, while UBQ11 decreased (Sun & Callis, 1997). Specific ubiquitin alleles were also shown to be upregulated in specific organs after heat stress in maize (Liu *et al.*, 1996), and a ubiquitin-conjugating enzyme has been shown to be upregulated by heat shock in tomato plants (Feussner *et al.*, 1997). Similarly, ubiquitin pathway genes showed increased transcript levels in heat-stressed tobacco (Rizhsky *et al.*, 2002). An E3 ubiquitin ligase, AtCHIP, is heat-induced in Arabidopsis, but surprisingly, over-expression of this gene resulted in increased heat sensitivity (Yan *et al.*, 2003). Small ubiquitin-like modifier proteins (SUMO) 1 and 2 – but not SUMO 3 – have been shown accumulate to high levels in heat-stressed Arabidopsis, and less so in HSP70 overexpressing plants. This suggests that induction of the ubiquitin pathway responds to increased need for removal of denatured proteins (Kurepa *et al.*, 2003). These data together indicate that upregulation of the ubiquitin pathway is an aspect of heat stress responses, but potentially is less important in plants than in yeast, possibly because they are protected through other mechanisms. Heat induction of components of the ubiquitin pathway may in some way ameliorate heat damage done in the absence of HSPs, and there is likely to be crosstalk between ubiquitin and HSP regulatory pathways.

The Arabidopsis *etr1* mutant, which is defective in ethylene signaling, has been shown to be sensitive to direct heat treatments (Larkindale & Knight, 2002), implicating ethylene as a component in heat stress signaling. Studies in other species have shown ACC oxidases, which are involved in the production of ethylene, are upregulated by heat (potato, Nie *et al.*, 2002). In contrast, ACC synthase levels were reduced at high temperature in the tropical fruit tree *Hancornia speciosa* (De Pereira-Netto & McCown, 1999). There have been no reports as to what effect this stress hormone has on plants under heat stress conditions, and there is no evidence in the literature of any relationship between ethylene signals and HSP synthesis. However, there are significant data suggesting that ethylene is important in protection against oxidative stress, and this may explain its role in heat stress (Varshney & Rout, 1998; Kato *et al.*, 2000; Bortier *et al.*, 2001; Moeder *et al.*, 2002; Nie *et al.*, 2002; Manning *et al.*, 2003).

Studies of touch-induced (TCH) genes have shown that they are regulated by a number of different stresses, including heat, but there has been no research to determine whether TCH genes are necessary for survival of high-temperature stress. It was, however, determined that TCH genes, unlike APX, are not regulated through the same pathways as HSPs. TCH gene expression is calcium dependent and the genes do not contain canonical 5' HSEs (Braam, 1992).

Microarray analysis of heat-regulated transcripts is in its early stages. An initial study in tobacco examined the transcripts of 170 cDNAs from plants drought-stressed with or without simultaneous heat stress (Rizhsky *et al.*, 2002). Many unique genes, not upregulated by heat or drought alone, were shown to be upregulated by a combination of the two stresses. The same authors have now used complete genome arrays from Arabidopsis to examine transcript changes in response to heat (38°C, 6 h), drought (70% relative water content) or heat plus drought. At the single timepoint tested, they found 262 increased transcripts in response to heat and 1075 increased in response to drought (Rizhsky *et al.*, 2004). Interestingly, of these genes only 29 represented overlap between the two different stresses. However, as in the tobacco study, many transcripts were increased only in response to the combined stresses (454). These data emphasize that understanding stress and utilizing any of this knowledge for crop improvement will require consideration of multiple, interacting environmental variables.

Microarray studies are also expanding our knowledge of potential functions required for heat stress tolerance. In our laboratory, we have identified over 100 'non-HSP' genes whose increased level of expression is associated with the acquisition of thermotolerance (Larkindale & Vierling, in preparation). It will be interesting to determine how these genes affect plant survival of high temperature treatments. Thus, diverse functions may be postulated to be important for either surviving heat shock or for the acquisition of thermotolerance, depending on the condition of the plant and its environment.

5.5.3   Other heat-protective responses

In addition to the genes induced at high temperature, the repertoire of heat-induced protective mechanisms is likely to include the accumulation of various small molecules thought to maintain proteins and membranes in functionally intact states. The role of such 'compatible solutes' in heat stress is best understood in yeast, where the nonreducing disaccharide trehalose accumulates in response to elevated temperatures and stabilizes cells against heat-induced viability loss (Elbein *et al.*, 2003). Although plants possess the enzymatic machinery for trehalose synthesis (Vogel *et al.*, 1998; Elbein *et al.*, 2003), there is no evidence for significant accumulation of this compound during heat stress. Overall, studies of metabolite changes during heat stress are only just beginning. Rizhsky *et al.* (2004) examined changes in polar metabolites in Arabidopsis after 6 h of 38°C stress with or without simultaneous drought treatment. Very few major changes were observed under heat stress alone, although lacitol, fucose and melibiose were increased over the control. Interestingly, heat and drought stresses together resulted in major differences, many of which were not seen with drought alone, including a dramatic increase in sucrose as well as elevated melibiose, maltose, gulose and fucose. The change in sucrose no doubt reflects dramatic changes in metabolic regulation, and how the other changes reflect metabolism versus adaptive accumulation of solutes remains to be determined. Also notable was the suppression of the drought-induced increase in proline when plants were also heat stressed. In fact, it appeared that high levels of proline were detrimental to heat-stressed plants, and active suppression of proline accumulation might occur during heat stress.

Membrane damage is also clearly a factor in temperature stress. Sharkey *et al.* (2001) found that the small hydrocarbon isoprene protects photosynthesis during brief periods of temperature stress in species that produce this hydrocarbon (e.g. kudzu and red oak). Similar protection resulted when a species that does not synthesize isoprene (*Phaseolus vulgaris*) was supplied the molecule exogenously. A mechanism for the protective properties of isoprene (and other volatiles such as monoterpenes, Loreto *et al.*, 1998) remains unknown, with Sharkey *et al.* (2001) invoking a 'general hydrocarbon effect' arising from the double bonds present in the molecule.

Mutants and transgenic plants with altered levels of lipid desaturation have been shown to have differential heat tolerance. Increased lipid saturation levels have been shown to correlate with increased heat tolerance of photosynthesis (Murakami *et al.*, 2000), while lipid saturation levels have been shown to increase during the acquisition of thermotolerance in bentgrass (Larkindale & Huang, 2004b). Increasing membrane fluidity has also been shown to induce some heat-induced transcripts in the absence of heat (Sangwan *et al.*, 2002), adding further evidence for interactions between heat-induced membrane changes and thermotolerance. Whether physical properties of the membrane phase play a significant role in rapid heat responses remains an intriguing possibility, with membrane perturbing reagents influencing responses to heat stress at the protein and physiological levels as discussed in Section 5.4.5 concerning sHSPs (Sangwan *et al.*, 2002; Török *et al.*, 2003).

5.5.4 Mutants defective in heat tolerance

A number of different mutants have been tested and shown to have reduced or increased ability to tolerate high temperature treatments (Table 5.3). The wide range of functions represented by these mutants further emphasizes the complexity of the heat response. In addition, several of these mutants have been found to accumulate normal levels of Hsp101 and/or sHSPs, consistent with the importance of other factors in heat tolerance. It should also be noted that a number of different assays for heat tolerance were used to assess the phenotypic response of these mutants to heat, such that it is difficult to make direct comparisons between experiments. It is clear that different mutants show different degrees of heat-sensitivity depending on the age of the plant and other stresses imposed simultaneously (Hong & Vierling, 2000, 2001; Hong *et al.*, 2003; Larkindale & Knight, 2002).

Several of the mutants with heat stress phenotypes are involved in specific signaling pathways: *NahG* (a transgenic – see Table 5.3) and *npr1* are involved in salicylic acid (SA) signaling (Lawton *et al.*, 1995; Cao *et al.*, 1997), *abo-*, *abi1* and *2* are involved in ABA signaling (Meyer *et al.*, 1994; Rodriguez *et al.*, 1998; Rai *et al.*, 2001), *etr1* in ethylene signaling (Chang *et al.*, 1993), and various lipid desaturase mutants are associated with membrane fluidity, a property proposed to act as a sensor through which plants and other organisms assess temperature stress (Vigh *et al.*, 1998; Orvar *et al.*, 2000; Browse & Xin, 2001; Sangwan *et al.*, 2002; reviewed in Mikami & Murata, 2003). The effects of these signals on processes involved in heat tolerance have only begun to be investigated, and what genes are induced or proteins are activated by these pathways remains unknown. As mentioned above, several of these mutants accumulate one or more HSPs normally, so it is probable that these signaling molecules are involved in other, non-HSP pathways (Hong *et al.*, 2003; Clarke *et al.*, 2004; Larkindale & Vierling, unpublished data). Signaling mutants are discussed in further sections.

Other mutants in Table 5.3 are defective in a variety of functions, including an enzyme in the GABA pathway, a chlorophyll-deficient mutant, a ubiquitin ligase, a glucosinolate deficient plant and a DNA helicase. At present, little is known about why these plants fail to thrive at high temperatures, although our recent analysis of *uvh6* suggests it fails to express a whole subset of non-HSP genes associated with thermotolerance (Larkindale & Vierling, in preparation). Further, considerable work will be required to determine the critical functions supplied by each of these genes under stress conditions.

We have taken a genetic approach to understanding acquired thermotolerance by directly isolating mutants defective in this process (Hong & Vierling, 2000). A similar approach has been taken by Burke *et al.* (2000). Based on an assay for acquired thermotolerance of hypocotyl elongation of 2.5-day-old dark grown seedlings, we identified seven loci, designated *hot1* through *7*, which are necessary for high temperature acclimation (Hong & Vierling, 2000; Hong *et al.*, 2003, unpublished). The *hot1* locus, which encodes Hsp101 as described earlier, is the only one of these genes that defines an HSP. The other loci do not map close to HSP or HSF genes, and all but *hot3* accumulate Hsp101 and sHsps at normal levels. Therefore, these other loci are expected to encode genes necessary for the acquisition of thermotolerance, but not associated with HSP induction. Another important phenotype of the HOT mutants is that several do not show defects in acquired thermotolerance when assayed at growth stages other than the 2.5-day-old seedling stage at which the mutation was identified (Hong *et al.*, 2003, unpublished). This fact most likely illustrates that there are changing requirements for heat tolerance at different growth stages, and/or that certain HOT genes become functionally redundant depending on growth stage. Further, there is no evidence that *hot1, 2, 3* or *4* are any more affected than wild type by chronic heat exposure. This observation provides direct genetic evidence for differences in processes required for acclimation to acute and chronic heat stress.

5.5.5   Transgenic plants with altered heat tolerance

An idea of what processes are involved in protection of cells from heat stress is also gained from studies in which constitutive expression of specific proteins has been shown to enhance heat tolerance. In addition to the studies already discussed concerning expression of sHsps/chaperones (Table 5.3), and manipulation of HSF gene expression (see Table 5.5), other transgenic plants with varying degrees of heat tolerance have been produced (Table 5.4). Surprisingly, however, such experiments have been quite limited compared to the experiments aimed at engineering drought, salt or cold stress tolerance.

Artificial introduction of high levels of the compatible solute glycine betaine into Arabidopsis through transformation with a bacterial choline oxidase gene engineered to target the protein to the chloroplast was found to significantly enhance germination rates and seedling growth at elevated temperatures (Alia *et al.*, 1998). Glycine betaine is implicated not only as an osmolyte, but also as a direct protectant of proteins and membranes (Sakamoto & Murata, 2002). The latter function might be more important in the case of heat stress. No measurements of physiological processes were reported for these plants, however, so at present no mechanistic interpretation is possible.

Altering fatty acid composition of lipids to increase high temperature stability of the photosynthetic membrane has also been shown to increase heat tolerance (Murakami *et al.*, 2000). Maintaining an intact photosynthetic system allows the

Table 5.4 Other transgenic plants showing altered thermotolerance

Gene inserted (species)	Construct	Transformed species	Effect	Reference
Ascorbate peroxidase (barley)	35S:sense	Arabidopsis	Increased thermotolerance	Shi *et al.* (2001)
ChyB: β-carotene hydroxylase (Arabidopsis)	35S:sense	Arabidopsis	Increased xanthophyll pool, increased heat tolerance	Davison *et al.* (2002)
CodA (choline oxidase) (*Arthrobacter globiformis*)	35S:sense	Arabidopsis	Increased glycine betaine, increased thermotolerance of seeds and young seedlings. Less Hsp70 induced	Alia *et al.* (1998)
FAD7: omega-3 desaturase (Arabidopsis)	35S:sense (cosuppression)	Arabidopsis	Cosuppression increased heat tolerance of photosynthesis	Murakami *et al.* (2000)
ANP: MAPKKK (Arabidopsis)	35S:sense	Arabidopsis	Initiated a MAPK cascade and stress gene induction; increased heat tolerance	Kovtun *et al.* (2000)

plant to continue to generate ATP under stress, and limits photo-oxidation due to the release of free radicals. This both reduces the level of oxidative stress experienced by the plant and allows the production of energy necessary for cellular repair during recovery from stress. Membrane fluidity alterations may also change the perception of the stress through lipid signaling, thus changing the response of protective mechanisms. Murakami *et al.* (2000) reduced trienoic fatty acids in the chloroplasts of transformed tobacco plants, and also studied the Arabidopsis *fad7/fad8* mutant that has similar fatty acid changes. As measured by O_2 evolution, photosynthesis did appear to be stabilized against short high temperature treatments (5 min at 40°C or 45°C) in the tobacco plants. Notably, plants could grow both under long-term chronic stress or survive short-term acute stress better than wild type. Further discussion of how altering lipids may effect stress tolerance is covered by Iba (2002).

There is much more limited data on the protective effects against heat of constitutive expression of other genes. Constitutive expression of an H_2O_2-responsive mitogen-activated protein kinase kinase kinase (MAPKKK) in tobacco (ANP1/NPK1) was found to protect plants against the lethality of 45 min at 48°C (Kovtun *et al.* 2000). The same plants showed tolerance to freezing and salt and were reported to otherwise exhibit normal growth. Interestingly, an NPK1-related transcript was significantly elevated by heat in studies of Rizhsky *et al.* (2004). A report by Shi *et al.* (2001) found a modest increase in heat tolerance of Arabidopsis plants constitutively expressing the barley APX1 gene. Twenty-day-old Arabidopsis plants overexpressing this gene were placed at

Table 5.5 HSF transgenics and mutants

Gene (species)	Construct or mutation	Transformed species	Effect	Reference
HSFA1 (tomato)	35S:sense	Tomato	Increased tolerance	Mishra *et al.* (2002)
HSFA1 (tomato)	35S:sense (cosuppression)	Tomato	Cosuppression resulted in plants highly sensitive to heat, reduced HSP synthesis	Mishra *et al.* (2002)
HSFA1a (Arabidopsis)	35S:sense	Arabidopsis	Increased thermotolerance	Lee *et al.* (1995)
HSFA1a (Arabidopsis)	Fusion construct with negative regulator EN-HSF1	Arabidopsis	Dominant negative reduced thermotolerance	Wunderlich *et al.* (2003)
HSFA1b (Arabidopsis)	35S:sense	Arabidopsis	Increased thermotolerance	Prändl *et al.* (1998)
HSFB1 (Arabidopsis)	35S:sense	Arabidopsis	No effect	Prändl *et al.* (1998)
HSFA1a (Arabidopsis)	T-DNA insertion mutant	Not applicable	No effect on HSP induction	Lohman *et al.* (2004)
HSFA1b (Arabidopsis)	T-DNA insertion mutant	Not applicable	No effect on HSP induction	Lohman *et al.* (2004)
HSFA1a/HSFA1b (Arabidopsis)	Double T-DNA insertion mutant	Not applicable	Reduced early HSP induction	Lohman *et al.* (2004)
Spl7 (*rice*)	HSF mutant, results in lesions of leaves	Not applicable	Forms lesions under high light/temperature	Yamanouchi *et al.* (2002)

35°C for 5 days and then observed after an additional 5 days at 23°C. Transgenic plants had 85% green tissue, as opposed to nontransgenics with 75% green tissue after a 5 day treatment at 35°C, presumably due to better control of oxidation stress. Other measurements of heat stress tolerance were not performed, but these results show that overexpression of this non-HSP gene can induce some form of heat tolerance. Doubling the pool of xanthophyll cycle intermediates by overexpression of β-carotene hydroxylase (Davison *et al.*, 2002) protected plants from heat stress under high light conditions, but tests of high light or heat stress alone were not reported. They concluded that the carotenoids were acting by general protection from oxidative stress.

Taken together, these data suggest that some form of heat tolerance can be induced through protection of several different systems damaged during heat stress. This is consistent with the interpretation that the acquisition of thermotolerance involves induction of a number of parallel systems that protect the plant against different types of heat-induced damage. The relative importance of these systems depending on the exact stress conditions and the age/developmental state of the plant remains to be determined.

5.6 Signaling pathways involved in response to heat

5.6.1 Heat shock transcription factors

HSP genes are transcribed in all eukaryotes through a similar process. Specific conserved transcription factors, called HSFs, recognize and bind to conserved *cis* elements (the HSE) in the promoter of HSP and other genes (Nover *et al.*, 2001). HSEs consist of the palindromic sequence 5'-AGAAnnTTCT-3'. In most species HSFs are constitutively expressed, and the proteins are found in the cytoplasm prior to stress. These pre-existing HSFs are believed to be held in an inactive state by association with Hsp70 and Hsp90 (Schöffl *et al.*, 1998). HSF activation is proposed to involve reduction in the amount of available Hsp70 and Hsp90 because these chaperones become saturated with heat-denatured substrates. However, other signaling components (in addition to denatured proteins) may act upstream of HSFs and merit further investigation. Once the plant has sensed an increase in temperature, HSFs go from a monomeric state in the cytoplasm to a trimeric state in the cell nucleus where they can then bind the HSEs (Lee *et al.*, 1995). HSF binding recruits other transcription components, resulting in gene expression within minutes of the increase in temperature.

Although this basic system is universal to eukaryotic cells, it is highly complicated in plants. Unlike animals and yeasts, which may have four or fewer HSFs, plants have been shown to have multiple copies of these genes: tomato has at least 17 and Arabidopsis has 21 different HSF genes (Nover *et al.*, 2001). These genes have been classified into three groups (classes A, B and C), which are discriminated by features of their flexible linkers and oligomerization domains (Nover *et al.*, 2001). Many of the HSFs are heat inducible suggesting that the specific HSF involved in transcription of a particular gene may vary depending on the timing and intensity of the stress. In general, as is summarized in Table 5.5, overexpression of plant HSFs can increase plant thermotolerance, but gene knockouts of individual HSFs tested so far have had little effect on survival of high temperature. This indicates that the HSF network is redundant, perhaps reflecting the importance of this high temperature response.

In Arabidopsis, HSFA1a and A1b appear to control the early response of many genes to heat (Lohmann *et al.*, 2004), but are nevertheless nonessential for heat tolerance. Other Arabidopsis HSFs are apparently responsible for the induction of genes expressed later, potentially including heat-inducible HSFs. Interestingly, in tomato, one particular HSF, HSFA1, has been proposed to be the 'master regulator' of the heat shock response. If this gene is suppressed, normal HSP production does not occur and the plant is extremely sensitive to high temperatures (Mishra *et al.*, 2002). Different HSFs have also been shown to act synergistically. HSFB1 in tomato does not act as a transcriptional activator by itself, but when expressed along with HSFA1 the expression of reporter genes is much higher than if HSFA1 alone is expressed. Thus, plants appear to have a remarkable ability to finely control the expression of heat-induced genes through the HSF system.

5.6.2   Other signaling pathways

Beyond HSFs, to date there have been no conclusive reports as to other signaling components involved in the upregulation of HSPs. A number of studies have used pretreatment of plants with a variety of signaling molecules to determine if these treatments could enhance thermotolerance (Dat *et al.*, 1998a, 1998b; Gong *et al.*, 1998; Jiang & Huang, 2001; Larkindale & Knight, 2002). These studies suggest that abscisic acid (ABA), SA, calcium ions, ethylene (or ACC, an ethylene precursor) and AOS are all potential signaling molecules associated with tolerance of heat stress. Pretreatment with any of these substances induces some degree of thermotolerance in plants. Sphingolipids, phosphatidic acid and changes in membrane fluidity, as well as several kinases and phosphatases have also been hypothesized to be involved in heat signaling and sensing (Sagwan *et al.*, 2002; Link *et al.*, 2002; Alfonso *et al.*, 2001; Mishkind, unpublished).

Mutant plants deficient in these signaling pathways provide a useful tool to investigate their potential involvement in the heat stress response. As discussed above, a number of mutants in signaling pathways are indeed sensitive to heat stress (Table 5.3). Treatment with calcium inhibitors also reduces plant thermotolerance (Larkindale & Knight, 2002). However, none of these mutants shows any defect in their ability to accumulate HSPs (Larkindale & Vierling, in preparation). These data indicate that plant responses to heat involve multiple signaling molecules that are not part of the HSP/HSF response pathway. This is further evidence for the existence of multiple parallel pathways that together induce the changes in the plant necessary for thermotolerance.

5.6.3   Abscisic acid

ABA has been shown to induce some degree of thermotolerance in a range of plant species (Robertson *et al.*, 1994; Gong *et al.*, 1998; Larkindale & Knight, 2002; Larkindale & Huang, 2004a). Plants treated with ABA show increased survival after a direct heat stress, but the improvement is not as great as that achieved by pretreatment with moderate temperature (acquired thermotolerance). Some ABA-signaling mutants show diminished ability both to tolerate high temperatures and to acquire thermotolerance, including mutants in the protein phosphatases *abi1* and *abi2* (Larkindale & Knight, 2002; Larkindale & Vierling, in preparation). In contrast, the ABA-insensitive mutant *abi3*, where the affected gene encodes a transcription factor involved in HSP expression in seeds (Rojas *et al.*, 1999; Wehmeyer & Vierling, 2000), is not defective in heat tolerance (Larkindale & Vierling, in preparation).

There are contrasting reports with respect to ABA and the induction of HSPs. A wheat Hsp100 family member was shown to be induced by ABA (Campbell *et al.*, 2001), while Hsp101 in maize was not (Nieto-Sotelo *et al.*, 1999). Sunflower sHSPs may be induced by ABA (Coca *et al.*, 1996), but wild-type ABA levels are not required for HSP induction in tomato (Bray, 1991). Work in our laboratory indicates *Arabidopsis* ABA mutants that are unable to acquire thermotolerance all accumulate HSPs to wild-type levels (Larkindale & Vierling, in preparation). This suggests that ABA plays a part in some pathway distinct from the induction of HSPs, but which is nonetheless essential to the acquisition of thermotolerance.

5.6.4 Salicylic acid

Like ABA, pretreatment with SA results in improved heat tolerance in a number of species (Dat *et al.*, 1998b; Larkindale & Knight, 2002; Clarke *et al.*, 2004; Larkindale & Huang, 2004a), and plants unable to utilize SA signaling show limited ability to tolerate heat stress (Larkindale & Knight, 2002; Clarke *et al.*, 2004; Larkindale & Vierling, in preparation). As with ABA, there is limited evidence that SA is associated with HSP production. At some concentrations SA pretreatment potentiated HSP synthesis in tomato plants, while at higher concentrations it induced Hsp70s in that species (Cronje & Bornman, 1999). In mammals, SA has been shown to initiate HSP synthesis (Jurivich *et al.*, 1992), although after the HSF has bound to the promoter of the HSP genes transcription is not initiated (Jurivich *et al.*, 1995). Data from our own work suggest that plants with reduced SA signaling accumulate HSPs normally, and treatment with an SA concentration sufficient to induce thermotolerance does not induce HSP synthesis in Arabidopsis. Therefore, the function of SA in thermotolerance has yet to be determined. Again it may represent an additional parallel, important signaling pathway of the heat response.

5.6.5 Calcium

The role of calcium in heat stress in plants has been debated for some time. There is little doubt that pretreatment in calcium chloride enhances a plant's ability to tolerate high temperatures, and that calcium inhibitors limit plant survival (Braam, 1992; Jiang & Huang, 2001; Larkindale & Knight, 2002; Larkindale & Huang, 2004a). However, only two groups have made direct measurements of calcium signals under heat stress in plants, and they have drawn opposite conclusions. In Arabidopsis, Larkindale and Knight (2002) failed to find any significant increase in cytosolic calcium during heating of 10-day-old seedlings, but saw significant increase in calcium immediately on the initiation of cooling. In contrast, in wheat, Liu *et al.* (2003) saw significant increases in calcium during heating and correlated this with decreased HSP synthesis in plants treated with calcium signaling inhibitors. Calcium and calmodulin treatments were also shown to induce maize HSF to HSE binding *in vitro*, while calcium and calmodulin inhibitors prevented binding at high temperatures (Li *et al.*, 2004). In Arabidopsis neither calcium inhibitor treatment nor addition of calcium chloride affected HSP synthesis during heat stress (Braam, 1992; Larkindale, 2001), nor did such treatments affect HSP induction in sugar beet (Kuznetsov *et al.*, 1998). These contrasting results may reflect a fundamental difference in the responses of monocots and dicots to high temperature, or may reflect differences in the heat treatments or other experimental parameters.

5.6.6   *Active oxygen species*

Hydrogen peroxide is produced early on in response to heat stress (Dat *et al.*, 1998b; Vacca *et al.*, 2004; Larkindale, 2001). While it has long been known that heat indirectly causes oxidative damage, presumably through the accumulation of AOS, the potential for these species to act as a signal has only recently been considered. However, H_2O_2 does induce some HSPs in some species, including rice (Lee *et al.*, 2000), and in mammalian cells H_2O_2 activates HSFs (Calabrese *et al.*, 2001). APX knockout and overexpressing plants ectopically express HSPs, along with a wide range of other stress components (Shi *et al.*, 2001; Pnueli *et al.*, 2003). In addition, prior treatment with H_2O_2 or menadione can increase plant thermotolerance (Larkindale, 2001; Larkindale & Huang, 2004a), suggesting a role for AOS in heat stress signaling. There also may be an association between the oxidative burst seen at the initiation of heating (Dat *et al.*, 1998b) and the accumulation of HSPs.

5.6.7   *Ethylene*

There has been limited study of ethylene as a potential signaling molecule during heat stress in plants. However, there is some evidence that this hormone may play a role in thermotolerance. ACC oxidases, involved in ethylene biosynthesis, have been shown to be heat upregulated (Nie *et al.*, 2002). Furthermore, pretreatment with ethylene induces some degree of thermotolerance (Larkindale & Knight, 2002; Larkindale & Huang, 2004a), and the *etr1* ethylene signaling mutant is sensitive to heat (Larkindale & Knight, 2002) (Table 5.3). Ethylene production has also been shown to be induced by heat treatments (Dziubinska *et al.*, 2003), and an ethylene response transcriptional coactivator was upregulated by drought and heat combined (Rizhsky *et al.*, 2002). Aminooxyacetic acid, an ethylene synthesis inhibitor, inhibits ethylene production and heat-induced apoptosis in tobacco plants (Yamada *et al.*, 2001, suggesting a further link between ethylene signaling and heat-induced responses. There is no evidence that this signal is associated with HSP induction, however, as HSPs are induced normally in *etr1* plants (Larkindale & Vierling, in preparation). This suggests whatever role ethylene plays in the response to heat stress is not through the induction of HSPs, but through some other stress pathway. This may be through apoptotic pathways (Yamada *et al.*, 2001), or through the suppression of heat-induced oxidative stress (Varshney & Rout, 1998; Kato *et al.*, 2000; Bortier *et al.*, 2001; Moeder *et al.*, 2002; Nie *et al.*, 2002; Manning *et al.*, 2003).

5.6.8   Signaling lipids

Despite intense interest in the role of lipid signaling in a wide variety of stress responses in plants (Meijer & Munnik, 2003; Worrall *et al.*, 2003), these pathways remain unexamined in the context of temperature stress. This situation stands in contrast to work in animals and yeast, where various lipid pathways have been shown to be engaged shortly after a heat shock. Sphingolipid-based pathways are implicated in both yeast and mammalian cells where heat stress activates serine palmitoyltransferase in response to heat stress. In yeast, this activity is required for trehalose accumulation and heat hardiness (Jenkins, 2003). Heat stress also induces ceramide in mammalian cells, which activates pathways that lead to apoptosis. Phospholipid-based signaling has also been implicated in mouse embryonic fibroblasts; reduced phospholipase C-1 activity reduces heat-induced protein kinase C (PKC) activation, which significantly compromises cell viability at elevated temperatures. It is interesting to note that this has no effect on HSP70 synthesis (Bai *et al.*, 2002). Preliminary experiments in tobacco suspension cultures and Arabidopsis seedlings, however, indicate that both PLC and PLD are activated immediately upon an increase in temperature (Mishkind *et al.*, in preparation). This suggests that there is a role for some form of lipid signaling in response to heat stress in plants.

5.6.9   Kinases and phosphatases

Little is known about whether MAP kinase cascades are involved in plant heat signaling. A specific MAPK was identified in Arabidopsis, which is upregulated by heat and by changes in membrane fluidity (Sangwan *et al.*, 2002). Another heat upregulated kinase was identified in tomato and was subsequently shown to phosphorylate HSFA3 (Link *et al.*, 2002). It is not known, however, if either of these is essential for heat tolerance, as the genes associated with the kinases have yet to be identified. Kovtun *et al.* (2000) have shown that H_2O_2 activates a specific Arabidopsis MAPKKK called ANP1. Activation of ANP1 initiates a kinase cascade including AtMPK3 and AtMPK6 that ultimately leads to increased transcription of Hsp18.2, a class II cytosolic sHsp. The effect of ANP1 activation on transcription of other HSP genes was not assessed, nor were HSP protein levels measured, but this result is consistent with other data showing some HSPs are induced by H_2O_2 (Lee *et al.*, 2000). Tobacco plants that constitutively expressed ANP1 showed increased resistance to heat treatment (48°C, 45 min) compared to wild type, as discussed earlier. However, induction of Hsp18.2 transcription by H_2O_2 and the ANP1 pathway is not as strong as heat-induced expression of HSPs; Hsp18.2 was induced only fivefold, while heat stress induction is easily 100-fold. This suggests that ANP1 HSP induction is not the sole mechanism through which Hsp18.2 is induced during heat shock. Thus, ANP1-mediated heat stress protection may result primarily from mechanisms separate from HSPs, such as limiting damage by reactive oxygen intermediates. Whether ANP1 is also activated by heat shock is not clear, and therefore, it is not possible to determine if this pathway normally contributes to tolerance to heat stress, or if the heat tolerance is just a by-product of inducing the oxidative stress response.

It is interesting to note that *S. cerevisiae* mutants defective in PKC fail to develop thermotolerance, but appear fully capable of inducing other proteins regulated by heat stress, including HSPs (Kamada *et al.*, 1995). The PKC pathway, which includes MAPK, MEK and MEKK kinases can also be induced by agents that perturb the plasma membrane, and PKC overexpression results in some increased ability to acquire thermotolerance in yeast. To date, no homologous heat-responsive system has been found in plants.

In terms of phosphatases, again there is weak evidence for specific genes functioning in heat stress signaling. In rice, a type 2A phosphatase was transcriptionally upregulated throughout the plant by heat, while another was downregulated in the stem under the same stress (Yu *et al.*, 2003). A specific Arabidopsis PP2A gene transcript, *AtB′γ*, has been shown to accumulate in response to heat stress (Haynes *et al.*, 1999), suggesting that PP2A hetero-trimers containing this subunit could be involved in stress response mechanisms in plants. On the contrary, transcripts from the *AtB′α*, *AtB′β*, *AtB′δ* and *AtB′ε* genes do not respond to such treatment (Latorre *et al.*, 1997; Haynes *et al.*, 1999; Terol *et al.*, 2002). The targets of these phosphatases are currently unknown, and again, it is not known if these proteins are required for plant survival at high temperature.

Altogether then, there is evidence of heat-inducible kinases and phosphat-ases, and evidence of kinases that affect HSP induction, and even some indirect evidence (in yeast) of kinases essential for thermotolerance. There is little evidence, however, directly linking any specific kinase or phosphatase to an essential function in the ability of a plant to survive heat stress. Further work is required in this area to determine which of these genes are essential for plant survival at high temperatures.

5.7 Genetic variation in heat tolerance

5.7.1 Agricultural/horticultural plants

Although studies have documented heritable differences in heat tolerance in both crop and horticultural species, these studies are difficult to compare because of the variety of species examined and the many different phenotypes quantified (yield, leaf firing, pollen viability, electrolyte leakage). Physiological assays have also revealed variations specific to acquired thermotolerance among cultivars of barley (Maestri *et al.*, 2002), potato (Ahn *et al.*, 2004), rice (Gesch *et al.*, 2003), pearl millet and sorghum (Howarth *et al.*, 1997), but follow-up quantitative genetic analysis of this variation has not been reported. Thus, there is little consensus on the numbers or types of genes that may be responsible for different heat tolerance traits (Marshall, 1982; Blum, 1988; Hall, 1990, 2001). The difficulties in breeding for heat tolerance have also limited genetic analysis in this area. Temperature cannot be controlled in the field, so breeders are at the mercy of the prevailing atmospheric conditions in most studies. Unless trial fields are irrigated, water availability will also severely affect plant success at high temperature.

A number of researchers have tried to investigate the relationship of genotypic differences in heat tolerance in relationship to differences in HSP quantity, quality or timing of synthesis. However, trying to link the variation in expression of a small set of genes to the differences in genotype is a difficult and imprecise approach. In the majority of cases, differences in HSP expression could not be correlated with differences in heat tolerance, and at best, suggestive correlations were observed with one or few HSPs (see Klueva *et al.*, 2001, for review). For example, a wheat line missing part of chromosome 1 in one copy of its genome was found to be highly heat tolerant, and acquired thermotolerance to 48°C after a lower temperature conditioning pretreatment than the parental line. HSP accumulation could also be observed at temperatures 4°C lower than in the parental line, which correlated with the induction of acquired thermotolerance (O'Mahony & Burke, 2000). Another study found that tissue culture-generated variants of creeping bentgrass (*Agrostis stolonifera*) selected for different levels of thermotolerance exhibited distinct expression profiles of chloroplast-localized sHSPs (Wang & Luthe, 2003).

Among the best-studied relationship of a heat-induced protein to differences in plant heat (and drought) tolerance comes from work of Ristic and coworkers (Ristic *et al.*, 1998; Bhadula *et al.*, 1998, 2001; Momcilovic & Ristic, 2004). The protein involved is an isoform of chloroplast elongation factor Tu, which, though not a classical HSP, appears to have some chaperone activity. These data suggest the interesting possibility that protection of chloroplast protein synthesis capability is an important aspect of heat tolerance.

5.7.2   *Natural variation in heat tolerance*

Studies of naturally occurring variation in heat stress responses complement forward and reverse genetic approaches aimed at uncovering the nature of the cellular machinery responsible for enhanced tolerance to high temperatures. Although ecotypic variation in heat tolerance has long been recognized, the numbers and types of genes involved remain unknown. Recently, some work has been directed at correlating HSP expression with natural variation in heat tolerance. Barua *et al.* (2003) concluded that levels of the chloroplast sHSP correlated with variation in thermotolerance of photosynthetic electron transport. Similarly, desert and coastal populations of *Encelia californica* were shown to accumulate sHSPs at different rates and also display differential heat sensitivity in photosynthetic electron transport (Knight & Ackerly, 2003). As more gene mapping is done in natural populations, many new opportunities to better understand stress responses and their overlap will arise.

Coupling natural variation with quantitative trait locus (QTL) mapping would appear well suited to this type of multigenic complex response, but so far is an underexplored approach for dissecting heat stress responses. Using this approach in rice or Arabidopsis would allow definition of major effect loci and their overlap depending on exact stress conditions, and offer the possibility to identify and clone critical genes. In Arabidopsis, accession variation in the ability to grow at high temperature (Langridge & Griffing, 1959) and differences in enzyme kinetic properties of accessions from contrasting temperature habitats have been reported (Simon *et al.*, 1983). Recently, QTL analysis has been applied to a study of tolerance of seeds to various types of stress at germination, including high temperatures during imbibition (Clerkx *et al.*, 2004). Interestingly, one locus identified as having an effect on heat tolerance overlapped with a locus associated with salt and ABA resistance, as well as with controlled deterioration and germination speed. This suggests that overall seed quality is correlated with tolerance to heat and other stresses. Notably, none of the QTL identified overlapped the Hsp101 gene, which when deleted leads to severe sensitivity to heat stress at imbibition (Hong & Vierling, 2000). This result emphasizes the fact that QTL mapping can uncover novel loci that were not previously identified by mutant screens. Further work of this type is certainly warranted.

5.8   Summary

In nature, different plant species, populations and even individuals may employ very different strategies to optimize growth and reproduction in the face of high temperature stress. Plants have evolved morphological, life history, physiological and cellular strategies not only to cope with temperature, but also to avoid high temperature damage. To date, the most successful strategies for enhancing agriculture in high temperature environments has primarily involved avoidance mechanisms, most often altering the timing of reproduction to achieve yield gains in hot climates (Mahan *et al.*, 1995; Slafer, 2003). Further advances in our understanding of responses to heat and our ability to manipulate heat tolerance will greatly benefit from application of genomics and proteomics techniques coupled with additional genetic analysis.

Acknowledgments

Research cited from the laboratory of E. Vierling has been supported by grants from the NIH, NSF, DOE and USDA.

References

Agarwal, M., Katiyar-Agarwal, S. & Grover, A. (2002) Plant Hsp100 proteins: structure, function and regulation. *Plant Science*, **163**, 397–405.

Agarwal, M., Katiyar-Agarwal, S., Sahi, C., Gallie, D. R. & Grover, A. (2001) *Arabidopsis thaliana* Hsp100 proteins: kith and kin. *Cell Stress & Chaperones*, **6**, 219–224.

Ahn, Y. J., Claussen, K. & Zimmerman, J. L. (2004) Genotypic difference in the heat-shock response and thermotolerance in four potato cultivars. *Plant Science*, **166**(4), 901–911.

Alfonso, M., Yruela, I., Almarcegui, S., Torrado, E., Perez, M. A. & Picorel, R. (2001) Unusual tolerance to high temperatures in a new herbicide-resistant D1 mutant from *Glycine max* (L.) *Merr.* cell cultures deficient in fatty acid desaturation. *Planta*, **212**(4), 573–582.

Alia, H. H., Sakamoto, A. & Murata, N. (1998) Enhancement of the tolerance of Arabidopsis to high temperatures by genetic engineering of the synthesis of glycinebetaine. *Plant Journal*, **16**(2), 155–161.

Alvim, F. C., Carolin, S. M., Cascarado, J. C., Nunes, C. C., Martinez, C. A., Otoni, W. C. & Fontes, E. P. (2001) Enhanced accumulation of BiP in transgenic plants confers tolerance to water stress. *Plant Physiology*, **126**, 1042–1054.

Apuya, N. R., Yadegari, R., Fischer, R. L., Harada, J. J., Zimmerman, J. L. & Goldberg, R. B. (2001) The Arabidopsis embryo mutant *schlepperless* has a defect in the chaperonin-60alpha gene. *Plant Physiology*, **126**, 717–730.

Asadulghani, S. Y. & Nakamoto, H. (2003) Light plays a key role in the modulation of heat shock response in the cyanobacterium *Synechocystis* sp PCC 6803. *Biochemical and Biophysical Research Communications*, **306**, 872–879.

Bai, X. C., Liu, A. L., Deng, F., Zou, Z. P., Bai, J., Ji, Q. S. & Luo, S. Q. (2002) Phospholipase C-1 is required for survival in heat stress: involvement of protein kinase C-dependent Bcl-2 phosphorylation. *Journal of Biochemistry (Tokyo)*, **131**(2), 207–212.

Barua, D., Downs, C. A. & Heckathorn, S. A. (2003) Variation in chloroplast small heat-shock protein function is a major determinant of variation in thermotolerance of photosynthetic electron transport among ecotypes of *Chenopodium album*. *Functional Plant Biology*, **30**(10), 1071–1079.

Basha, E., Lee, G. J., Demeler, B. & Vierling, E. (2004) Chaperone activity of cytosolic small heat shock proteins from wheat. *European Journal of Biochemistry*, **271**, 1426–1436.

Berardini, T. Z., Bollman, K., Sun, H. & Poethig, R. S. (2001) Regulation of vegetative phase change in *Arabidopsis thaliana* by cyclophilin 40. *Science*, **291**(5512), 2405–2407.

Bernacchi, C. J., Portis, A. R., Nakano, H., von Caemmerer, S. & Long, S. P. (2002) Temperature response of mesophyll conductance. Implications for the determination of Rubisco enzyme kinetics and for limitations to photosynthesis *in vivo*. *Plant Physiology*, **130**(4), 1992–1998.

Bhadula, S. K., Elthon, T. E., Habben, J. E., Helentjaris, T. G., Jiao, S. & Ristic, Z. (2001) Heat-stress induced synthesis of chloroplast protein synthesis elongation factor (EF-Tu) in a heat-tolerant maize line. *Planta*, **212**(3), 359–366.

Bhadula, S. K., Yang, G. P., Sterzinger, A. & Ristic, Z. (1998) Synthesis of a family of 45 kD heat shock proteins in a drought and heat resistant line of maize under controlled and field conditions. *Journal of Plant Physiology*, **152**, 104–111.

Blum, A. (1988) *Plant Breeding for Stress Environments*. CRC Press, Boca Raton, FL.

Borrelli, M. J., Stafford, D. M., Karczewski, L. A., Rausch, C. M., Lee, Y. J. & Corry, P. M. (1996) Thermotolerance expression in mitotic CHO cells without increased translation of heat shock proteins. *Journal of Cell Physiology*, **169**(3), 420–428.

Bortier, K., Dekelver, G., De Temmerman, L. & Ceulemans, R. (2001) Stem injection of *Populus nigra* with EDU to study ozone effects under field conditions. *Environmental Pollution*, **111**(2), 199–208.

Boston, R. S., Viitanen, P. V. & Vierling, E. (1996) Molecular chaperones and protein folding in plants. *Plant Moleclar Biology*, **32**(1–2), 191–222.

Bouche, N., Fait, A., Bouchez, D., Moller, S. G. & Fromm, H. (2003) Mitochondrial succinic-semialdehyde dehydrogenase of the gamma-aminobutyrate shunt is required to restrict levels of reactive oxygen intermediates in plants. *Proceedings of the National Academy of Sciences*, **100**(11), 6843–6848.

Braam, J. (1992) Regulated expression of the calmodulin-related TCH genes in cultured Arabidopsis cells: induction by calcium and heat shock. *Proceedings of the National Academy of Sciences*, **89**(8), 3213–3216.

Bray, E. A. (1991) Wild type ABA levels are not required for heat shock protein induction in tomato. *Plant Physiology*, **97**(2), 817–820.

Browse, J. & Xin, Z. (2001) Temperature sensing and cold acclimation. *Current Opinion in Plant Biology*, **4**(3), 241–246.

Burke, J. J., O'Mahony, P. J. & Oliver, M. J. (2000) Isolation of Arabidopsis mutants lacking components of acquired thermotolerance. *Plant Physiology*, **123**(2), 575–588.

Calabrese, V., Scapagnini, G., Catalano, C., Bates, T. E., Dinotta, F., Micali, G. & Giuffrida Stella, A. M. (2001) Induction of heat shock protein synthesis in human skin fibroblasts in response to oxidative stress: regulation by a natural antioxidant from rosemary extract. *International Journal of Tissue Reactions*, **23**(2), 51–58.

Campbell, J. L., Klueva, N. Y., Zheng, H. G., Nieto-Sotelo, J., Ho, T. D. & Nguyen, H. T. (2001) Cloning of new members of heat shock protein HSP101 gene family in wheat (*Triticum aestivum* (L.) *Moench*) inducible by heat, dehydration, and ABA. *Biochimica et Biophysica Acta*, **1517**(2), 270–277.

Cao, D., Froehlich, J. E., Zhang, H. & Cheng, C. L. (2003) The chlorate-resistant and photomorphogenesis-defective mutant cr88 encodes a chloroplast-targeted HSP90. *Plant Journal*, **33**(1), 107–118.

Cao, H., Glazebrook, J., Clarke, J. D., Volko, S. & Dong, X. (1997) The Arabidopsis NPR1 gene that controls systemic acquired resistance encodes a novel protein containing ankyrin repeats. *Cell*, **88**(1), 57–63.

Carlson, R. E. (1990) Heat stress, plant-available soil moisture, and corn yields in Iowa: a short- and long-term view. *Journal of Production Agriculture*, **3**, 293–297.

Chang, C., Kwok, S. F., Bleecker, A. B. & Meyerowitz, E. M. (1993) Arabidopsis ethylene-response gene ETR1: similarity of product to two-component regulators. *Science*, **262** (5133), 539–544.

Chen, Q., Lauzon, L. M., DeRocher, A. E. & Vierling, E. (1990) Accumulation, stability, and localization of a major chloroplast heat-shock protein. *Journal of Cell Biology*, **110**, 1873–1883.

Clarke, S. M., Mur, L. A., Wood, J. E. & Scott, I. M. (2004) Salicylic acid dependent signaling promotes basal thermotolerance but is not essential for acquired thermotolerance in *Arabidopsis thaliana*. *Plant Journal*, **38**(3), 432–447.

Clerkx, E. J. M., El-Lithy, M. E., Vierling, E., Ruys, G. J., Blankestijn-De Vries, H., Groot, S. P. C., Vreugenhil, D. & Koornneef, M. (2004) Analysis of natural allelic variation of Arabidopsis seed germination and seed longevity traits between the accessions Landsberg *erecta* and Shakdara, using a new recombinant inbred line population. *Plant Physiology*, **135**, 432–443.

Coca, M. A., Almoguera, C., Thomas, T. L. & Jordano, J. (1996) Differential regulation of small heat-shock genes in plants: analysis of a water-stress-inducible and developmentally activated sunflower promoter. *Plant Molecular Biology*, **31**(4), 863–876.

Commuri, P. D. & Jones, R. J. (2001) High temperatures during endosperm cell division in maize: a genotypic comparison under *in vitro* and field conditions. *Crop Science*, **41**, 1122–1130.

Crafts-Brandner, S. J. & Salvucci, M. E. (2000) Rubisco activase constrains the photosynthetic potential of leaves at high temperature and CO_2. *Proceedings of the National Academy of Sciences*, **97**(24), 13430–13435.

Cronje, M. J. & Bornman, L. (1999) Salicylic acid influences Hsp70/Hsc70 expression in *Lycopersicon esculentum*: dose- and time-dependent induction or potentiation. *Biochemical and Biophysical Research Communications*, **265**(2), 422–427.

Cross, R. H., McKay, S. A. B., McHughen, A. G. & Bonham-Smith, P. C. (2003) Heat-stress effects on reproduction and seed set in *Linum usitatissimum* L. (flax). *Plant Cell and Environment*, **26**, 1013–1020.

Dale, R. J. (1983) Temperature perturbations in the midwestern and southeastern United States important for corn production. In C. D. Raper & P. J. Kramer (eds) *Crop Reactions to Water and Temperature Stresses in Humid, Temperate Climates*. Westview Press, Boulder, CO, pp. 21–32.

Dat, J. F., Foyer, C. H. & Scott, I. M. (1998a) Changes in salicylic acid and antioxidants during induced thermotolerance in mustard seedlings. *Plant Physiology*, **118**(4), 1455–1461.

Dat, J. F., Lopez-Delgado, H., Foyer, C. H. & Scott, I. M. (1998b) Parallel changes in H_2O_2 and catalase during thermotolerance induced by salicylic acid or heat acclimation in mustard seedlings. *Plant Physiology*, **116**(4), 1351–1357.

Davison, P. A., Hunter, C. N. & Horton, P. (2002) Overexpression of beta-carotene hydroxylase enhances stress tolerance in Arabidopsis. *Nature*, **418**(6894), 203–206.

Deaton, J., Sun, J., Holzenburg, A., Struck, D. K., Berry, J. & Young, R. (2004) Functional bacteriorhodopsin is efficiently solubilized and delivered to membranes by the chaperonin GroEL. *Proceedings of the National Academy of Sciences of the United States of America*, **101**, 2281–2286.

De Pereira-Netto, A. B. & McCown, B. H. (1999) Thermally induced changes in shoot morphology of *Hancornia speciosa* microcultures: evidence of mediation by ethylene. *Tree Physiology*, **19**(11), 733–740.

DeRocher, A. E., Helm, K. W., Lauzon, L. M. & Vierling, E. (1991) Expression of a conserved family of cytoplasmic low molecular weight heat shock proteins during heat stress and recovery. *Plant Physiology*, **96**, 1038–1047.

Dziubinska, H., Filek, M., Koscielniak, J. & Trebacz, K. (2003) Variation and action potentials evoked by thermal stimuli accompany enhancement of ethylene emission in distant non-stimulated leaves of *Vicia faba* minor seedlings. *Journal of Plant Physiology*, **160**(10), 1203–1210.

Elbein, A. D., Pan, Y. T., Pastuszak, I. & Carroll, D. (2003) New insights on trehalose: a multifunctional molecule. *Glycobiology*, **13**(4), 17R–27R.

Feussner, K., Feussner, I., Leopold, I. & Wasternack, C. (1997) Isolation of a cDNA coding for an ubiquitin-conjugating enzyme UBC1 of tomato – The first stress-induced UBC of higher plants. *FEBS Letters*, **409**(2), 211–215.

Friant, S., Meier, K. D. & Riezman, H. (2003) Increased ubiquitin-dependent degradation can replace the essential requirement for heat shock protein induction. *EMBO Journal*, **22**(15), 3783–3791.

Fu, S., Meeley, R. & Scanlon, M. J. (2002) Empty pericarp2 encodes a negative regulator of the heat shock response and is required for maize embryogenesis. *Plant Cell*, **14**(12), 3119–3132.

Gesch, R. W., Kang, I.-H., Gallo-Meagher, M., Vu, J. C. V., Boote, K. J., Allen, L. H. & Bowes, G. (2003) Rubisco expression in rice leaves is related to genotypic variation of photosynthesis under elevated growth CO_2 and temperature. *Plant Cell and Environment*, **26**, 1941–1950.

Giese, K. & Vierling, E. (2002) Changes in oligomerization are essential for the chaperone activity of a small heat shock protein *in vivo* and *in vitro*. *Journal of Biological Chemistry*, **277**, 46310–46318.

Gong, M., Li, Y. J. & Chen, S. Z. (1998) Abscisic acid-induced thermotolerance in maize seedlings is mediated by calcium and associated with antioxidant systems. *Journal of Plant Physiology*, **153**(3–4), 488–496.

Hall, A. E. (1990) Breeding for heat tolerance – An approach based on whole-plant physiology. *HortScience*, **25**, 17–19.

Hall, A. E. (2001) *Crop Responses to the Environment*. CRC Press, Boca Raton, FL.

Härndahl, U. & Sundby, C. (2001) Does the chloroplast small heat shock protein protect photosystem II during heat stress *in vitro*? *Physiologia Plantarum*, **111**, 273–275.

Härndahl, U., Hall, R. B., Osteryoung, K. W., Vierling, E., Bornman, J. F. & Sundby, C. (1999) The chloroplast small heat shock protein undergoes oxidation-dependent conformational changes and may protect plants from oxidative stress. *Cell Stress & Chaperones*, **4**(2), 129–138.

Hawthorn, L. R., Kerr, L. B. & Campbell, W. F. (1966) Relation between temperature of developing pods and seeds and scalded seeds of garden pea. *Journal of the American Society for Horticultural Science*, **88**, 437–440.

Haynes, J. G., Hartung, A. J., Hendershot, J. D., Passingham, R. S. & Rundle, S. J. (1999) Molecular characterization of the B′ regulatory subunit gene family of *Arabidopsis* protein phosphatase 2A. *European Journal of Biochemistry*, **260**, 127–136.

Heckathorn, S. A., Downs, C. A., Sharkey, T. D. & Coleman, J. S. (1998) The small, methionine-rich chloroplast heat-shock protein protects photosystem II electron transport during heat stress. *Plant Physiology*, **116**(1), 439–444.

Hershko, D. D., Robb, B. W., Luo, G. J., Paxton, J. H. & Hasselgren, P. O. (2003) Interleukin-6 induces thermotolerance in cultured Caco-2 cells independent of the heat shock response. *Cytokine*, **21**(1), 1–9.

Hill, J. E. & Hemmingsen, S. M. (2001) *Arabidopsis thaliana* type I and II chaperonins. *Cell Stress & Chaperones*, **6**, 190–200.

Hong, S. W. & Vierling, E. (2000) Mutants of *Arabidopsis thaliana* defective in the acquisition of tolerance to high temperature stress. *Proceedings of the National Academy of Sciences of the United States of America*, **97**, 4392–4397.

Hong, S. W. & Vierling, E. (2001) Hsp101 is necessary for heat tolerance but dispensable for development and germination in the absence of stress. *Plant Journal*, **27**, 25–35.

Hong, S. W., Lee, U. & Vierling, E. (2003) Arabidopsis *hot* mutants define multiple functions required for acclimation to high temperatures. *Plant Physiology*, **132**(2), 757–767.

Howarth, C. J., Pollock, C. J. & Peacock, J. M. (1997) Development of laboratory-based methods for assessing seedling thermotolerance in pearl millet. *New Phytologist*, **137**, 129–139.

Hubert, D. A., Tornero, P., Belkhadir, Y., Krishna, P., Takahashi, A., Shirasu, K. & Dangl, J. L. (2003) Cytosolic HSP90 associates with and modulates the Arabidopsis RPM1 disease resistance protein. *EMBO Journal*, **22**, 5679–5689.

Hurkman, W. J., McCue, K. F., Altenbach, S. B., Korn, A., Tanaka, C. K., Kotathari, K. M., Johnson, E. L., Bechtel, D. B., Wilson, J. D., Anderson, O. D. & DuPont, F. M. (2003) Effect of temperature on expression of genes encoding enzymes for starch biosynthesis in developing wheat endosperm. *Plant Science*, **164**, 873–881.

Iba, K. (2002) Acclimative response to temperature stress in higher plants: approaches of gene engineering from temperature tolerance. *Annual Review of Plant Biology*, **53**, 225–245.

Ishikawa, A., Tanaka, H., Nakai, M. & Asahi, T. (2003) Deletion of a chaperonin 60 beta gene leads to cell death in the Arabidopsis lesion initiation 1 mutant. *Plant and Cell Physiology*, **44**(3), 255–261.

Jenkins, G. M. (2003) The emerging role for sphingolipids in the eukaryotic heat shock response. *Cellular and Molecular Life Sciences*, **60**, 701–710.

Jenkins, M. E., Suzuki, T. C. & Mount, D. W. (1997) Evidence that heat and ultraviolet radiation activate a common stress-response program in plants that is altered in the *uvh6* mutant of *Arabidopsis thaliana*. *Plant Physiology*, **115**(4), 1351–1358.

Jeong, M. J., Park, S. C. & Byun, M. O. (2001) Improvement of salt tolerance in transgenic potato plants by glyceraldehyde-3 phosphate dehydrogenase gene transfer. *Molecules and Cells*, **12**(2), 185–189.

Jiang, Y. & Huang, B. (2001) Effects of calcium on antioxidant activities and water relations associated with heat tolerance in two cool-season grasses. *Journal of Experimental Botany*, **52**(355), 341–349.

Jurivich, D. A., Pachetti, C., Qiu, L. & Welk, J. F. (1995) Salicylate triggers heat shock factor differently than heat. *Journal of Biological Chemistry*, **270**(41), 24489–24495.

Jurivich, D. A., Sistonen, L., Kroes, R. A. & Morimoto, R. I. (1992) Effect of sodium salicylate on the human heat shock response. *Science*, **255**(5049), 1243–1245.

Kamada, Y., Jung, U. S., Piotrowski, J. & Levin, D. E. (1995) The protein kinase C-activated MAP kinase pathway of *Saccharomyces cerevisiae* mediates a novel aspect of the heat shock response. *Genes & Development*, **9**, 1559–1571.

Katiyar-Agarwal, S., Agarwal, M. & Grover, A. (2003) Heat-tolerant basmati rice engineered by over-expression of hsp101. *Plant Molecular Biology*, **51**(5), 677–686.

Kato, M., Hayakawa, Y., Hyodo, H., Ikoma, Y. & Yano, M. (2000) Wound-induced ethylene synthesis and expression and formation of 1-aminocyclopropane-1-carboxylate (ACC) synthase, ACC oxidase, phenylalanine ammonia-lyase, and peroxidase in wounded mesocarp tissue of *Cucurbita maxima*. *Plant and Cell Physiology*, **41**(4), 440–447.

Keeler, S. J., Boettger, C. M., Haynes, J. G., Kuches, K. A., Johnson, M. M., Thureen, D. L., Keeler, C. L. & Kitto, S. L. (2000) Acquired thermotolerance and expression of the HSP100/ClpB genes of lima bean. *Plant Physiology*, **123**, 1121–1132.

Keeling, P. L., Bacon, P. J. & Holt, D. C. (1993) Elevated temperature reduces starch deposition in wheat endosperm by reducing the activity of soluble starch synthase. *Planta*, **191**, 342–348.

Kelley, W. L. (1999) Molecular chaperones: how J domains turn on Hsp70s. *Current Biology*, **9**, R305–R308.

Kim, S. Y., Hong, C. B. & Lee, I. (2001) Heat shock causes stage-specific male sterility in *Arabidopsis thaliana*. *Journal of Plant Research*, **114**, 301–307.

Klueva, N. Y., Maestri, E., Marmiroli, N. & Nguyen, H. T. (2001) Mechanisms of thermotolerance in crops. In A. S. Basra (ed.) *Crop Responses and Adaptations to Temperature Stress*. Food Products Press, Binghamton, New York, pp. 177–217.

Knight, C. A. & Ackerly, D. D. (2003) Small heat shock protein responses of a closely related pair of desert and coastal *Encelia*. *International Journal of Plant Science*, **164**(1), 53–60.

Kobata, T. & Uemuki, N. (2004) High temperatures during the grain-filling period do not reduce the potential grain dry matter increase of rice. *Agronomy Journal*, **96**, 406–414.

Kovacs, E., van der Vies, S. M., Glatz, A., Torok, Z., Varvasovszki, V., Horvath, I. & Vigh, L. (2001) The chaperonins of *Synechocystis* PCC 6803 differ in heat inducibility and chaperone activity. *Biochemical and Biophysical Research Communications*, **289**, 908–915.

Kovtun, Y., Chiu, W.-L., Tena, G. & Sheen, J. (2000) Functional analysis of oxidative stress-activated mitogen-activated protein kinase cascade in plants. *Proceedings of the National Academy of Sciences*, **97**, 2940–2945.

Kozlowski, T. T. & Pallardy, S. G. (2002) Acclimation and adaptive responses of woody plants to environmental stresses. *Botanical Review*, **68**(2), 270–334.

Krishna, P. & Gloor, G. (2001) The Hsp90 family of proteins in *Arabidopsis thaliana*. *Cell Stress & Chaperones* **6**, 238–246.

Külz, D. (2003) Evolution of the cellular stress proteome: from monophyletic origin to ubiquitous function. *Journal of Experimental Biology*, **206**, 3119–3124.

Kurepa, J., Walker, J. M., Smalle, J., Gosink, M. M., Davis, S. J., Durham, T. L., Sung, D. Y. & Vierstra, R. D. (2003) The small ubiquitin-like modifier (SUMO) protein modification system in Arabidopsis. Accumulation of SUMO1 and -2 conjugates is increased by stress. *Journal of Biological Chemistry*, **278**(9), 6862–6872.

Kuznetsov, V. V., Andreev, I. M. & Trofimova, M. S. (1998) The synthesis of HSPs in sugar beet suspension culture cells under hyperthermia exhibits differential sensitivity to calcium. *Biochemistry and Molecular Biology International*, **45**(2), 269–278.

Langridge, J. & Griffing, B. (1959) A study of high temperature lesions in *Arabidopsis thaliana*. *Australian Journal of Biological Sciences*, **12**, 117–135.

Larkindale J. (2001) Cell signaling in response to heat shock in *Arabidosis thaliana*. D. Phil. dissertation. New College, Oxford University, UK.

Larkindale, J. & Huang, B. (2004a) Thermotolerance and antioxidant systems in *Agrostis stolonifera*: involvement of salicylic acid, abscisic acid, calcium, hydrogen peroxide, and ethylene. *Journal of Plant Physiology*, **161**(4), 405–413.

Larkindale, J. & Huang, B. (2004b) Changes of lipid composition and saturation level in leaves and roots for heat-stressed and heat-acclimated creeping bentgrass (*Agrostis stolonifera*). *Environmental and Experimental Botany*, **51**(1), 57–67.

Larkindale, J. & Knight, M. R. (2002) Protection against heat stress-induced oxidative damage in Arabidopsis involves calcium, abscisic acid, ethylene, and salicylic acid. *Plant Physiology*, **128**(2), 682–695.

Latorre, K. A., Harris, D. M. & Rundle, S. J. (1997) Differential expression of three Arabidopsis genes encoding the B′ regulatory subunit of protein phosphatase 2A. *European Journal of Biochemistry*, **245**, 156–163.

Law, R. D. & Crafts-Brandner, S. J. (2001) High temperature stress increases the expression of wheat leaf ribulose-1,5-bisphosphate carboxylase/oxygenase activase protein. *Archives of Biochemistry and Biophysics*, **386**(2), 261–267.

Law, R. D., Crafts-Brandner, S. J. & Salvucci, M. E. (2001) Heat stress induces the synthesis of a new form of ribulose-1,5-bisphosphate carboxylase/oxygenase activase in cotton leaves. *Planta*, **214**(1), 117–125.

Lawton, K., Weymann, K., Friedrich, L., Vernooij, B., Uknes, S. & Ryals, J. (1995) Systemic acquired resistance in Arabidopsis requires salicylic acid but not ethylene. *Molecular Plant – Microbe Interactions*, **8**(6), 863–870.

Leborgne-Castel, N., Jelitto-Van Dooren, E. P. W. M., Crofts, A. J. & Denecke, J. (1999) Overexpression of BiP in tobacco alleviates endoplasmic reticulum stress. *Plant Cell*, **11**, 459–469.

Lee, B. H., Won, S. H., Lee, H. S., Miyao, M., Chung, W. I., Kim, I. J. & Jo, J. (2000) Expression of the chloroplast-localized small heat shock protein by oxidative stress in rice. *Gene*, **245**(2), 283–290.

Lee, J. H. & Schöffl, F. (1996) An Hsp70 antisense gene affects the expression of HSP70/HSC70, the regulation of HSF, and the acquisition of thermotolerance in transgenic *Arabidopsis thaliana*. *Molecular and General Genetics*, **252**(1–2), 11–19.

Lee, J. H., Hubel, A. & Schöffl, F. (1995) Derepression of the activity of genetically engineered heat shock factor causes constitutive synthesis of heat shock proteins and increased thermotolerance in transgenic Arabidopsis. *Plant Journal*, **8**(4), 603–612.

Lee, S. M. & Park, J. W. (1998) Thermosensitive phenotype of yeast mutant lacking thioredoxin peroxidase. *Archives of Biochemistry and Biophysics*, **359**(1), 99–106.

Lee, S., Sowa, M. E., Watanabe, Y. H., Sigler, P. B., Chiu, W., Yoshida, M. & Tsai, F. T. F. (2003) The structure of ClpB: a molecular chaperone that rescues proteins from an aggregated state. *Cell*, **115**, 229–239.

Lee, Y. R., Nagao, R. T. & Key, J. L. (1994) A soybean 101-kD heat shock protein complements a yeast HSP104 deletion mutant in acquiring thermotolerance. *Plant Cell*, **6**(12), 1889–1897.

Li, B., Liu, H. T., Sun, D. Y. & Zhou, R. G. (2004) Ca (2+) and calmodulin modulate DNA-binding activity of maize heat shock transcription factor in vitro. *Plant and Cell Physiology*, **45**(5), 627–634.

Lin, B.-L., Wang, J.-S., Liu, H.-C., Chen, R.-W., Meyer, Y., Bakarat, A. & Delseny, M. (2001) Genomic analysis of the Hsp70 superfamily in *Arabidopsis thaliana*. *Cell Stress & Chaperones*, **6**, 201–208.

Link, V., Sinha, A. K., Vashista, P., Hofmann, M. G., Proels, R. K., Ehness, R. & Roitsch, T. (2002) A heat-activated MAP kinase in tomato: a possible regulator of the heat stress response. *FEBS Letters*, **531**(2), 179–183.

Liu, H. T., Li, B., Shang Z. L., Li, X. Z., Mu, R. L., Sun, D. Y. & Zhou, R. G. (2003). Calmodulin is involved in heat shock signal transduction in wheat. *Plant Physiology*, **132**(3), 1186–1195.

Liu, L., Maillet, D. S., Frappier, J. R., d'Ailly, K., Walden, D. B. & Atkinson, B. G. (1996) Characterization, chromosomal mapping, and expression of different ubiquitin fusion protein genes in tissues from control and heat-shocked maize seedlings. *Biochemistry and Cell Biology*, **74**(1), 9–19.

Liu, Y., Burch-Smith, T., Schiff, M., Feng, S. & Dinesh-Kimar, P. (2004) Molecular chaperone Hsp90 associates with resistance protein N and its signaling proteins SGT1 and Rar1 to modulate an innate immune response in plants. *Journal of Biological Chemistry*, **279**, 2101–2108.

Lohmann, C., Eggers-Schumacher, G., Wunderlich, M. & Schöffl, F. (2004) Two different heat shock transcription factors regulate immediate early expression of stress genes in Arabidopsis. *Molecular Genetics and Genomics*, **271**(1), 11–21.

Loreto, F., Förster, A., Dürr, M., Csiky, O. & Seufert, G. (1998) On the monoterpene emission under heat stress and on the increased thermotolerance of leaves of *Quercus ilex* L. fumigated with selected monoterpenes. *Plant Cell and Environment*, **21**, 101–107.

Ludwig-Müller, J., Krishna, P. & Forreiter, C. (2000) A glucosinolate mutant of Arabidopsis is thermosensitive and defective in cytosolic Hsp90 expression after heat stress. *Plant Physiology*, **123**(3), 949–958.

Lund, P. A., Large, A. T. & Kapatai, G. (2003) The chaperonins: perspectives from the archaea. *Biochemical Society Transactions*, **31**(Pt 3), 681–685.

Lupas, A. N. & Martin, J. (2002) AAA proteins. *Current Opinion in Structural Biology*, **12**, 746–753.

Lu, R., Malcuit, I., Moffett, P., Ruiz, M. T., Peart, J., Wu, A.-J., Rathjen, J. P., Bendahmane, A., Day, L. & Baulcombe, D. C. (2003) High throughput virus-induced gene silencing implicates heat shock proein 90 in plant disease resistance. *EMBO Journal*, **22**, 5690–5699.

Macario, A. J., Lange, M., Ahring, B. K. & De Macario, E. C. (1999) Stress genes and proteins in the archaea. *Microbiology and Molecular Biology Reviews*, **63**, 923–967.

Maestri, E., Klueva, N., Perrotta, C., Gulli, M., Nguyen, H. T. & Marmiroli, N. (2002) Molecular genetics of heat tolerance and heat shock proteins in cereals. *Plant Molecular Biology*, **48**, 667–681.

Mahan, J. R., McMichael, B. L. & Wanjura, D. F. (1995) Methods for reducing the adverse effects of temperature stress on plants: a review. *Environmental and Experimental Botany*, **35**, 251–258.

Malik, M. K., Slovin, J. P., Hwang, C. H. & Zimmerman, J. L. (1999) Modified expression of a carrot small heat shock protein gene, Hsp17.7, results in increased or decreased thermotolerance. *Plant Journal*, **20**, 89–99.

Manning, W. J., Flagler, R. B. & Frenkel, M. A. (2003) Assessing plant response to ambient ozone: growth of ozone-sensitive loblolly pine seedlings treated with ethylenediurea or sodium erythorbate. *Environmental Pollution*, **126**(1), 73–81.

Marshall, H. G. (1982) Breeding for tolerance to heat and cold. In M. N. Christiansen, C. F. Lewis (eds) *Breeding Plants for Less Favorable Environments*. John Wiley & Sons, New York, pp. 47–70.

Martineau, J. R., Specht, J. E., Williams, J. H. & Sullivan, C. Y. (1979) Temperature tolerance in soybeans. I. Evaluation of a technique for assessing cellular membrane thermostability. *Crop Science*, **19**, 75–81.

Maurizi, M. R. & Xia, D. (2004) Protein binding and disruption by Clp/Hsp100 chaperones. *Structure*, **12**, 175–183.

Mayer, M. P., Brehmer, D., Gässler, C. S. & Bukau, B. (2002) Hsp70 chaperone machines. *Advances in Protein Chemistry*, **59**, 1–44.

Meijer, H. J. G. & Munnik, T. (2003) Phospholipid-based signaling in plants. *Annual Review of Plant Biology*, **54**, 265–306.

Meyer, K., Leube, M. P. & Grill, E. (1994) A protein phosphatase 2C involved in ABA signal transduction in *Arabidopsis thaliana*. *Science*, **264**(5164), 1452–1455.

Mikami, K. & Murata, N. (2003) Membrane fluidity and the perception of environmental signals in cyanobacteria and plants. *Progress in Lipid Research*, **42**(6), 527–543.

Mishra, S. K., Tripp, J., Winkelhaus, S., Tschiersch, B., Theres, K., Nover, L. & Scharf, K. D. (2002) In the complex family of heat stress transcription factors, HsfA1 has a unique role as master regulator of thermotolerance in tomato. *Genes and Development*, **16**(12), 1555–1567.

Moeder, W., Barry, C. S., Tauriainen, A. A., Betz, C., Tuomainen, J., Utriainen, M., Grierson, D., Sandermann, H., Langebartels, C. & Kangasjarvi, J. (2002) Ethylene synthesis regulated by biphasic induction of 1-aminocyclopropane-1-carboxylic acid synthase and 1-aminocyclopropane-1-carboxylic acid oxidase genes is required for hydrogen peroxide accumulation and cell death in ozone-exposed tomato. *Plant Physiology*, **130**(4), 1918–1926.

Mogk, A., Deuerling, E., Vorderwülbecke, S., Vierling, E. & Bukau, B. (2003a) Small heat shock proteins, ClpB and the DnaK system form a functional triade in reversing protein aggregation. *Molecular Microbiology*, **50**, 585–595.

Mogk, A., Schlieker, C., Friedrich, K. L., Schöfeld, H. J., Vierling, E. & Bukau, B. (2003b) Refolding of substrates bound to small HSPs relies on a disaggregation reaction mediated most efficiently by ClpB/DnaK. *Journal of Biological Chemistry*, **278**, 31033–31042.

Momicilovic, I. & Ristic, Z. (2004) Localization and abundance of chloroplast protein synthesis elongation factor (EF-Tu) and heat stability of chloroplast stromal proteins in maize. *Plant Science*, **166**(1), 81–88.

Murakami, Y., Tsuyama, M., Kobayashi, Y., Kodama, H. & Iba, K. (2000) Trienoic fatty acids and plant tolerance of high temperature. *Science*, **287**(5452), 476–479.

Neuwald, A. F., Aravind, L., Spouge, J. L. & Koonin, E. V. (1999) AAA$^+$: a class of chaperone-like ATPases associated with the assembly, operation, and disassembly of protein complexes. *Genome Research*, **9**, 27–43.

Nieto-Sotelo, J., Kannan, K. B., Martinez, L. M. & Segal, C. (1999) Characterization of a maize heat-shock protein 101 gene, HSP101, encoding a ClpB/Hsp100 protein homologue. *Gene*, **230**(2), 187–195.

Nieto-Sotelo, J., Martinez, L. M., Ponce, G., Cassab, G. I., Alagon, A., Meeley, R. B., Ribaut, J. M., & Yang, R. (2002) Maize HSP101 plays important roles in both induced and basal thermotolerance and primary root growth. *Plant Cell*, **14**(7), 1621–1633.

Nie, X., Singh, R. P. & Tai, G. C. (2002). Molecular characterization and expression analysis of 1-aminocyclopropane-1-carboxylate oxidase homologs from potato under abiotic and biotic stresses. *Genome*, **45**(5), 905–913.

Nobel, P. S. (1991) *Physicochemical and Environmental Plant Physiology*. Academic Press, San Deigo, CA.

Nover, L., Bharti, K., Doring, P., Mishra, S. K., Ganguli, A. & Scharf, K. D. (2001) Arabidopsis and the heat stress transcription factor world: how many heat stress transcription factors do we need? *Cell Stress & Chaperones*, **6**(3), 177–189.

O'Mahony, P. & Burke, J. (2000) A ditelosomic line of 'Chinese Spring' wheat with augmented acquired thermotolerance. *Plant Science*, **158**(1–2), 147–154.

Ortiz, C. & Cardemil, L. (2001) Heat-shock responses in two leguminous plants: a comparative study. *Journal of Experimental Botany*, **52**(361), 1711–1719.

Orvar, B. L., Sangwan, V., Omann, F. & Dhindsa, R. S. (2000) Early steps in cold sensing by plant cells: the role of actin cytoskeleton and membrane fluidity. *Plant Journal*, **23**(6), 785–794.

Panchuk, I. I., Volkov, R. A. & Schöffl, F. (2002) Heat stress- and heat shock transcription factor-dependent expression and activity of ascorbate peroxidase in Arabidopsis. *Plant Physiology*, **129**(2), 838–853.

Park, S. M. & Hong, C. B. (2002) Class I small heat-shock protein gives thermotolerance in tobacco. *Journal of Plant Physiology*, **159**, 25–30.

Patterson, B. D. & Graham, D. (1987) Temperature and metabolism. In *The Biochemistry of Plants*, Vol. 12. Academic Press, New York, pp. 153–199.

Peacock, J. M., Soman, P., Jayachandran, R., Rani, A. U., Howarth, C. J. & Thomas, A. (1993) Effects of high soil surface temperature on seedling survival in pearl millet. *Experimental Agriculture*, **29**, 215–225.

Pnueli, L., Liang, H., Rozenberg, M. & Mittler, R. (2003) Growth suppression, altered stomatal responses, and augmented induction of heat shock proteins in cytosolic ascorbate peroxidase (Apx1)-deficient Arabidopsis plants. *Plant Journal*, **34**(2), 187–203.

Prändl, R., Hinderhofer, K., Eggers-Schumacher, G. & Schöffl, F. (1998) HSF3, a new heat shock factor from *Arabidopsis thaliana*, derepresses the heat shock response and confers thermotolerance when over-expressed in transgenic plants. *Molecular and General Genetics*, **258**(3), 269–278.

Prasad, T. K. & Stewart, C. R. (1992) cDNA clones encoding *Arabidopsis thaliana* and *Zea mays* mitochondrial chaperonin HSP60 and gene expression during seed germination and heat shock. *Plant Molecular Biology*, 873–885.

Qi, Y., Sun, Y., Xu, L., Xu, Y. & Huang, H. (2004) ERECTA is required for protection against heat-stress in the AS1/AS2 pathway to regulate adaxial–abaxial leaf polarity in Arabidopsis. *Planta*, **219**(2), 270–276.

Queitsch, C., Hong, S. W., Vierling, E. & Lindquist, S. (2000) Heat shock protein 101 plays a crucial role in thermotolerance in Arabidopsis. *Plant Cell*, **12**(4), 479–492.

Rai, S. P., Luthra, R., Gupta, M. M. & Kumar, S. (2001) Pleiotropic morphological and abiotic stress resistance phenotypes of the hyper-abscisic acid producing Abo-mutant in the periwinkle *Catharanthus roseus*. *Journal of Biosciences*, **26**(1), 57–70.

Raper, C. D. & Kramer, D. M. (1983) *Crop Reactions to Water and Temperature Stresses in Humid, Temperature Climates*. Westview Press, Boulder, CO, pp. 1–373.

Riezman, H. (2004) Why do cells require heat shock proteins to survive heat stress? *Cell Cycle*, **3**(1), 61–63.

Ristic, Z., Yang, G. P., Martin, B. & Fullerton, S. (1998) Evidence of association between specific heat-shock protein(s) and the drought and heat tolerance phenotype in maize. *Journal of Plant Physiology*, **153**, 497–505.

Rizhsky, L., Liang, H. & Mittler, R. (2002) The combined effect of drought stress and heat shock on gene expression in tobacco. *Plant Physiology*, **130**(3), 1143–1151.

Rhizhsky, L., Liang, H., Shuman, J., Shulaev, V., Davletova, S. & Mittler, R. (2004) When defense pathways collide. The response of Arabidopsis to a combination of drought and heat stress. *Plant Physiology*, **134**, 1683–1696.

Robertson, A. J., Ishikawa, M., Gusta, L. V. & MacKenzie, S. L. (1994) Abscisic acid-induced heat tolerance in *Bromus inermis* Leyss cell-suspension cultures. Heat-stable, abscisic acid-responsive polypeptides in combination with sucrose confer enhanced thermostability. *Plant Physiology*, **105**(1), 181–190.

Rodriguez, P. L., Benning, G. & Grill, E. (1998) ABI2, a second protein phosphatase 2C involved in abscisic acid signal transduction in Arabidopsis. *FEBS Letters*, **421**(3), 185–190.

Rojas, A., Almoguera, C. & Jordano, J. (1999) Transcriptional activation of a heat shock gene promoter in sunflower embryos: synergism between ABI3 and heat shock factors. *Plant Journal*, **20**(5), 601–610.

Rokka, A., Zhang, L. & Aro, E. M. (2001) Rubisco activase: an enzyme with a temperature-dependent dual function? *Plant Journal*, **25**(4), 463–471.

Sakamoto, A. & Murata, N. (2002) The role of glycine betaine in the protection of plants from stress: clues from transgenic plants. *Plant Cell and Environment*, **25**, 163–171.

Salvucci, M. E. & Crafts-Brandner, S. J. (2004) Inhibition of photosynthesis by heat stress: the activation state of Rubisco as a limiting factor in photosynthesis. *Physiologia Plantarum*, **120**, 179–186.

Sangster, T. A., Lindquist, S. & Queitsch, C. (2004) Under cover: causes, effects and implications of Hsp90-mediated genetic capacitance. *BioEssays*, **26**, 348–362.

Sangwan, V., Orvar, B. L., Beyerly, J., Hirt, H. & Dhindsa, R. S. (2002) Opposite changes in membrane fluidity mimic cold and heat stress activation of distinct plant MAP kinase pathways. *Plant Journal*, **31**(5), 629–638.

Sanmiya, K., Suzuki, K., Egawa, Y. & Shono, M. (2004) Mitochondrial small heat-shock protein enhances thermotolerance in tobacco plants. *FEBS Letters*, 16 **557**(1–3), 265–268.

Sato, S., Peet, M. M. & Thomas, J. F. (2002) Determining critical pre- and post-anthesis periods and physiological processes in *Lycopersicon esculentum* Mill. Exposed to moderately elevated temperatures. *Journal of Experimental Botany*, **53**, 1187–1195.

Scharf, K.-D., Siddique, M. & Vierling, E. (2001) The expanding family of *Arabidopsis thaliana* small heat stress proteins and a new family of proteins containing alpha-crystallin domains (Acd proteins). *Cell Stress & Chaperones*, **6**(3), 225–237.

Schirmer, E. C., Glover, J. R., Singer, M. A. & Lindquist, S. (1996) HSP100/Clp proteins: a common mechanism explains diverse functions. *Trends in Biochemical Sciences*, **21**, 289–296.

Schirmer, E. C., Lindquist, S. & Vierling, E. (1994) An Arabidopsis heat shock protein complements a thermotolerance defect in yeast. *Plant Cell*, **6**, 1899–1909.

Schöffl, F., Prändl, R. & Reindl, A. (1998) Regulation of the heat-shock response. *Plant Physiology*, **117**, 1135–1141.

Schroda, M., Vallon, O., Wollman, F. A. & Beck, C. F. (1999) A chloroplast-targeted heat shock protein 70 (HSP70) contributes to the photoprotection and repair of photosystem II during and after photoinhibition. *Plant Cell*, **11**(6), 1165–1178.

Sharkey, T. D., Chen, X. & Yeh, S. (2001) Isoprene increases thermotolerance of fosmidomycin-fed leaves. *Plant Physiology*, **125**(4), 2001–2006.

Shi, W. M., Muramoto, Y., Ueda, A. & Takabe, T. (2001) Cloning of peroxisomal ascorbate peroxidase gene from barley and enhanced thermotolerance by overexpression in *Arabidopsis thaliana*. *Gene*, **273**(1), 23–27.

Simon, J.-P., Potvin, C. & Blanchard, M.-H. (1983) Thermal adaptation and acclimation of higher plants at the enzyme level: kinetic properties of NAD malate dehydrogenase and glutamate oxaloacetate trasnaminase in two genotypes of *Arabidopsis thaliana* (Brassicaceae). *Oecologia*, **60**, 143–148.

Slafer, G. A. (2003) Genetic basis of yield as viewed from a crop physiologist's perspective. *Annals of Applied Biology*, **124**(2), 117–128.

Smith, B. J. & Yaffe, M. P. (1991) Uncoupling thermotolerance from the induction of heat shock proteins. *Proceedings of the National Academy of Sciences*, **88**(24), 11091–11094.

Soto, A., Allona, I., Collada, C., Guevara, M.-A., Casado, R., Rodriguez-Cerezo, E., Aragoncillo, C. & Gomez, L. (1999a) Heterologous expression of a plant small heat-shock protein enhances *Escherichia coli* viability under heat and cold stress. *Plant Physiology*, **120**, 521–528.

Soto, T., Fernandez, J., Vicente-Soler, J., Cansado, J. & Gacto, M. (1999b) Accumulation of trehalose by overexpression of tps1, coding for trehalose-6-phosphate synthase, causes increased resistance to multiple stresses in the fission yeast *Schizosaccharomyces pombe*. *Applied and Environmental Microbiology*, **65**(5), 2020–2024.

Storozhenko, S., De Pauw, P., Van Montagu, M., Inze, D. & Kushnir, S. (1998) The heat-shock element is a functional component of the Arabidopsis APX1 gene promoter. *Plant Physiology*, **118**(3), 1005–1014.

Sun, C. W. & Callis, J. (1997) Independent modulation of *Arabidopsis thaliana* polyubiquitin mRNAs in different organs and in response to environmental changes. *Plant Journal*, **11**(5), 1017–1027.

Sung, D.-Y. & Guy, C. L. (2003) Physiological and molecular assessment of altered expression of Hsc70-1 in Arabidopsis. Evidence for pleiotropic consequences. *Plant Physiology*, **132**(2), 979–987.

Sung, D.-Y., Kaplan, F., Lee, K.-J. & Guy, C. L. (2003) Acquired tolerance to temperature extremes. *Trends in Plant Science*, **8**(4), 179–187.

Sung, D.-Y., Vierling, E. & Guy, C. L. (2001) Comprehensive expression profile analysis of the Arabidopsis hsp70 gene family. *Plant Physiology*, **126**, 789–800.

Sun, W., Bernard, C., van de Cotte, B., Van Montagu, M. & Verbruggen, N. (2001) *At-HSP17.6A*, encoding a small heat-shock protein in *Arabidopsis*, can enhance osmotolerance upon overexpression. *Plant Journal*, **27**, 407–415.

Swan, T. M. & Watson, K. (1999) Stress tolerance in a yeast lipid mutant: membrane lipids influence tolerance to heat and ethanol independently of heat shock proteins and trehalose. *Canadian Journal of Microbiology*, **45**(6), 472–479.

Takahashi, A., Casais, C., Ichimura, K. & Shirasu, K. (2003) HSP90 interacts with RAR1 and SGT1 and is essential for RPS2-mediated disease resistance in *Arabidopsis*. *Proceedings of the National Academy of Sciences*, **100**, 11777–11782.

Tanaka, Y., Nishiyama, Y. & Murata, N. (2000) Acclimation of the photosynthetic machinery to high temperature in *Chlamydomonas reinhardtii* requires synthesis *de novo* of proteins encoded by the nuclear and chloroplast genomes. *Plant Physiology*, **124**(1), 441–449.

Taub, D. R., Seemann, J. R. & Coleman, J. S. (2000) Growth in elevated CO_2 protects photosynthesis against high-temperature damage. *Plant Cell and Environment*, **23**, 649–656.

Terol, J., Bargues, M., Carrasco, P., Perez-Alonso, M. & Paricio, N. (2002) Molecular characterization and evolution of the protein phosphatase 2A B′ regulatory subunit family in plants. *Plant Physiology*, **129**(2), 808–822.

Thompson, L. M. (1975) Weather variability, climatic change, and grain production. *Science*, **188**, 535–541.

Török, Z., Goloubinoff, P., Tsvetkova, N. M., Glatz, A., Balogh, G., Varvasovszki, V., Los, D. A., Vierling, E., Crowe, J. H. & Vigh, L. (2001) HSP17 is an amphitropic protein that stabilizes heat-stressed membranes and binds denatured proteins for subsequent chaperone-mediated refolding. *Proceedings of the National Academy of Sciences*, **98**, 3098–3103.

Török, Z., Tsvetkova, N. M., Balogh, G., Horváth, I., Nagy, E., Penzes, Z., Hargitai, J., Bensaude, O., Csermely, P., Crowe, J. H., Maresca, B. & Vigh, L. (2003) Heat shock protein coinducers with no effect on protein denaturation specifically modulate the membrane lipid phase. *Proceedings of the National Academy of Sciences*, **100**(6), 3131–3136.

Trent, J. D., Kagawa, H. K., Paavola, C. D., McMillan, R. A., Howard, J., Jahnke, L., Lavin, C., Embaye, T. & Henze, C. E. (2003) Intracellular localization of a group II chaperonin indicates a membrane-related function. *Proceedings of the National Academy of Sciences*, **100**, 15589–15594.

Tsvetkova, N. M., Horváth, I., Török, Z., Wolkers, W. F., Balogi, Z., Shigapova, N., Crowe, L. M., Tablin, F., Vierling, E., Crowe, J. H. & Vigh, L. (2002) Small heat-shock proteins regulate membrane lipid polymorphism. *Proceedings of the National Academy of Sciences*, **99**, 13504–13509.

Turner, N. C. & Kramer, P. J. (1980) *Adaptation of Plants to Water and High Temperature Stress*. John Wiley & Sons, New York.

Vacca, R. A., de Pinto, M. C., Valenti, D., Passarella, S., Marra, E. & De Gara, L. (2004) Production of reactive oxygen species, alteration of cytosolic ascorbate peroxidase, and impairment of mitochondrial metabolism are early events in heat shock-induced programmed cell death in tobacco bright-yellow 2 cells. *Plant Physiology*, **134**(3), 1100–1112.

van Montfort, R. L. M., Basha, E., Friedrich, K. L., Slingsby, C. & Vierling, E. (2001) Crystal structure and assembly of a eukaryotic small heat shock protein. *Nature Structural Biology*, **8**, 1025–1030.

van Montfort, R. L. M., Slingsby, C. & Vierling, E. (2002) Structure and function of the small heat shock protein/-crystallin family of molecular chaperones. In A. L. Horwich (ed.) *Protein Folding in the Cell*, Vol. 59. Academic Press, New York, pp. 105–156.

Varshney, C. K. & Rout, C. (1998) Ethylene diurea (EDU) protection against ozone injury in tomato plants at Delhi. *Bulletin of Environmental Contamination and Toxicology*, **61**(2), 188–193.

Vierling, E. (1991) The roles of heat shock proteins in plants. *Annual Review of Plant Physiology and Plant Molecular Biology*, **42**, 579–620.

Vigh, L., Maresca, B. & Harwood, J. L. (1998) Does the membrane's physical state control the expression of heat shock and other genes? *Trends in Biochemical Sciences*, **23**, 369–374.

Vogel, G., Aeschbacher, R. A., Boller, T. & Wiemken, A. (1998) Trehalose-6-phosphate phosphatases from *Arabidopsis thaliana*: identification by functional complementation of the yeast *tps2* mutant. *Plant Journal*, **13**, 673–683.

Wang, D. & Luthe, D. S. (2003) Heat sensitivity in a bentgrass variant. Failure to accumulate a chloroplast heat shock protein isoform implicated in heat tolerance. *Plant Physiology*, **133**(1), 319–327.

Wang, W., Vinocur, B. & Altman, A. (2003) Plant responses to drought, salinity and extreme temperatures: toward genetic engineering for stress tolerance. *Planta*, **218**, 1–14.

Wang, W., Vincour, B., Shoseyov, O. & Altman, A. (2004) Role of plant heat shock proteins and molecular chaperones in the abiotic stress response. *Trends in Plant Science*, **9**, 244–252.

Waters, E. R. (2003) Molecular adaptation and the origin of land plants. *Molecular Phylogenetics and Evolution*, **29**(3), 456–463.

Waters, E. R. & Vierling, E. (1999) The diversification of plant cytosolic small heat shock proteins preceded the divergence of mosses. *Molecular Biology and Evolution*, **16**, 127–139.

Waters, E. R., Lee, G. J. & Vierling, E. (1996) Evolution, structure and function of the small heat shock proteins in plants. *Journal of Experimental Botany*, **47**, 325–338.

Wehmeyer, N. & Vierling, E. (2000) The expression of small heat shock proteins in seeds responds to discrete developmental signals and suggests a general protective role in desiccation tolerance. *Plant Physiology*, **122**(4), 1099–1108.

Weis, E. & Berry, J. A. (1988) Plants and high temperature stress. *Symposia of the Society for Experimental Biology*, **42**, 329–346.

Wells, D. R., Tanguay, R. L., Le, H. & Gallie, D. R. (1998) HSP101 functions as a specific translational regulatory protein whose activity is regulated by nutrient status. *Genes & Development*, **12**, 3236–3245.

Whitham, S. A., Anderberg, R. J., Chisholm, S. T. & Carrington, J. C. (2000) Arabidopsis RTM2 gene is necessary for specific restriction of tobacco etch virus and encodes an unusual small heat shock-like protein. *Plant Cell*, **12**(4), 569–582.

Wise, R. R., Olson, A. J., Schrader, S. M. & Sharkey, T. D. (2004) Electron transport is the functional limitation of photosynthesis in field-grown Pima cotton plants at high temperature. *Plant Cell and Environment*, **27**(6), 717–724.

Worrall, D., Ng, C. K.-Y. & Hetherington, A. M. (2003) Sphingolipids, new players in plant signaling. *Trends in Plant Science*, **8**, 317–320.

Wunderlich, M., Werr, W. & Schöffl, F. (2003) Generation of dominant-negative effects on the heat shock response in *Arabidopsis thaliana* by transgenic expression of a chimaeric HSF1 protein fusion construct. *Plant Journal*, **35**(4), 442–451.

Yamada, T., Marubashi, W., Nakamura, T., Niwa, M. (2001) Possible involvement of auxin-induced ethylene in an apoptotic cell death during temperature-sensitive lethality expressed by hybrid between *Nicotiana glutinosa* and *N. repanda*. *Plant Cell Physiology*, **42**, 923–930.

Yamanouchi, U., Yano, M., Lin, H., Ashikari, M. & Yamada, K. (2002) A rice spotted leaf gene, Spl7, encodes a heat stress transcription factor protein. *Proceedings of the National Academy of Sciences*, **99**(11), 7530–7535.

Yan, J., Wang, J., Li, Q., Hwang, J. R., Patterson, C. & Zhang, H. (2003) AtCHIP, a U-box-containing E3 ubiquitin ligase, plays a critical role in temperature stress tolerance in Arabidopsis. *Plant Physiology*, **132**(2), 861–869.

Yeh, C. H., Chang, P. L., Yeh, K.-W., Lin, W.-C., Chen, Y.-M. & Lin, C.-Y. (1997) Expression of a gene encoding a 16.9-kDa heat-shock protein, Oshsp16.9, in *Escherichia coli* enhances thermotolerance. *Proceedings of the National Academy of Sciences*, **94**, 10967–10972.

Young, L. W., Wilen, R. W. & Bonham-Smith, P. C. (2004) High temperature stress of *Brassica napus* during flowering reduces micro- and megagametophyte fertility, induces fruit abortion, and disrupts seed production. *Journal of Experimental Botany*, **55**(396), 485–495.

Yu, R. M., Zhou, Y., Xu, Z. F., Chye, M. L. & Kong, R. Y. (2003). Two genes encoding protein phosphatase 2A catalytic subunits are differentially expressed in rice. *Plant Molecular Biology*, **51**(3), 295–311.

Zabaleta, E., Assad, N., Oropeza, A., Salerno, G. & Herrera-Estrella, L. (1994a) Expression of one of the members of the Arabidopsis chaperonin 60 beta gene family is developmentally regulated and wound-repressible. *Plant Molecular Biology*, **24**, 195–202.

Zabaleta, E., Oropeza, A., Assad, N., Mandel, A., Salerno, G. & Herrera-Estrella, L. (1994b) Antisense expression of chaperonin 60 in transgenic tobacco plants leads to abnormal phenotypes and altered distribution of photoassimilates. *Plant Journal*, **6**, 425–432.

Zahedi, M., Sharma, R. & Jenner, C. F. (2003) Effects of high temperature on grain growth and on the metabolites and enzymes in the starch-synthesis pathway in the grains of two wheat cultivars differing in their responses to temperature. *Functional Plant Biology*, **30**, 291–300.

Zatsepina, O. G., Velikodvorskaia, V. V., Molodtsov, V. B., Garbuz, D., Lerman, D. N., Bettencourt, B. R., Feder, M. E. & Evgenev, M. B. (2001) A *Drosophila melanogaster* strain from sub-equatorial Africa has exceptional thermotolerance but decreased Hsp70 expression. *Journal of Experimental Biology*, **204**(Pt 11), 1869–1881.

6 Adaptive responses in plants to nonoptimal soil pH

V. Ramírez-Rodríguez, J. López-Bucio and L. Herrera-Estrella

6.1 Introduction

Soil is a multiphasic system containing nutrients in varying degrees of solubility. In natural ecosystems, nutrients originate from inorganic and organic materials and only a small fraction of these elements are dissolved in the liquid-phase. The soil solution is therefore the immediate source of plant nutrients. Nutrient behavior depends on a large number of soil properties: texture, structure, water content and pH. The optimal pH for soils is in the neutral range (6–7.5). However, under unfavorable conditions pH value can increase or decrease depending on environmental factors and human activities. Examples of these situations occur in alkaline and acid soils, which represent important marginal land resources for agriculture. Alkaline soils have an intrinsically low P, Fe and Mo availability, which are imposed by the high abundance of cations. In acid soils, plants suffer principally from the high content of toxic aluminum and low phosphorus availability (Von-Uexkull & Muttert, 1995). Considering that nearly 4000 million hectares of the global land comprises acid soils and that alkaline soils cover over 25% of the Earth's surface, these soils represent a major agronomic problem due to their reduced potential for crop production (Marschner, 1995).

Several studies indicate that grain production worldwide should increase in the coming years to sustain the food needs of the growing human population. The world population is expected to reach 10 billion by 2050. Alkaline and acid soils represent one of the very few options for agricultural expansion. During the two last decades, myriad studies have been conducted to understand the physiological and biochemical adaptive strategies used by plants to grow in acid and alkaline soils. Only recently, developments in molecular biology have resulted in identification, isolation and molecular characterization of genes involved in metal tolerance and nutrient scavenging, which have important implications for improving plant growth by means of transgenic approaches. For instance, several microbial and plant genes have been used to enhance root exudation of organic acids and metal-binding molecules such as phytosiderophores and phytochelatins (PCs). Root exudates chemically modify the rhizosphere, allowing more nutrients to be extracted by plants. Recently, both enhanced nutrient uptake and metal tolerance in plants have been achieved by manipulation of metal-binding compounds such as citrate,

PCs, metallothioneins (MTs), phytosiderophores and ferritin, or by the expression of metal transporter proteins. In this chapter, we review current knowledge on the plant's adaptive responses to acid and alkaline soils, with particular emphasis on low nutrient availability and metal tolerance. The potential of transgenic approaches to develop crops that can be more efficiently grown in these marginal lands is also discussed.

6.2 Soil pH

Understanding soil chemistry is essential for proper crop management aimed at obtaining optimal plant productivity. Soil pH is a critical factor that influences many soil properties and determines soil fertility. In aqueous (liquid) solutions, an acid is a substance that donates hydrogen ions (H^+) to another molecule (Tisdale *et al.*, 1993). Soil pH is a measure of the number of protons in the soil solution. The actual concentration of hydrogen ions in the soil solution would be quite small. For example, a soil with pH 4.0 has a hydrogen ion concentration in the soil water of just 0.0001 moles per liter. Because the pH scale employs the use of logarithms, each whole number change (for example from 5.0 to 4.0) represents a tenfold increase in the concentration of H^+ ions. Therefore, when the amount of hydrogen ions increases, pH decreases. Soils with pH values around 7 have a hydrogen ion concentration 100 times lower than soils with pH 5. Table 6.1 shows the descriptive ranges for pH in soils.

6.3 Soil acidification

Soil formation is a natural process that starts when rock surfaces are first colonized by algae and lichens. Removal of cations and retention of anions

Table 6.1 Descriptive ranges for pH in soils

Ultra acid	< 3.5
Extremely acid	3.5–4.4
Very strongly acid	4.5–5.0
Strongly acid	5.1–5.5
Moderately acid	5.6–6.0
Slightly acid	5.6–6.0
Neutral	6.6–7.3
Slightly alkaline	7.4–7.8
Moderately alkaline	7.9–8.4
Strongly alkaline	8.5–9.0
Very strongly alkaline	9.0

from soil particles by environmental or human factors results in soil acidification. In near-neutral soils rich in weatherable minerals, soil acidification is relatively rapid and mainly associated with CO_2 deprotonation. In acid soils, the process of acidification is slower and due to cation assimilation and organic acid deprotonation. The principal H^+ sources are largely derived from the carbon and nitrogen cycles. In natural ecosystems, soils become gradually more acidic with time. Important factors that have contributed to an increased soil acidification rate include human activities such as agriculture, excessive use of fertilizers, mining and industry, which have led to an accelerated release of metals into the soil, causing serious environmental problems (Breemen et al., 1983).

6.4 Acid soils

Acidity degrees vary considerably due to environmental factors during soil formation; the climate, parent material and vegetation are especially important in determining pH. Acid soils range from moderately acidic (5.5–4.5) to extremely acidic (<4.5). They are characterized by low cation-exchange capacity (CEC) and low base saturation. Table 6.2 shows the World Reference Base (WRB) soil classification system. According to this classification, acid soils are divided into acrisols, arenosols, cambisols, histosols, ferralsols, luvisols, planosols, podzols and fluvisols. Most of these groups occur in the tropics and subtropics as well as in soils from temperate climate regions. Acrisols and ferralsols are most common in old land surfaces in humid tropical climates and podzols are typical soils of the northern coniferous forests but may also occur in the tropics.

Acid soils hamper agricultural productivity because of P, Ca, Mg and Mo deficiency and aluminum toxicity (Von-Uexkull & Muttert, 1995). In general, soils with pH <5 and > 60% Al saturation of the cation-exchange capacity (CEC) generally contain phytotoxic concentrations of Al^{3+}. High concentrations of Al^{3+} affect root growth, uptake and translocation of nutrients, cell division, respiration, nitrogen mobilization and glucose phosphorylation. Therefore, in the context of agricultural soils, it is necessary to make a distinction between acid soils having Al toxicity and acid soils without Al toxicity because of the differing management regimes needed for these soils (Kidd & Proctor, 2001).

High acidity considerably affects the physical, chemical and biological properties of soils. The low calcium concentrations in the soil solution restrict biological activity and soil structure stability. The major production constraint on strongly acid mineral soils is, however, the possible occurrence of high aluminum concentrations that have strong toxic effects on plants. In particular, many tropical soils that are extremely weathered contain high amounts of

Table 6.2   Global land extension of acids and calcareous soils

Soil	Europe	North America	South and C. America	Africa	Australia	Asia
Acids soils						
Acrisols	4170[*]	114 813	341 161	92 728	32 482	174 541
Arenosols	3806	25 512	118 967	462 401	193 233	12 889
Ferrasols	0	0	423 353	31924	0	–
Histosols	32 824	93 462	9245	12 270	1167	101 933
Podzols	213 624	220 770	5522	11 331	8459	22 423
Calscisols	56 657	114 720	24 318	171 237	113 905	117 270

[*]Extension is in thousand of ha.

exchangeable aluminum. The initial effect of aluminum toxicity in plant growth is a reduction in root elongation. Later, root systems become brown in color and branching is reduced. The symptoms of aluminum toxicity are similar to those of phosphorus or calcium deficiency (Kochian, 1995). When aluminum is not present, as in organic soils, their cation-buffering capacity is lost, the pH can fall to well below 4 and H^+ ions dominate the composition of the soil solution, then competing with other cations for root absorption sites, interfering with ion transport and uptake by roots (Kidd & Proctor, 2001).

6.5   Calcareous soils

Calcareous soils are rich in calcium carbonate and occur mainly in the semiarid subtropics of both hemispheres. They are characterized by the presence of calcium carbonate in the parent material and by a calcic horizon, a layer of secondary accumulation of carbonates (usually Ca or Mg), that contains excess concentrations of calcium carbonate. In the WRB soil classification system, calcareous soils occur mainly in the reference soil group of calcisols.

The accumulation of calcium carbonate in the soil layers occurs when the carbonate concentration in the soil solution remains high. Accumulation starts in the fine- and medium-sized pores at the surface of contact between the soil particles. This accumulation may be rather concentrated in a narrow zone or more dispersed, depending upon the quantity and frequency of rainfall, topography, soil texture and vegetation. In some soils, calcium carbonate deposits are concentrated into layers that can become very hard and impermeable to water (also called 'caliche'). The caliche layers are formed by rainfall (at nearly constant annual rates) leaching of salts to a particular depth in the soil at which the water content is so low that the carbonates precipitate. Caliche layers are also formed by salt moving upward from a water table (caused by irrigation) and precipitating near the top of the capillary fringe.

Calcium carbonate is especially abundant in regions where calcareous parent material occurs. These regions are characterized by sparse natural vegetation of xerophytic shrubs and ephemeral grasses. The total extent of calcisols is estimated in 800 million hectares worldwide, mainly concentrated in arid or Mediterranean climates. However, this is difficult to estimate since many calcisols occur together with solonchaks that are actually salinized calcisols or with other carbonate-enriched soils, which are not classified as calcisols.

Calcareous soils tend to be low in organic matter and available nitrogen. Their high pH also results in low phosphate availability (due to the formation of unavailable calcium phosphates such as apatite) and sometimes in reduced micronutrient (e.g. zinc and iron) availability. There may also be problems of potassium and magnesium nutrition as a result of the nutritional imbalance between these elements and calcium.

Calcareous soils can contain $CaCO_3$ and $NaCO_3$, both relatively insoluble, which lead to a saturation of the Ca and Na exchange sites. In this case, soil pH is determined by the dissociation of Na and Ca hydroxides. Soil pH can vary from 7 to 8.3 when only $CaCO_3$ is present and from 8.5 to 10 when $NaCO_3$ is a major soil component.

6.6   Plant responses to soil stress

The many facets of plant responses to environmental stress may include morphological and cytological changes, alteration in metabolic pathways, and transcriptional regulation of genes (Meyerowitz & Somerville, 1994). Plants growing on acid and calcareous soils are continuously exposed to either mineral deficiencies or metal toxicity. Therefore, these plants must be able to increase the solubility of nutrients and adjust selective ion uptake. In general, plants increase nutrient solubilization and uptake by excreting compounds such as sugars, organic acids, secondary metabolites and enzymes that alter the chemistry of the rhizosphere (a narrow zone of soil surrounding the root). The secretion of substances by roots varies according to the species, the soil conditions and the nutritional status of the plant. On the other hand, activation of carriers and transporters plays a role in metal uptake selectivity. In the following section, we focus our discussion on the mechanisms that plants have evolved to cope with metal ion toxicity and phosphorus and iron deficiencies, which are the most prevalent stresses that plants encounter in acid and calcareous soil.

6.7   Plant responses to heavy metals

The transport of heavy metals into roots is increased in acid soils. At high concentrations, plant micronutrients such as Fe, Mn, Cu and Zn and other metals such as Al, Cd, Pb and Ni cause phytotoxic effects, which include reduction of root growth, reduction in chlorophyll content and photosynthesis, inhibition of enzyme activities and damage to chloroplasts and mitochondria. Thus, many plant species have developed genetic and physiological tolerance to survive in natural metal-rich soils or in those soils subjected to increasing heavy metal pollution. It is common to find species of Gramineae, Lamiaceae, Fabaceae and Brassicaceae widely distributed throughout heavy-metal-contaminated ecosystems worldwide (Herrera-Estrella *et al.*, 1999).

Heavy-metal-tolerant plants have two basic strategies to respond to metal toxicity. They can either be excluders, in which differential metal uptake or transport between root and shoot leads to a more or less constant low level of metal ions in shoots over a wide range of external concentrations, or be accumulators, where metals are taken up from soil, concentrated in specific plant parts and transformed into physiologically tolerable forms.

In the excluder strategy, metals can be maintained outside the root through root exudation of chelating substances or by the activation of membrane transporters that pump out metal ions back into the soil.

6.8   Aluminum tolerance by exclusion

The aluminum cation Al^{3+} is toxic to many plants at micromolar concentrations. Several plant species have evolved mechanisms that enable them to grow on acid soils where toxic concentrations of Al^{3+} can limit plant growth. Recently, it has been suggested that Al tolerance can be due to enhanced rhizosphere alkalinization from roots, increased Al efflux across the plasma membrane and by exclusion of Al outside the cell membrane (Kochian, 1995).

Several plant species are known to secrete organic acids from their roots in response to Al. Citrate, oxalate and malate are some of the commonly released organic acid anions that can form sufficiently strong complexes with Al^{3+} to protect plant roots. Increased Al resistance correlates with greater rates of organic acid exudation. In snapbean, maize and *Cassia tora*, increased citrate secretion leads to Al tolerance, whereas in some Al-tolerant wheat varieties, Al tolerance depends on Al-inducible release of malate. Root exudation of oxalate in buckwheat also correlates with Al tolerance (Ryan *et al.*, 2001). Two mechanisms of organic acid secretion have been identified in plants. In the first mechanism, no discernible delay is observed between the addition of Al and the onset of organic acid release. This rapid response suggests that Al activates pre-existing proteins and that induction of protein synthesis is apparently not required. In this case, Al might simply activate a transporter on the plasma membrane to initiate organic anion efflux (Delhaize *et al.*, 1993; Zheng *et al.*, 1998). In the second mechanism, organic acid secretion is delayed for several hours after exposure to Al^{3+}. This type of response might indicate that induction of protein synthesis is required. The synthesized proteins could be involved in organic acid metabolism or in the transport of organic acid anions (Ma *et al.*, 1997a, 1997b). Figure 6.1 shows two models for Al-stimulated secretion of organic acids by roots.

Although many types of organic acids are found in root cells, only a few of these are specifically secreted in response to Al exposure (Ma, 2000). This suggests that specific transport systems for organic acid anions exist on the plasma membrane. In wheat and maize, this transport system has been identified as an anion channel (Ryan *et al.*, 1997; Kollmeier, 2001). Anion channels are membrane-bound transport proteins that allow the passive flow of anions down their electrochemical gradient. Patch-clamp studies on protoplasts prepared from wheat roots showed that Al^{3+} activates an anion channel in the plasma membrane that is permeable to malate and chloride (Ryan *et al.*, 1997; Zhang *et al.*, 2001). Evidence from studies on maize has shown that Al is able to activate the channel in isolated plasma membrane patches, indicating

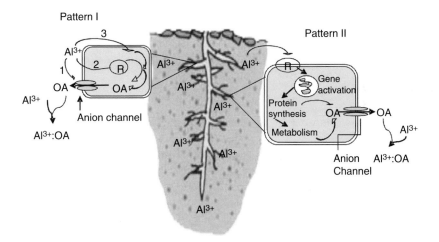

Figure 6.1 Models for aluminum (Al) stimulated secretion of organic acids (OA) from plant roots. Pattern I: Al activates an anion channel on the plasma membrane that is permeable to organic acid anions. This stimulation could occur in one of three ways: (1) Al^{3+} interacts directly with the channel protein to trigger its opening; (2) Al^{3+} interacts with a specific receptor (R) on the membrane surface or with the membrane itself to initiate a secondary messenger cascade that activates the channel; and (3) Al^{3+} enters the cytoplasm and activates the channel directly, or indirectly via secondary messengers. Pattern II: Al interacts with the cell, perhaps via a receptor protein (R) on the plasma membrane, to activate the transcription of genes that encode proteins involved in the metabolism of organic acids and their transport across the plasma membrane. Organic acid anions form a stable complex with Al, thereby keeping Al^{3+} outside of root cells.

that secondary messengers are either not required or are membrane-bound (Pineros & Kochian, 2001). Some cultivars of wheat have evolved mechanisms that minimize the harmful effects of Al. The nature of these mechanisms has been investigated using near-isogenic lines that differ in Al tolerance at a single dominant locus designated as *Alt1*. The *Alt1* locus cosegregates with an Al-activated malate efflux from root apices (Delhaize *et al.*, 1993). The *Alt1* gene was recently cloned (Sasaki *et al.*, 2004). It encodes a membrane protein, which is constitutively expressed in greater level in root apices of the Al-tolerant wheat line. Heterologous expression of *Alt1* in *Xenopus* oocytes, rice and cultured tobacco cells conferred an Al-activated malate efflux. This information suggests that *Alt1* encodes an aluminum-activated malate transporter. It would be interesting to test whether the expression on *Alt1* in Al-sensitive crop and pasture species can increase plant production in acid soils.

Several groups have attempted to manipulate citrate metabolism using genetic engineering to increase organic acid efflux. Transgenic tobacco lines expressing a citrate synthase gene from the bacterium *Pseudomonas aeruginosa* showed increases in internal concentrations of citrate, increased citrate secretion and enhanced Al tolerance (De la Fuente *et al.*, 1997). Overexpression of the carrot mitochondrial citrate synthase in *Arabidopsis* resulted in increased citrate synthase activity, increased citrate concentrations and a 60% increase in citrate efflux. These changes were associated with a small enhancement in Al tolerance (Koyama *et al.*, 2000). The relationship between increased citrate efflux and a greater capacity to synthesize citrate in transgenic plants suggests that changes in metabolism might also drive citrate secretion (see below for further details).

6.9 Aluminum tolerance by internal accumulation

It is well known that some Al-tolerant plant species can accumulate high concentrations of Al without showing symptoms of Al toxicity. Remarkably, Al stimulates the growth of *Melastoma malabathricum*, a tropical rainforest species known to accumulate Al (Watanabe *et al.*, 1998). Buckwheat leaves accumulate over 400 mg/kg dry weight when grown on acid soils. *Hydrangea* plants can accumulate high concentrations of Al (>3000 mg/kg dry weight) in leaves over several months of growth (Ma *et al.*, 1997a, 1997b). Recent evidence has shown that these Al-accumulator species detoxify internal Al^{3+} by forming Al–organic acid complexes. Complexes of Al citrate (1:1) in *Hydrangea* leaves and Al oxalate (1:3) in buckwheat have been identified by Al nuclear magnetic resonance. A current hypothesis is that chelation of Al by organic acids effectively reduces the activity of Al^{3+} in the cytosol, preventing the formation of complexes of Al and essential cellular components (Watanabe *et al.*, 1998).

6.10 Metal hyperaccumulators

Some plants have the extraordinary ability to accumulate high concentrations of heavy metals in their tissues. These plants hyperaccumulate zinc, nickel or cadmium without suffering cellular damage. Hyperaccumulators come from a number of diverse plant taxa, although the majority belongs to the family Brassicaceae (Baker et al., 1994). One Brassica species, Alyssum lesbiacum, can be grown in Ni^{2+}-rich soil with only moderate growth reduction. Ni^{2+} is rapidly transported into the plant, where it accumulates to over 3% of the dry weight in aboveground tissues (Krämer et al., 1996). Hyperaccumulation of heavy metals is a complex phenomenon. Likely, it involves several steps, including metal transport across plasma membranes of root cells, xylem loading and translocation, detoxification and sequestration of metals at the whole plant and cellular level (Lombi et al., 2002). Figure 6.2 focus on the mechanisms used by plants to take up and accumulate metals.

Some Ni hyperaccumulators respond to Ni^{2+} exposure with a large dose-dependent increase in histidine concentrations in the xylem. Thus, chelation with certain ligands, for example histidine, nicotianamine (Pich et al., 1994) and citrate (Senden et al., 1995), appears to route metals primarily to the xylem. In the hyperaccumulator species Thlaspi goesingense, histidine may also be involved in the mobilization and transport of Ni^{2+} and Zn^{2+} from the rhizosphere into the roots (Meagher, 2000).

Little is known about the mechanisms of metal translocation into the shoot. Organic acid transport in the transpiration stream has been correlated with the transport of micronutrients. Formation of metal–citrate complexes results in more zinc passing through the excised stems of Pinus radiata, and copper in Papyrus stems (Senden et al., 1995). These complexes are efficiently transported in the xylem and once in the shoot, metals such as Cd, Ni and Zn occur preferentially in epidermal and mesophyll cells, depending on the plant species. In these cells, metal ions are detoxified by compartmentalization into the vacuole, where they are probably complexed with organic acids.

Other mechanisms for sequestering metals ions in plant tissues involve two classes of cysteine-rich peptides, the metallothioneins (MTs) and phytochelatins (PCs). Metals such as Cd, Co, Cu, Hg and Ni are sequestered by binding to the organic sulfur (R–SH) in the cysteine residues of these peptides (Kägi & Schaffer, 1998; Schat et al., 2002). In vitro, MTs form metal–ligand complexes; their specificity correlates with the thiolate series for cation binding in the order Cu > Cd > Pb > Zn (Kägi & Schaffer, 1998).

Plants have a complex family of MT genes encoding peptides that are generally composed of 60–80 amino acids and contain 9–16 cysteine residues (Zhou & Goldsbrough, 1995). The MTs are primarily thought to chaperone nutrient metals to their various necessary roles (e.g. insertion into an enzymatic center during protein folding). MTs can also protect plants from the effects

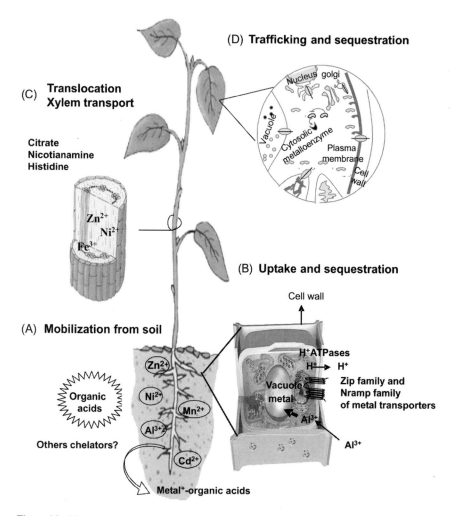

Figure 6.2 Mechanisms involved in metal accumulation by plants. (A) Root exudates and acidification of the rhizosphere aid in metal mobilization in soil. (B) Uptake of both hydrated metal ions and metal–chelate complexes is mediated by plasma membrane proteins. Into the cell, metals are chelated by organic acids and excess metal is transported to vacuole, where it is accumulated. (C) Metals are transported from the root to the shoot via the xylem in the form of metal–chelate complexes. (D) In the leaf apoplast, metals can move cell-to-cell through plasmodesmata. Intracellular distribution of essential transition metals is mediated by specific metallochaperones and transporters localized in endomembranes. Abbreviations and symbols: CW, cell wall; M, metal; filled circles, chelators; filled ovals, transporters; bean-shaped structures, metallochaperones.

of toxic metal ions and aid in their accumulation. For example, transgenic expression of the 32 amino acid metal-binding domain of the mouse MT in tobacco confers moderate levels of Cd resistance and accumulation (Pan *et al.,*

1994). MT–metal complexes can be glutathionated (Brouwer *et al.*, 1993), suggesting that these complexes might be transported into vacuoles for long-term sequestration.

PCs are nonribosomally synthesized peptides. PCs form ligand complexes with nutrient and toxic metals and aid transport into vacuoles (Salt & Rauser, 1995), where the metals are sequestered. Mutants in the synthesis of PCs or their precursors are hypersensitive to Cd and many other sulfur-reactive metals, demonstrating the role of PCs in protecting plants from toxic metals (Howden *et al.*, 1995). Further evidence of this role comes from the expression of a bacterial glutathione synthetase (GS) in *Brassica juncea*. Transgenic *B. juncea* plants have greater GSH and PC concentrations and increased Cd^{2+} tolerance and accumulation than untransformed controls (Zhu *et al.*, 1999).

6.11 Plant responses to mineral deficiency

6.11.1 Phosphorus deficiency

Few unfertilized soils release P fast enough to support the high growth rates of crop plant species. In many agricultural systems in which application of P to the soil is necessary, the recovery of applied P by crop plants is very low in the growing season. Over 80% of the P applied as fertilizer can become immobile and unavailable for plant uptake because of adsorption, precipitation or conversion to the organic form (Holford, 1997).

Soil P is found mainly in two different pools, the inorganic orthophosphate (Pi) form, which is the most readily available for plant uptake and the organic form, of which phytic acid (inositol hexaphosphate) is usually a major component. Depending on the soil characteristics, mineral forms of P such as Ca–P, Al–P and Fe–P can be found in high amounts in the soil (Schachtman *et al.*, 1998). Therefore, although P can be quite abundant in many soils, it is largely unavailable for uptake. As such, crop yield on 30–40% of the world's arable land is limited by P availability. The acid-weathered soils of the tropics and subtropics and alkaline soils of arid zones are particularly prone to P deficiency. Application of P-containing fertilizers is usually the recommended treatment for enhancing soil P availability and stimulating crop yields (Von-Uexkull & Muttert, 1995).

Plants have evolved two broad strategies for P acquisition and use in nutrient-limiting environments:

(1) those aimed at optimizing the use of P and
(2) those directed toward enhanced acquisition or uptake.

Processes that optimize the use of P involve decreased growth rate, increased growth per unit of P uptake, remobilization of internal P and modifications in carbon metabolism (Raghothama, 1999; Uhde-Stone *et al.*, 2003). Processes that lead to enhanced uptake include increased production and secretion of phosphatases, exudation of organic acids, greater root growth along with modified root architecture, expansion of root surface area by prolific development of root hairs and enhanced expression of Pi transporters (Raghothama, 1999; Abel *et al.*, 2002).

Genotypic differences in P assimilation efficiency among crops greatly depend on their ability to solubilize and acquire P from soil. Organic acid excretion has been correlated with the capacity of some plant species to solubilize phosphate from low-solubility compounds such as Al–P, Ca–P and rock phosphate. Increments in organic acid exudation are observed in rape (*Brassica napus*) and white lupin (*Lupinus albus*) as part of their responses to stress caused by P deficiency (Dinkelaker *et al.*, 1989; Hoffland *et al.*, 1989).

In white lupin, P deprivation induces the formation of cluster roots. These cluster roots, also termed proteoid roots, excrete large quantities of citrate, allowing the plant to enhance the solubilization of phosphate from low-solubility P sources (Dinkelaker *et al.*, 1995).

In addition to the correlation found between organic acid exudation and the enhanced capacity of certain plants to acquire P, organic acids such as citrate, malate, oxalate, tartrate, malonate and lactate are commonly used by soil chemists to extract phosphate from insoluble compounds present in the soil. Of all the organic acids mentioned above, citrate is the most effective P-solubilizing compound (Staunton & Leprince, 1996). In the soil, most organic acids are produced by decomposition of organic matter or by microbial activity. Citrate and malate form stable complexes with cations such as Ca, Al and Fe. This correlates with their high P-solubilizing capacity. Traina *et al.* (1987) observed a rapid release of phosphate from an acid soil in the presence of citrate. Similarly, greenhouse studies by Bolan *et al.* (1994) show that exudation of citrate, malate and oxalate into the soil by roots of *Lotus pedunculatus* was sufficient to solubilize phosphate from rock phosphate. Recently, Luo and coworkers evaluated the role of organic acid exudation from plant roots in Al tolerance and P uptake in rape and tomato under field conditions. They showed that both Al tolerance and P uptake were higher in rape than in tomato and that these enhancements depended on the greater exudation of citrate from rape roots. These results confirm than under field conditions, citrate exudation is an effective mechanism for Al detoxification and P acquisition (Luo *et al.*, 1999).

6.11.2   Improving P efficiency in transgenic plants

Since the exudation of organic acids by roots has been strongly related to Al tolerance and P solubilization in various plant species, several attempts have been made in the past 5 years to develop transgenic plants that produce and release increased amounts of organic acids. Constitutive expression of the citrate synthase gene from *P. aeruginosa* in the cytoplasm of transgenic tobacco plants enhanced synthesis and excretion of citrate. The increase in citrate synthesis and excretion in these transgenic plants was reported to result in an increased tolerance to Al toxicity (De la Fuente *et al.*, 1997). Further analysis of these plants showed that they had elevated performance to grow and reproduce in calcareous soil with low P content (López-Bucio *et al.*, 2002).

Although there is a report suggesting that the increased expression of enzymes involved in organic acid synthesis does not result in the accumulation of organic acids or its excretion (Delhaize *et al.*, 2001), results from several other research groups have confirmed that this strategy is valid. Koyama *et al.* (2000), using transgenic Arabidopsis plants that constitutively express a mito-chondrial citrate synthase from carrot, reported that transgenic Arabidopsis plants having an elevated citrate efflux from roots had a greater capacity to absorb P from acid soil containing Al–P compounds (Koyama *et al.*, 2000). More recently, an Arabidopsis mitochondrial CS gene was expressed in canola using an *Agrobacterium tumefaciens* system (Anoop *et al.*, 2003). Increased levels of CS activity and greater tolerance to phytotoxic concentrations of Al were reported for these canola transgenic lines when compared to untrans-formed controls. These transgenic lines showed enhanced levels of cellular shoot citrate and a twofold increase in citrate exudation when exposed to 150 μM Al (Anoop *et al.*, 2003). Because citrate is an efficient chelator of P, the transgenic canola plants could also show improved assimilation of P, although this has not yet been tested. Other studies have also suggested that the strategy of increasing the expression of enzymes involved in organic acid biosynthesis can be used to enhance the synthesis and exudation of organic ions. For instance, transgenic alfalfa overexpressing a nodule-specific malate dehydrogenase showed enhanced exudation of organic ions from their roots, which correlated with enhanced Al tolerance and P uptake under field condi-tions (Tesfaye *et al.*, 2001). Taken together, the information available suggests that modulation of different enzymes involved in organic acid synthesis and turnover, such as citrate synthase and malate dehydrogenase, could be con-sidered for gene manipulation to improve Al tolerance and P nutrition.

In many agricultural soils, organic matter is a major P source. It is estimated that in tropical soils up to 80% of P is present in the organic form, of which phytate (inositol hexaphosphate) is a major component. To use phytate, some plants are able to excrete enzymes (including phosphatases and phytases) that hydrolyze organic P, making it available in the rhizosphere. To test the hypothesis that enzymatic degradation of phytate could release substantial Pi for plant growth, Richardson *et al.* (2001) generated transgenic Arabidopsis plants expressing an *Aspergillus niger* phytase. These plants were able to secrete a substantial amount of phytase into their growth medium, which enabled them to obtain Pi from phytate. A 20-fold increase in total root phytase activity in transgenic lines improved P nutrition, such that the growth of these plants in a medium with phytate was equivalent to control plants supplied with a readily available source of soluble phosphate (Richardson et al., 2001). This work demonstrates that in addition to the secretion of organic acids by roots, the secretion of the enzyme phytase represents an important target for engineering plants for growth on P-limited soils.

To improve P uptake in plants, P transporters may also play a critical role. In cell culture experiments, expression of an Arabidopsis high-affinity P transporter gene enhanced biomass production and P uptake capacity of transgenic tobacco cells under P-limited conditions (Mitsukowa et al., 1997). However, since Pi has a low mobility constant in the soil, the real potential of over-expressing P transporters to develop P-efficient transgenic plants remains to be determined. Cellular and molecular approaches are now being employed to isolate and characterize more nutrient ion transporters in model plants such as Arabidopsis, tobacco, tomato and Medicago truncatula. A major future challenge is to improve crop plants with greater expression of nutrient ion transporters to test their nutrient acquisition capacity in low-fertility soils.

6.11.3   Plant responses to iron deficiency

Iron deficiency is a major problem for crop production worldwide but particularly in calcareous soils. The main symptom of Fe deficiency in plants is chlorosis in young leaves. Although abundant in the Earth's crust, iron is extremely insoluble in oxygenated environments, being mainly present as oxyhydrates with low bioavailability. In aerated systems, the concentrations of ionic Fe^{3+} and Fe^{2+} are extremely low (10^{-10} M or lower). Chelates of Fe^{3+} and occasionally of Fe^{2+} are therefore the dominant forms of iron in the soil solution.

As a further constraint, high levels of bicarbonate often decrease the mobility of Fe species far below that required by plants. Because of its fundamental role in many vital processes, organisms have evolved mechanisms that transform iron into forms that can be used by cells. In plants, most of the components involved in iron acquisition and uptake have been characterized at the molecular level during the past few years and microarray analysis of iron-inducible genes has provided insights into global changes in the metabolism of plants in response to their nutritional status (see below for further details). Uptake of iron according to the plant's need requires mechanisms that can sense and respond to changes in iron levels, either within an individual cell or in its immediate environment. In higher plants, uptake rates are correlated with the requirement of the shoot rather than with the iron concentration of the cells that mediate uptake. Such behavior is indicative of the involvement of a diffusible signal communicating the iron status between the shoot and the roots (Grusak, 1995).

In spite of being extremely insoluble at neutral and basic pHs, ferric iron (Fe^{3+}) is the predominant form of this nutrient in soil. Strategies for iron acquisition in plants involve the reduction of ferric iron to its ferrous form (Fe^{2+}). Figure 6.3 shows the ways used by plants to acquire iron. Plants solubilize and absorb Fe^{2+} using one of two strategies. One of these strategies is used by grasses and involves the biosynthesis and secretion of mugineic acid (MA) into the rhizosphere (Mori, 1999). MAs are peptide-like compounds known as phytosiderophores. The first step in the synthesis of MAs involves the formation of nicotianamine (NA) from three molecules of S-adenosylmethionine by NA synthase. Then, NA aminotransferase transfers an amino group to NA, forming an unstable intermediate that is reduced to

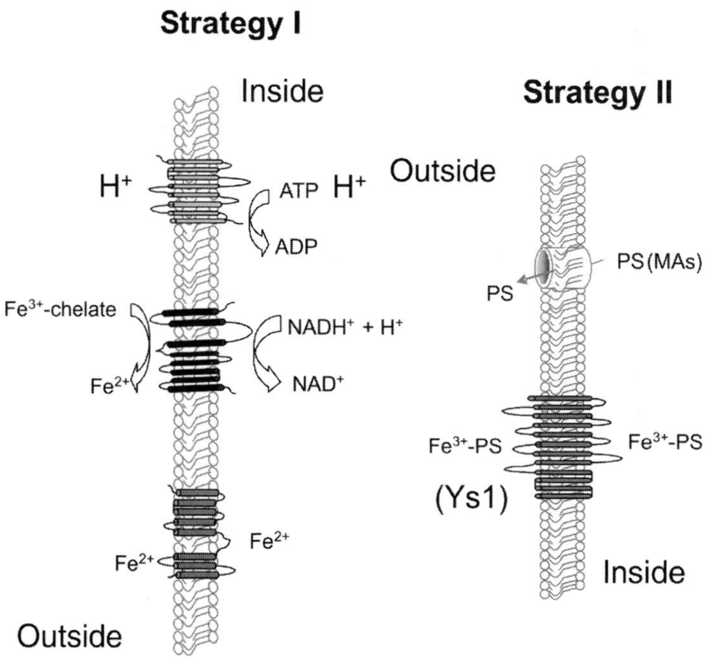

Figure 6.3 Strategies of iron uptake. Strategy I is carried out by all higher plants, except the Gramineae, which use strategy II. The model plants *Arabidopsis thaliana*, *Lycopersicum sculentum* (tomato) and *Pisum sativum* (pea) are all strategy I plants. They improve iron uptake by three reactions: (i) the excretion of protons via P-type ATPase to acidify the surrounding solution, thus increasing Fe^{3+} solubility; (ii) reduction of Fe^{3+} by a Fe^{3+}-chelate reductase to the more soluble Fe^{2+} form and (iii) plasmalemma transport of Fe^{2+} by iron transporters. All three components of this strategy increase their activities during iron deficiency. Strategy II plants comprise the grasses such as wheat (*Triticum aestivum*), barley (*Hordeum vulgare*), rice (*Oryza sativa*) and maize (*Zea mays*). Their roots release phytosiderophores (PSs) that chelate Fe^{3+} in the rhizosphere. Specific transporter proteins then import the Fe^{3+}–PS complexes into the cell. Iron deficiency upregulates the underlying genes.

deoxymugineic acid, from which other MAs can be synthesized. Recently, successful identification of MA biosynthetic genes and cloning of the MA-uptake system has been reported (Higuchi *et al.*, 1999; Mori, 1999). Figure 6.4 shows the pathway for MA biosynthesis.

The other strategy, present in dicotyledonous and nongraminaceous mono-cotyledonous plants, involves Fe uptake in a way similar to that in *Sacchar-omyces cerevisiae*, including:

(1) acidification of the rhizosphere;
(2) reduction of Fe^{3+} to Fe^{2+} and
(3) transport of Fe^{2+} across the root plasma membrane.

In the past few years, knowledge about the molecular basis of iron transport systems in plants has greatly increased. The *FRO2* gene from Arabidopsis, encoding a Fe^{3+} chelate reductase, was cloned based on its homology to the yeast *FRE1* and *FRE2* reductase genes (Robinson *et al.*, 1999). In addition,

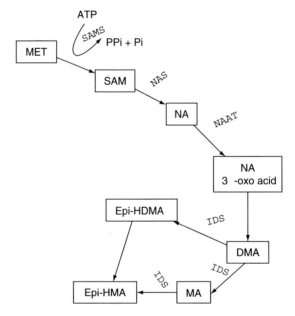

Figure 6.4 Phytosiderophore biosynthesis. Mugineic acid (MA) biosynthesis starts from the condensation of three molecules of methionine (me) to form *S*-adenosyl methionine (SAM) that are integrated into nicotianamine (NA) in one enzymatic step catalyzed by nicotianamine synthase. NA synthase has been cloned from tomato, barley and rice and its expression in roots is strongly upregulated by iron availability in strategy II plants (Higuchi *et al.*, 1999). Deamination of NA by nicotianamine aminotransferase leads to deoxyMA production, which is then hydroxylated by the IDS3 protein to form MA (Takahashi *et al.*, 1999). Both, MA and deoxyMA, may be further hydroxylated by the IDS2 protein to form further MA derivatives (Mori, 1999).

functional complementation of the *fet3/fet4* yeast mutant, defective in both high- and low-affinity iron transport systems, allowed the cloning of a new type of metal transporter of *Arabidopsis* named *IRT1*. This transporter mediates Fe^{2+} transport when expressed in yeast, and its expression is upregulated under iron-deficient conditions in roots (Eide *et al.*, 1996).

Other systems for iron uptake have been recently described. These include the *AtNramp1*, *AtNramp3* and *AtNramp4* genes, which encode a family of divalent cation transporters. These genes are involved in iron homeostasis, and their expression is inducible by iron starvation in plants. In addition, a gene encoding the major Fe^{3+} uptake system of grasses has been cloned. This corresponds to the maize *yellow stripe 1* gene (*ys1*), which encodes a membrane protein for which eight homologues exist in *Arabidopsis* (Curie *et al.*, 2000, 2001). Another family of potential iron transporters from plants is the ZIP family, which comprises over 25 members ubiquitously present in organisms from *Trypanosome* to humans and plants. Common features shared by ZIP proteins include eight putative transmembrane domains and an intracellular loop rich in histidine, potentially involved in metal binding (Eng *et al.*, 1998). The Arabidopsis *IRT2* gene belongs to the ZIP family of metal transporters and encodes a root-periphery iron transporter. *IRT2* expression in yeast suppresses the growth defect of iron transport yeast mutants and enhances iron uptake and accumulation. *IRT2* expression is specifically detected in the external cell layers of the root system, and is upregulated by iron deficiency (Vert *et al.*, 2001).

To begin elucidating the molecular mechanisms that regulate plant responses to low Fe availability, several mutants have been identified in Arabidopsis and other plant species. For instance, a tomato mutant (*fer*) that develops severe chlorosis on normal soils and lacks the typical iron-deprivation responses of the wild-type parent, including the formation of ectopic root hairs, and the increased excretion of protons and ferric reductase activity has been isolated (Ling *et al.*, 1996). Ling *et al.* (2002) used a map-based approach to clone the *fer* gene and demonstrate that it encodes a protein containing a basic helix–loop–helix domain, a protein domain commonly present in eukaryotic transcription factors. The *fer* gene is expressed at the root tip independently of the iron status of the plant (Ling *et al.*, 2002). These results suggest that FER may control root physiology and development at the transcriptional level in response to iron supply.

6.12 Morphological responses to mineral deficiency

6.12.1 Effects of iron availability on transfer cell formation

Transfer cells are parenchyma cells with conspicuous ingrowths of secondary wall material that protrude into the cell lumina and increase the cell surface.

Under ordinary conditions, transfer cells are found in areas with high solute fluxes in a variety of vegetative and reproductive tissues. In roots, transfer cells occur in specialized tissues such as the pericycle adjacent to nodules of leguminous plants, in vascular parenchyma and at sites where lateral roots join the parent root (McDonald *et al.*, 1996).

In strategy I plants, rhizodermal transfer cells have been classified based on response to Fe deficiency. The formation of transfer cells in the root epidermis is correlated with the stage of differentiation of the stele. In most species, the transfer-cell-bearing zone is located primarily in developed root hair zones characterized by mature xylem vessels, thus allowing an efficient long-distance transport of Fe (Landsberg, 1994). From their structural character-istics and anatomical distribution, the formation of transfer cells is seen as a mechanism for increasing the rate of solute transport at the apoplast–symplast boundary (Schmidt, 1999).

6.12.2   *Effects of nutrient availability on root hair formation*

Root hairs are long tubular-shaped outgrowths from root epidermal cells that increase the total absorptive surface of the root system and participate in nutrient and water uptake. One of the most important effects of nutrient deficiency on root development is the induction of some epidermal cell layers to form long root hairs. Conditions of low P and Fe availability induce the formation of roots with an abundance of long root hairs (López-Bucio *et al.*, 2003). The elongation of root hairs is regulated by P availability in a dose-dependent manner. Recent work in Arabidopsis has shown that P deprivation not only affects root hair elongation but also produces up to a fivefold increase in root hair density. This effect is due to an increase in the number of epidermal cells that differentiate into trichoblasts (Ma *et al.*, 2001). When iron is limiting, root hair formation and elongation also increase. The extra root hairs that result from limiting iron availability are often located in positions that are occupied by nonhair cells under normal conditions. Although similar, the changes in root hair morphology in response to P and Fe have been found to be under the control of different signaling pathways (Schmidt & Schikora, 2001). Plant hormones, mainly auxins, cytokinins and ethylene, appear to be the key factors in these nutrient-mediated root hair alterations (Schmidt *et al.*, 2000).

6.12.3   *Effects of nutrient availability on root branching*

The production of lateral roots generates the final architecture of the root system. Root branching is a basic structural trait in most plants, offering important opportunities for soil exploration. A number of plant species that are well adapted to grow on infertile soils with low nutrient availability exploit better the soil environment through changes in root branching patterns. In particular, white lupin (*L. albus*) plants grow and proliferate in soils with limited P availability through the formation of cluster roots. Cluster roots, also termed proteoid roots, consist of groups of small lateral roots that arise from the perycicle and are specialized in P uptake. Cluster roots develop more extensively when lupin plants are exposed to P-limiting conditions. These roots have determinate growth, that is, their root meristematic cells divide only for a limited period and then differentiate. After just a few days of growth, proteoid roots become exhausted and form large numbers of root hairs. The increased P-uptake capacity of cluster roots compared to roots that have normal growth is provided by their increased absorptive surface, increased exudation of organic acids and phosphatase and possibly greater expression of P transporters (Neumann & Martinoia, 2002).

In Arabidopsis, the formation of lateral roots is greatly influenced by phosphate and nitrate availability. The root architecture of Arabidopsis plants that have been grown in low P medium resembles that of the cluster roots of lupins: lateral roots arise in close proximity to each other, have determinate growth and are densely covered by root hairs (Williamson *et al.*, 2001). It has been reported that low concentrations of P favored lateral root growth over primary root growth. When quantified, the number of lateral roots was up to five times greater in plants grown in a limiting (1 μM) P concentration than in plants supplied with optimal (1 mM) P (Williamson *et al.*, 2001; López-Bucio *et al.*, 2002). Anatomical and biochemical analyses have shown that low-P-grown mature roots lack a normal apex and have increased expression of P transporter genes. Taken together, these results suggest that in Arabidopsis the low-P-induced determinate root growth is directed toward increasing the P-uptake capacity of the plant's root system by modulating root architecture and by inducing the expression of genes that are involved in P uptake (López-Bucio *et al.*, 2003). Changes in nitrate and phosphate availability have been found to have contrasting effects on lateral root formation and elongation. In Arabidopsis, increasing nitrate availability reduces primary root elongation, whereas an increase in P supply has the opposite effect.

6.13 Functional genomics for the discovery of genes involved in mineral nutrition

The term functional genomics refers to the development and application of genomewide experimental approaches to assess gene function by making use of the information provided by the sequencing of plant genomes (Bouchez & Hofte, 1998). In particular, the availability of the complete sequence of two angiosperm genomes, the dicot *Arabidopsis thaliana* and the monocot rice (*Oryza sativa*), has helped in the discovery of new genes (Martienssen &

McCombie, 2001; Goff *et al.*, 2002). A rapid way to establish an inventory of expressed genes is by determining partial sequences of complimentary DNA (cDNA) called expressed sequence tags (ESTs). In this approach, sequences of 300–500 base pairs are determined from one or both ends of randomly chosen cDNA clones. Thousands of sequences from any particular plant species can thus be determined with a limited investment. The complete genome sequences of Arabidopsis and rice now make it possible to study the genomes of economically important crops by comparing the expressed genes (ESTs) in these crops to the complete genomes of higher plants.

With these approaches the focus of the analysis is shifted from individual components to whole plants, involving high-throughput methods for the study of large numbers of genes in parallel. Having identified a new sequence, indirect information on cellular or developmental function can be obtained by comparison to sequence databases, and from spatial and temporal expression patterns; for example, the presence of mRNA in different cell types during development or in different environments. Knocking out or overexpressing the gene permits the gene sequence to be linked to a phenotype from which a biological role may be deduced.

6.14 Application of functional genomics to iron and phosphorus nutrition

In mineral nutrition, the application of functional genomics has addressed the challenges of identifying and characterizing the components of signaling cascades that plants use to sense changes in both their internal mineral status and the rhizosphere mineral environment. With the goal of a better understanding of the molecular processes involved in the adaptation of plants to Fe deficiency, Thimm *et al.* (2001) investigated gene expression in response to Fe deficiency in Arabidopsis roots and shoots through the use of a cDNA array representing 6000 individual gene sequences. In this work, Arabidopsis seedlings were grown for 1, 3 and 7 days in the absence of Fe, and gene expression was investigated. An expression analysis of genes in glycolysis, the tricarboxylic acid cycle and the oxidative pentose phosphate pathway revealed an induction of several enzymes within 3 days of Fe limitation. In roots, these changes were found primarily in clones homologous to enzymes in fermentation including lactate dehydrogenase, pyruvate decarboxylase and alcohol dehydrogenase. In shoots, expression of hexokinase, Glc-6-phosphate isomerase, Fru-bisphosphate aldolase and triose phosphate isomerase was increased. In addition, genes encoding α-amylase, the phosphate translocator and the H^+-Suc symporter were induced. As for glycolysis, no changes in gene expression for enzymes in the citrate cycle were observed after 1 day of Fe-deficient growth. However, after 3 days the increased expression of isocitrate lyase, malate dehydrogenase and cytochrome c reductase/oxidase was noteworthy. At day 7, isocitrate dehydrogenase expression also increased. Taken together, these data confirm that Fe deficiency causes changes in gene expression in roots and shoots. The induction in expression of enzymes usually involved in anaerobic respiration was interpreted as an effect of Fe deficiency on energy production through oxidative phosphorylation. Changes in expression of genes involved in the citrate cycle may be involved in the strategies used by plants to cope with iron limitation that include acidification of the rhizosphere and the accumulation of citrate and malate in roots.

Wang *et al.* (2002) generated a cDNA array consisting of 1280 genes from tomato (*Lycopersicon esculentum*) roots for expression profiling in mineral nutrition. This array was used to search for genes induced by phosphorus, potassium and iron deficiencies. RNA gel-blot analysis was conducted to study the regulation of these genes in response to withholding P, K or Fe. From this research, a set of genes previously not associated with mineral nutrition were identified, such as transcription factors, a mitogen-activated protein (MAP) kinase, a MAP kinase kinase and 14-3-3 proteins. To investigate whether the internal nutritional status of the leaves produced a signal-triggering gene induction, both the transcription factor and MAP kinase expression was studied in roots of decapitated tomato plants. Both genes were induced as rapidly as 1 h after withholding the nutrients from roots of intact or decapitated plants, suggesting that the rapid response to the absence of P, K or Fe in the root environment is triggered either by a root-localized signal or because of root sensing of the surrounding medium. Interestingly, the authors found that expression of P, K and Fe transporters was also rapidly induced by deprivation of all three mineral nutrients. This finding suggests that there is a crosstalk between signaling pathways for plant responses to different mineral nutrients and that coordination and regulation of the uptake and transport of different minerals may occur in the root system.

To identify Arabidopsis genes whose expression increases in response to phosphate starvation, Hammond *et al.* (2003) grew plants hydroponically in a medium lacking P, and total RNA samples from shoot material were used to challenge Affymetrix GeneChips representing 8100 Arabidopsis genes. The results obtained from the microarray technology and confirmed by quantitative PCR showed that changes in gene expression under P limitation could be either common to a variety of environmental perturbations or specific to P starvation. The expression of 60 genes was transiently upregulated 2.5-fold 4 h after withdrawing P. Many of these genes were common to other environmental stresses and included five chitinases, four peroxidases and a PR-1-like protein. These genes have been reported to be involved in cell rescue and defense, which is consistent with the common horticultural observation that plants subjected to a mild P stress are less susceptible to pathogens. One peroxidase in particular (At3g49129) has been shown to be upregulated in response to aluminum stress (Richards *et al.*, 1998), which suggests that nutritional and heavy metal responses may have source-common response mechanisms. Other genes that were transiently upregulated by P starvation have been found to be regulated by oxidative stress, including a cytochrome P450, a C2H2-type zinc finger protein and a blue copper-binding protein. The expression of *SQD1*, a gene involved in the synthesis of sulfolipids, responded specifically to P starvation and was increased after 100 h of P-limited growth (Hammond *et al.*, 2003). Based on these results the authors proposed that monitoring plants – also termed 'smart' plants – can be genetically engineered by transformation with a construct containing the promoter of a gene upregulated specifically by P starvation upstream of a reporter gene such as β-glucuronidase *(GUS)*. This approach may accelerate the discovery of regulatory sequences that could direct the expression of transgenes in plants in a tissue- and nutrient-specific manner. More important, phosphate monitoring crops would allow precision management of P fertilization, thereby maintaining yields while reducing costs, conserving natural resources and preventing pollution, which represent major goals of modern agriculture.

References

Abel, S., Ticconi, C. A. & Delatorre, C. A. (2002) Phosphate sensing in higher plants. *Physiologia Plantarum*, **115**, 1–8.

Anoop, V. M., Basu, U., McCammon, M. T., McAlister-Henn, L. & Taylor, G. J. (2003) Modulation of citrate metabolism alters aluminum tolerance in yeast and transgenic canola overexpressing a mitochondrial citrate synthase. *Plant Physiology*, **132**, 2205–2217.

Baker, A. J. M., Reeves, R. D. & Hajar, A. S. M. (1994) Heavy-metal accumulation and tolerance in British populations of the metallophyte *Thlaspi caerulescens* (Brassicaceae). *New Phytologist*, **127**, 61–68.

Bolan, N. S., Naidu, R., Mahimairaja, S. & Baskaran, S. (1994) Influence of low-molecular weight organic acids on the solubilization of phosphates. *Biology and Fertility of Soils*, **18**, 311–319.

Bouchez, D. & Hofte, H. (1998) Functional genomics in plants. *Plant Physiology*, **118**, 725–732.

Breemen, V. N., Mulder, S. & Driscoll, C. T. (1983) Acidification and alkalinisation of soils. *Plant and Soil*, **75**, 283–308.

Brouwer, M., Hoexum-Brouwer, T. & Cashon R. E. (1993) A putative glutathione-binding site in Cd/Zn-metallothionein identified by equilibrium binding and molecular-modeling studies. *Biochemical Journal*, **294**, 219–225.

Curie, C., Alonso, J. M., Lee, J. M., Ecker, J. R. & Briat J. F. (2000) Involvement of NRAMP1 from *Arabidopsis thaliana* in iron transport. *Biochemical Journal*, **347**, 749–755.

Curie, C., Panaviene, Z., Loulergue, C., Dellaporta, S. L., Briat, J. F. & Walker E. L. (2001) Maize *yellow stripe1* (*ys1*) encodes a membrane protein directly involved in Fe(III) uptake. *Nature*, **409**, 346–349.

De la Fuente, J. M., Ramírez-Rodriguez, V., Cabrera-Ponce, J. L. & Herrera-Estrella L. (1997) Aluminum tolerance in transgenic plants by alteration of citrate synthesis. *Science*, **276**, 1566–1568.

Delhaize, E., Ryan, P. R. & Randall, P. J. (1993) Aluminum tolerance in wheat (*Triticum aestivum* L) II. Aluminum-stimulated excretion of malic acid from root apices. *Plant Physiology*, **103**, 695–702.

Delhaize E., Hebb D. M. & Ryan P. R. (2001) Expression of a *Pseudomonas aeruginosa* citrate synthase gene in tobacco is not associated with either citrate overproduction or efflux. *Plant Physiology*, **125**, 2059–2067.

Dinkelaker, B., Römheld, V. & Marschner, H. (1989) Citric acid excretion and precipitation of calcium citrate in the rhizosphere of white lupin (*Lupinus albus* L.). *Plant Cell and Environment*, **12**, 285–292.

Dinkelaker, B., Römheld, V. & Marschner, H. (1995) Distribution and function of proteoid roots and other root clusters. *Botanica Acta*, **108**, 183–200.

Dyson, T. (1999) World food trends and prospects to 2025. *Proceedings of the National Academy of Sciences of the United States of America*, **96**, 5929–5936.

Eide, D. J., Broderius, M., Fett, J. & Guerinot, M. L. (1966) A novel iron-regulated metal transporter from plants identified by functional expression in yeast. *Proceedings of the National Academy of Sciences of the United States of America*, **93**, 5624–5628.

Eng, B. H., Guerinot, M. L., Eide, D. & Sauer, M. H. Jr. (1998) Sequence analyses and phylogenetic characterization of the ZIP family of metal ion transport proteins. *Journal of Membrane Biology*, **166**, 1–7.

Goff, S. A., Ricke, D., Lan, T. H., Presting, G., Wang, R., Dunn, M., Glazebrook, J., Sessions, A., Oeller, P., Varma H., Hadley, D., Hutchison, D., Martin, C., Katagiri, F., Lange, B. M., Moughamer, T., Xia, Y., Budworth, P., Zhong, J., Miguel, T., Paszkowski, U., Zhang, S., Colbert, M., Sun, W. L., Chen, L., Cooper, B., Park, S., Wood, T. C., Mao, L., Quail, P., Wing, R., Dean, R., Yu, Y., Zharkikh, A., Shen, R., Sahasrabudhe, S., Thomas, A., Cannings, R., Gutin, A., Pruss, D., Reid, J., Tavtigian, S., Mitchell, J., Eldredge, G., Scholl, T., Miller, R. M., Bhatnagar, S., Adey, N., Rubano, T., Tusneem, N., Robinson, R., Feldhaus, J., T, Macalma., Oliphant, A. & Briggs, S. (2002) A draft sequence of the rice genome (*Oryza sativa* L. ssp. Japonica). *Science*, **296**, 92–100.

Grusak, M. A. (1995) Whole-root iron [III]-reductase activity throughout the life cycle of iron-grown *Pisum sativum* L. (*Fabaceae*). Relevance to the iron nutrition of developing seeds. *Planta*, **197**, 111–117.

Hammond, J. P., Bennett, M. J., Bowen, H. C., Broadley, M. R., Eastwood, D. C., May, S. T., Rahn, C., Swarup, R., Woolaway, K. E., & White, P. J. (2003) Changes in gene expression in Arabidopsis shoots during phosphate starvation and the potential for developing smart plants. *Plant Physiology*, **132**, 578–596.

Herrera-Estrella, L., Guevara-García, A. & López-Bucio, J. (1999) Heavy metal adaptation. *Encyclopedia of Life Sciences*. Macmillan Publishers, London, pp. 1–5.

Higuchi, K., Suzuki, K., Nakanishi, H., Yamaguchi, H., Nishizawa, N. K. & Mori, S. (1999) Cloning of nicotianamine synthase genes, novel genes involved in the biosynthesis of phytosiderophores. *Plant Physiology*, **119**, 471–480.

Hoffland, E., Findenegg, G. & Nelemans, J. A. (1989) Solubilization of rock phosphate by rape. Local root exudation of organic acids as response to P starvation. *Plant and Soil*, **113**, 161–165.

Holford, I. C. R. (1997) Soil phosphorus: its measurement and its uptake by plants. *Australian Journal of Soil Research*, **35**, 227–239.

Howden, R., Goldsbrough, P. B., Andersen, C. R. & Cobbett, C. S. (1995) Cadmium sensitive *cad1* mutants of *Arabidopsis thaliana* are phytochelatin deficient. *Plant Physiology*, **107**, 1059–1066.

Kägi, J. H. & Schaffer, A. (1988) Biochemistry of metallothionein. *Biochemistry*, **27**, 8509–8515.

Kidd, P. S. & Proctor, J. (2001) Why plants grow poorly on very acid soils: are ecologists missing the obvious? *Journal of Experimental Botany*, **52**, 791–799.

Kochian, L. V. (1995) Cellular mechanisms of aluminum toxicity and resistance in plants. *Annual Review of Plant Physiology and Plant Molecular Biology*, **46**, 237–260.

Kollmeier, M. (2001) Aluminum activates a citrate-permeable anion channel in the Al-sensitive zone of the maize root apex: a comparison between an Al-sensitive and an Al-tolerant cultivar. *Plant Physiology*, **126**, 397–410.

Koyama, H., Kawamura, A., Kihara, T., Hara T., Takita, E. & Shibata, D. (2000) Overexpression of mitochondrial citrate synthase in *Arabidopsis thaliana* improved growth on a phosphorus limited soil. *Plant and Cell Physiology*, **41**, 1030–1037.

Krämer, U., Cotter-Howells, J. D., Charnock, J. M., Baker, A. J. M. & Smith, J. A. C. (1996) Free histidine as a metal chelator in plants that accumulate nickel. *Nature*, **379**, 635–638.

Landsberg, E. C. (1994) Transfer cell formation in sugar beet roots induced by latent Fe deficiency. *Plant and Soil*, **165**, 197–205.

Ling, H.-Q., Bauer, P., Keller, B. & Ganal, M. (2002) The tomato *fer* gene encoding a bHLH protein controls iron uptake responses in roots. *Proceedings of the National Academy of Sciences of the United States of America*, **99**, 13938–13943.

Ling, H.-Q., Pich, A., Scholz, G. & Ganal, M. W. (1996) Genetic analysis of two tomato mutants affected in the regulation of iron metabolism. *Molecular and General Genetics*, **252**, 87–92.

Lombi, E., Tearall, K. L., Howarth, J. R., Zhao, F. J., Hawkesford, M. J. & Mcgrath, S. P. (2002) Influence of iron status on cadmium and zinc uptake by different ecotypes of the hyperaccumulator *Thlaspi caerulescens*. *Plant Physiology*, **128**, 1359–1367.

López-Bucio, J., Cruz-Ramírez, A. & Herrera-Estrella, L. (2003) The role of nutrient availability in regulating root architecture. *Current Opinion in Plant Biology*, **6**, 280–287.

López-Bucio, J., Hernández-Abreu, E., Sánchez-Calderón, L., Nieto-Jacobo, M. F., Simpson, J. & Herrera-Estrella, L. (2002) Phosphate availability alters architecture and causes changes in hormone sensitivity in the Arabidopsis root system. *Plant Physiology*, **129**, 244–256.

López-Bucio, J., Martínez de la Vega, O., Guevara-García, A. & Herrera-Estrella, L. (2000) Enhanced phosphorus uptake in transgenic tobacco plants that overproduce citrate. *Nature Biotechnology*, **18**, 450–453.

Luo, H. M., Watanabe, T., Shinano, T. & Tadano, T. (1999) Comparison of aluminum tolerance and phosphate absorption between rape (*Brassica napus* L.) and tomato (*Lycopersicum esculentum* Mill.) in relation to organic acid exudation. *Soil Science and Plant Nutrition*, **45**, 897–907.

Ma, J. (2000) Role of organic acids in detoxification of aluminum in higher plants. *Plant and Cell Physiology*, **41**, 383–390.

Ma, J. F., Zheng, S. J. & Matsumoto, H. (1997a) Specific secretion of citric acid induced by Al stress in *Cassia tora* L. *Plant and Cell Physiology*, **38**, 1019–1025.

Ma, Z., Bielenberg, D. G., Brown, K. M. & Lynch, J. P. (2001) Regulation of root hair density by phosphorus availability in *Arabidopsis thaliana*. *Plant Cell and Environment*, **24**, 459–467.

Ma, J. F., Hiradate, S., Nomoto, K., Iwashita, T., & Matsumoto, H. (1997b) Internal detoxification mechanism of Al in *Hydrangea*. Identification of Al form in the leaves. *Plant Physiology*, **113**, 1033–1039.

Marschner, H. (1995) *Mineral Nutrition of Higher Plants*. Academic Press, London.

Martienssen, R. & McCombie, W. R. (2001) The first plant genome. *Cell*, **105**, 571–574.

McDonald, R., Fieuw, S. & Patrick, J. W. (1996) Sugar uptake by the dermal transfer cells of developing cotyledons of *Vicia faba* L. *Planta*, **198**, 502–509.

Meagher, R. B. (2000) Phytoremediation of toxic elemental and organic pollutants. *Current Opinion in Plant Biology*, **3**, 153–162.

Meyerowitz, E. M. & Somerville, C. R. (1994) *Arabidopsis*. Cold Spring Harbor Press, New York.

Mitsukowa, N., Okamura, S., Shirano, Y., Sato, S., Kato T., Harashima, S. & Shibata, D. (1997) Over-expression of an *Arabidopsis thaliana* high-affinity phosphate transporter gene in tobacco cultured cells enhances cell growth under phosphate-limited conditions. *Proceedings of the National Academy of Sciences of the United States of America*, **94**, 7098–7102.

Mori, S. (1999) Iron acquisition by plants. *Current Opinion in Plant Biology*, **2**, 250–253.

Neumann, G. & Martinoia, E. (2002) Cluster roots – an underground adaptation for survival in extreme environments. *Trends in Plant Science*, **7**, 162–167.

Pan, A., Tie, F., Duau, Z., Yang, M., Wang, Z., Li, L., Chen, Z. & Ru, B. (1994) Alpha-domain of human metallothionein I (A) can bind to metal in transgenic tobacco plants. *Molecular and General Genetics*, **242**, 666–674.

Pich, A., Scholz, G. & Stephan, U. W. (1994) Iron-dependent changes of heavy metals, nicotianamine, and citrate in different plant organs and in the xylem exudate of two tomato genotypes. Nicotianamine as possible copper translocator. *Plant and Soil*, **165**, 189–196.

Pineros, M. A. & Kochian, L. V. (2001) A patch clamp study on the physiology of aluminum toxicity and aluminum tolerance in *Zea mays*: identification and characterization of Al^{3+}-induced anion channels. *Plant Physiology*, **125**, 292–305.

Raghothama, K. G. (1999) Phosphate acquisition. *Annual Review of Plant Physiology and Plant Molecular Biology*, **50**, 665–693.

Richards, K. D., Schott, E. J., Sharma, Y. K., Davis, K. R. & Gardner, R. C. (1998) Aluminum induces oxidative stress genes in *Arabidopsis thaliana*. *Plant Physiology*, **116**, 409–418.

Richardson, A. E., Hadobas, P. A. & Hayes, J. E. (2001) Extracellular secretion of *Aspergillus* phytase from *Arabidopsis* roots enable plants to obtain phosphorus from phytate. *Plant Journal*, **25**, 641–649.

Robinson, N. J., Procter, C. M., Connolly, E. L., & Guerinot, M. L. (1999) A ferric-chelate reductase for iron uptake from soils. *Nature*, **397**, 694–697

Ryan, P. *et al.* (2001) Function and mechanism of organic anion exudation from plant roots. *Annual Review of Plant Physiology and Plant Molecular Biology*, **52**, 527–560.

Ryan, P. R. *et al.* (1997) Aluminum activates an anion channel in the apical cells of wheat roots. *Proceedings of the National Academy of Sciences of the United States of America*, **94**, 6547–6552.

Salt, D. E. & Rauser, W. E. (1995) MgATP-dependent transport of phytochelatins across the tonoplast of oat roots. *Plant Physiology*, **107**, 1293–1301.

Sasaki, T., Yamamoto, Y., Ezaki, B., *et al.* (2004) A wheat gene encoding an aluminum-activated malate transporter. *Plant Journal*, **37**, 645–653.

Schachtman, D. P., Reid, R. J. & Ayling, S. L. (1998) Expression of the yeast FRE genes in transgenic tobacco. *Journal of Plant Physiology*, **118**, 51–58.

Schat, H., Llugany, M., Voijs, R., Hartley-Whitaker, J. & Bleeker, P. M. (2002) The role of phytochelatins in constitutive and adaptive heavy metal tolerance in hyperaccumulator and non-hyperaccumulator metallophytes. *Journal of Experimental Botany*, **53**, 2381–2392.

Schmidt, W. (1999) Mechanism and regulation of reduction-based iron uptake in plants. *New Phytologist*, **141**, 1–26.

Schmidt, W. & Schikora, A. (2001) Different pathways are involved in phosphate and iron stress-induced alterations of root epidermal cell development. *Plant Physiology*, **125**, 2078–2084.

Schmidt, W., Tittel, J. & Schikora, A. (2000) Role of hormones in the induction of iron deficiency responses in Arabidopsis roots. *Plant Physiology*, **122**, 1109–1118.

Senden, M. H. M. *et al.* (1995) Citric acid in tomato plant roots and its effect on cadmium uptake and distribution. *Plant and Soil*, **171**, 333–339.

Staunton, S. & Leprince, F. (1996) Effect of pH and some organic anions on the solubility of soil phosphate: implications for P bioavailability. *European Journal of Soil Science*, **47**, 231–239.

Takahashi, M., Yamagachi, H., Nakanishi, H., Shioiri, T., Nishizawa, N. K. & Mori, S. (1999) Cloning two genes for nicotianamine aminotransferase, a critical enzyme in iron acquisition (strategy II) in Graminaceous plants. *Plant Physiology*, **121**, 947–956.

Tesfaye, M., Temple, S. J., Allan, D. L., Vance, C. P. & Samac, D. A. (2001) Overexpression of malate dehydrogenase in transgenic alfalfa enhances organic acid synthesis and confers tolerance to aluminum. *Plant Physiology*, **127**, 1863–1874.

Thimm, O., Eissegmann, B., Kloska, S., Altmann, T. & Buckhout, T. J. (2001) Response of Arabidopsis to iron deficiency stress as revealed by microarray analysis. *Plant Physiology*, **127**, 1030–1043.

Tisdale, S. L., Nelson, W. L., Beaton, J. D. & Havlin, J. L. (1993) Soil acidity and basicity. In *Soil Fertility and Fertilizers*, 5th ed. Macmillan, New York, pp. 364–404.

Traina, S. J., Sposito, G., Bradford, G. R. & Kafkafi, U. (1987) Kinetic study of citrate effects on orthophosphate solubility in an acid montmorillonitic soil. *Soil Science Society of American Journal*, **51**,
368–375.

Uhde-Stone, C., Gilbert, G., Johnson, J., Litjens, R., Zinn, K. E., Temple, S. J., Vance, C. P. & Allan, D. L.
(2003) Acclimation of white lupin to phosphorus deficiency involves enhanced expression of genes
related to organic acid metabolism. *Plant and Soil*, **248**, 99–116.

Vert, G., Briat, J. F. & Curie, C. (2001) Arabidopsis IRT2 gene encodes a root-periphery iron transporter.
Plant Journal, **26**, 181–189.

Von-Uexkull, H. R. & Muttert, M. E. (1995) Global extent, development and economic impact of acid soils.
Plant and Soil, **171**, 1–15.

Wang, Y. H., Garvin, D. F. & Kochian L. V. (2002) Rapid induction of regulatory and transporter genes in
response to phosphorus, potassium, and iron deficiencies in tomato roots. Evidence for crosstalk in root/
rhizosphere-mediated signals. *Plant Physiology*, **130**, 1361–1370.

Watanabe, T. *et al.* (1998) Distribution and chemical speciation of aluminum in the Al accumulator plant,
Melastoma malabathricum L. *Plant and Soil*, **201**, 165–173.

Williamson, L., Ribrioux, S., Fitter, A. & Leyser, O. (2001) Phosphate availability regulates root system
architecture in Arabidopsis. *Plant Physiology*, **126**, 1–8.

Zhang, W. H., Ryan, P. R. & Tyerman, S. D. (2001) Aluminum activates malate-permeable channels in the
apical cells of wheat roots. *Plant Physiology*, **125**, 1459–1472.

Zheng, S. J., Ma, J. F. & Matsumoto, H. (1998) High aluminum resistance in buckweath. I Al-induced specific
secretion of oxalic acid from root tips. *Plant Physiology*, **117**, 745–751.

Zhou, J. & Goldsbrough, P. B. (1995) Structure, organization and expression of the metallothionein gene
family in Arabidopsis. *Molecular and General Genetics*, **248**, 318–328.

Zhu, Y. L., Pilon-Smits, E. H. A., Jouanin, I. & Terry, N. (1999) Overexpression of glutathione synthase in
Indian mustard enhances cadmium accumulation and tolerance. *Plant Physiology*, **119**, 73–79.

7   Plant response to herbicides

William E. Dyer and Stephen C. Weller

7.1   Introduction

Herbicides, in contrast to the other stresses discussed in this book, represent a deliberate stress imposed on plants by humans. Herbicides are chemicals used to manage unwanted plants in the ecosystem, plants that are commonly referred to as 'weeds'. Although there are many definitions for this term, we will use the following: 'a plant that forms populations that are able to enter habitats cultivated, markedly disturbed or occupied by man and potentially depress or displace the resident populations that are deliberately cultivated or are of ecological or aesthetic interest' (Navas, 1991). This definition recognizes the ecology and biology of weedy species, as well as the impact of weeds on human endeavors (Bridges, 1995). In general, weeds have substantial levels of genetic variability and plasticity that enable them to invade, reproduce and establish in new environments (Mack *et al.*, 2000). These same characteristics dictate that plant response to herbicides is extremely variable, while at the same time providing a genetic background in which the evolution of resistant individuals can occur at a relatively high frequency.

Where did herbicides come from and why have they become important for vegetation control? Most herbicides were discovered through greenhouse screening of mass-synthesized or off-the-shelf compounds for toxic activity against plants. This approach, while totally empirical, has been surprisingly successful in finding chemicals that affect specific biochemical processes in plants. Many selective herbicides, which kill members of certain plant families (weeds) while not injuring others (crops), have also been identified through this approach. More recently, sophisticated methods for herbicide discovery have been explored, including combinatorial chemistry and quantitative structure–activity relationship (QSAR) models that regress physicochemical parameters against herbicidal activities. Comparative molecular field analysis and comparative molecular similarity indices analysis (CoMSIA) methodologies, well-established ligand-based molecular design tools widely used by medicinal and pesticide chemists, are also being used for herbicide discovery (Bordas *et al.*, 2003). Single proton NMR profiling of metabolic products coupled with principal component analysis/cluster mapping was used to screen compounds for herbicidal activity (Hole *et al.*, 2000). This approach, which used a highly mechanized, high throughput screening (HTS) system, successfully confirmed known herbicides as well as accurately predicted the herbicidal activity of several unknown compounds. The technique was subsequently further refined and used to assign metabolite profiles to 400 plant extracts (Ott *et al.*, 2003). These approaches show significant promise, but there is no solid evidence yet that they will be more successful or efficient for new herbicide discovery than traditional screening methods.

Since their introduction, herbicides have been valuable tools to investigate plant physiology, metabolism and genetics. For example, much of our knowledge of photosystem II (PSII) is a result of studies on atrazine resistance and the investigation of the specific mechanisms involved (Naber & van Rensen, 1991). Also, detailed knowledge of herbicide enzyme targets such as 5-enolpyruvylshikimate-3-phosphate synthase (EPSPS) and acetolactate synthase (ALS) has been obtained through herbicide physiology studies. More recently, they have become useful tools as selectable markers in plant transformation and as specific metabolic probes in genomic studies and gene discovery.

Plant response to herbicides is highly variable, and depends on three major interrelated aspects of the target species and the herbicide's biochemical target. First, the 300 or so herbicides in use today represent more than 40 chemical families and target about 30 separate plant enzymes or pathways. Thus, unlike other abiotic stresses, they can impose many different physiological stresses on plants and their effects vary widely across plant families. Because of this variability, plant response (injury) to an individual herbicide can range from no response, to relatively minor short-term stresses from which it recovers, to severe lethal impacts. Second, certain weedy species that are only partially sensitive to herbicides' effects may experience acute or long-lasting stress and yet survive. Herbicides, like any other xenobiotic, trigger a multitude of plant defense responses, and plant families and individual species vary widely in their abilities to induce such pathways. And third, selective herbicides can induce stress on crop plants, as many times these species must induce catabolic systems to degrade the xenobiotic. Crops thus experience transient stress as the herbicide is metabolized or as inhibited enzymes are replaced via *de novo* synthesis.

Herbicides exert some of the strongest selection pressures ever faced by plants. They are designed to kill 100% of multiple species in a variety of ecosystems. Herbicides represent about 60% of all pesticides applied worldwide, with annual use estimated at 310 000 metric tons in Europe, North America and West Asia in 1997, the most recent figures available (FAO Statistics, 2004). Herbicides are used on millions of hectares of cropland, rangeland, forestry, turf and rights-of-way, targeting literally billions of individual plants. Many herbicides have long soil residues that extend their selection on plants and germinating seedlings throughout the growing season, and some persist for several years. Herbicides have only been in widespread use for 50 years, making them by far the evolutionarily newest form of stress experienced by plants. Nonetheless, the extremely strong selection pressures they impose have had profound effects on the plant species composition of agroecosystems. This selection has rapidly led to the development of herbicide-resistant weed biotypes, the study of which can be exploited as a paradigm of 'compressed evolution' of plant response to strong stress agents. Thus, in addition to their lethal and sublethal effects on target plants, herbicides simultaneously exert extremely strong selection for resistant biotypes.

Plants can become resistant to herbicides through three mechanisms: an altered target site, metabolic detoxification, or physiological avoidance such as sequestration or reduced translocation (reviewed by Powles & Holtum, 1994). Of these, the most common is selection for an altered target site (Sherman et al., 1996). Genetic variation and background mutations give rise to preexisting, herbicide-resistant alleles that can be selected in most plant populations (Gressel, 2002). Metabolic detoxification involves inducing (usually) or selecting for high constitutive levels of enzymes that can rapidly degrade the herbicide before irreversible damage occurs. Sequestering or otherwise localizing the herbicide to metabolically inactive sites is the least common method of achieving resistance. However, recent characterizations of glyphosate-resistant weed biotypes suggest that a combination of several mechanisms including decreased translocation may be responsible. Investigations of these resistance mechanisms and their variability among plant species can provide valuable insights into the many ways that plants respond to herbicide stress.

Much has been written about the relative fitness of herbicide-resistant and susceptible biotypes of weedy species (reviewed in Warwick & Black, 1994). The comparison is important because of its impact on our ability to predict the persistence and spread of resistant biotypes within and among agroecosystems. The question is also of interest because it has probative value in investigating the association between the selection of mutants and their performance ('Haldanian fitness') in the environment (reviewed in Bergelson & Purrington, 1996; Bergelson et al., 1996). Historically, the first cases of evolved resistance were to atrazine, a photosynthetic inhibitor (Ryan, 1970). These and other biotypes with resistance at the site of herbicide action suffered substantial fitness costs, because the mutation conferring resistance also reduced photosynthetic efficiency (Conard & Radosevich, 1979). However, in almost all cases involving other herbicide families, fitness differences between resistant and susceptible biotypes are small, variable and highly dependent on environmental conditions. Differences are often not detectable under field conditions. Obviously, resistant biotypes have an absolute fitness advantage in the presence of the herbicide. Similarly, it is clear that there must be some inherent fitness cost associated with the resistance mechanism, or the resistant type would be the wildtype (Gressel, 2002). It seems that there are few generalities that can be made about the degree of fitness costs associated with each resistant species/herbicide pair. However, study of these biotypes can often provide valuable insight into the pleiotropic and often unpredictable effects that accompany plant response to strong stressors.

Herbicides are usually not 100% effective due to application errors and environmental constraints, so all vegetation is not killed. Reduced rates and improper timing of herbicide applications can also cause plant stress without lethality. To further complicate their overall effects, herbicides are often applied in mixtures of two or more compounds that have different molecular targets and are toxic to different or overlapping spectra of plant species.

Our discussion of herbicide-induced stress will be organized along five general categories of herbicide effects:

(1) photosynthetic inhibitors,
(2) biosynthetic inhibitors,
(3) hormone disruptors,
(4) inducers of herbicide metabolism and
(5) genome effects.

Table 7.1 shows the biochemical targets and pathways of selected herbicides from the first three categories. In addition to the herbicides listed here, there are several with unknown biochemical targets.

7.2 Photosynthetic inhibitors

Two herbicide groups inhibit photosynthesis. The first group consists of triazines, ureas and uracils, which are soil-applied compounds widely used for about 30 years in a number of cropping systems and for industrial weed control. Other members of this group include foliar-applied herbicides such as bentazon, bromoxynil, desmedipham, phenmedipham and propanil. The triazine herbicide atrazine was the most widely used herbicide in the USA until about 1993, when its uses were significantly restricted due to environmental and health concerns (Loosli, 1995). Atrazine controls a wide range of weedy species in corn production, and its residual properties can provide season-long weed control. These herbicides target the D1 protein in PSII of the photosynthetic electron transport pathway. The second group consists of bipyrilidium herbicides, of which paraquat is the best known. Paraquat is still one of the most widely used herbicides in the world, although its use is highly restricted in the USA and other countries because of severe mammalian toxicity (Bromilow, 2004). Paraquat is applied to foliage and causes very rapid wilting and desiccation; thus it is used as a desiccant and defoliant in addition to its weed-management roles. The precise molecular target for this family is unknown, but they bind very near to ferredoxin in the photosystem I pathway.

The most logical result of photosynthetic inhibition is that affected plants would starve due to the loss of fixed carbon. While this effect would eventually

Table 7.1   Categories, biochemical targets and affected pathways of selected herbicides

Category	Enzyme or target	Pathway	Selected chemical families or herbicides
Photosynthetic inhibitors	D1 (*psbA*)	Photosystem II	Triazines, ureas, uracils, bentazon, bromoxynil and propanil
	Ferredoxin (near)	Photosystem I	Bipyridiliums
Biosynthetic inhibitors	Acetolactate synthase	Branched-chain amino acids	Sulfonylureas, imidazolinones, triazolopyrimidines, pyrimidinyloxybenzoates and sulfonylanilides
	Acetyl-CoA carboxylase	Fatty acids	Aryloxyphenoxypropionates and cyclohexanediones
	Alpha tubulin	Microtubule assembly	Dinitroananilines, pyridines, amides and DCPA
	Cellulose synthase	Cellulose biosynthesis	Nitriles and benzamide
	Dihydropteroate synthase	Folic acid	Asulam
	EPSPS	Aromatic amino acids	Glyphosate
	Long-chain fatty acid: CoA ligase	Fatty acid elongation	Carbamothioates and chloroacteamides
	Glutamine synthetase	Ammonia metabolism	Glufosinate
	4-Hydroxyphenylpyruvate dioxygenase	Quinones	Triketones
	Lycopene cyclase	Carotenoids	Clomazone
	Phytoene desaturase	Carotenoids	Pyridazinones
	Protoporphyrinogen oxidase	Chlorophyll and heme	Diphenylethers, oxidiazoles, N-phenylheterocycles and N-phenylphtalimide
Hormone disruptors	Auxin receptor(s)	Auxin action	Phenoxies, benzoics and picolinic acids
	ent-Kaurene synthase	Gibberellic acid	Paclobutrazole

happen, more rapid toxicity results from blocking electron transfer, preventing the conversion of light energy into electrochemical energy and resulting in the production of high-energy reactive oxygen species (ROS) such as superoxide and singlet oxygen. Singlet oxygen and other ROS are highly toxic compounds that cause acute membrane damage (Blokhina *et al.*, 2003). Although these molecules have very short half-lives and can therefore travel only submicron distances before decaying, they can nonetheless accumulate and cause significant local damage. If unchecked, ROS can be converted by superoxide dismutase (SOD) into hydrogen peroxide, which diffuses over much greater distances and causes further oxidative damage. Thus, the toxicity of photosynthetic inhibitor herbicides is mostly due to the lethal photooxidative stress that they impose, and much less to their interruption of carbon fixation.

Although ROS are highly toxic compounds, they are also known to fulfill important positive roles in physiological functions. For example, ROS are produced in response to a number of different physiological stimuli and they likely function as second messengers to directly or indirectly modulate many proteins (Mori & Schroeder, 2004). Thus, sublethal doses of photosynthetic inhibitor herbicides may generate low levels of ROS, which in turn could cause a multitude of downstream effects not related to toxicity. Such effects have not been investigated to our knowledge.

Atrazine inhibition of photosynthesis could be largely overcome by the concomitant feeding of sucrose to Arabidopsis seedlings (Sulmon *et al.*, 2004). The reversal was not related to a compensation of carbon lost through photosynthetic inhibition, but was more due to an induction of *psbA* mRNA and D1 protein levels. Overexpression of the D1 protein would effectively titrate the number of inhibited target sites, and this mechanism of achieving resistance has been reported for other herbicides. The protective effects of added sugars on photosynthetic machinery could also be related to the general protective effects of compatible solutes, such as the elimination of free radicals and stabilization of protein structure, as shown for drought and salt stresses (Garg *et al.*, 2002).

7.2.1 Resistance

Biotypes of over 60 weed species have developed resistance to PSII inhibitors (termed 'triazine-resistant' here unless noted otherwise) and over 25 biotypes are resistant to bipyridiliums. Resistance to PSII inhibitors is primarily due to a serine$_{264}$ to glycine substitution in the D1 binding site that blocks herbicide binding. In most resistant biotypes studied, resistance is associated with the reduced photosynthesis efficiency, making these biotypes less fit. The resistant D1 protein either transfers electrons more slowly to Q_B, the subsequent electron acceptor (Naber & van Rensen, 1991) or it is more sensitive to photoinhibition and thus has a greater turnover rate (Niyogi, 1999). Urea herbicides usually retain activity on triazine-resistant biotypes since they bind in a slightly different region of the D1 protein. Two weedy species, *Abutilon theophrasti* (velvetleaf) and *Lolium rigidum* (rigid ryegrass) exhibit resistance due to increased herbicide metabolism (Anderson & Gronwald, 1991; Burnet *et al.*, 1993). Resistance to urea herbicides has occurred in a few species and is due to a serine$_{264}$ to threonine substitution in the D1 protein (Masabni & Zandstra, 1999). Most cases of resistance have resulted from repeated herbicide applications over many years and the long soil residual period of most of these herbicides. Strategies to overcome resistance consist of alternating herbicides with different mechanisms of action or using mixtures of different herbicides.

There are two competing theories for the mechanism of resistance to bipyridilium herbicides. The first is that herbicide molecules are sequestered in cell walls before they can exert toxic effects (Fuerst & Vaughn, 1990). Alternatively, enhanced detoxification of ROS by elevated levels of SOD, ascorbate, peroxidase or glutathione reductase may confer resistance (Shaaltiel & Gressel, 1986). In either case, the evolution of resistance was unanticipated by some, since these herbicides have no soil activity and their selection intensity should therefore be reduced. However, producers compensated for residual activity by making repeated applications in orchards and nurseries, effectively creating intense selection pressure and essentially forcing the evolution of resistant biotypes (Turcsanyi et al., 1994).

7.3 Biosynthetic inhibitors

7.3.1 Branched-chain amino acid synthesis inhibitors

Table 7.1 lists the sulfonylurea, imidazolinone, triazolopyrimidine sulfonanilide and pyrimidyloxy benzoate chemical families that inhibit ALS, also known as acetohydroxyacid synthase (EC 4.1.3.18) (LaRossa & Schloss, 1984). This is the first enzyme in the biosynthetic pathway for the branched-chain amino acids leucine, isoleucine and valine. ALS catalyzes the condensation of either two molecules of pyruvate to form acetolactate in the leucine and valine pathway, or one molecule of pyruvate with one molecule of α-ketobutyrate to form 2-aceto-2-hydroxybutyrate as the first step in isoleucine biosynthesis. Much of our detailed knowledge about this enzyme comes from studies of these herbicides and their interactions with ALS (Duggleby & Pang, 2000).

Inhibition of ALS leads to the depletion of the branched-chain amino acids, and plant death is thought to result from eventual protein starvation (Singh, 1999). Although this seems to be a fairly straightforward mechanism of toxicity, there are other, more subtle physiological effects associated with these herbicides. First, ALS inhibition led to the accumulation of the α-ketoacids 2-oxobutyrate (2-oxobutanoate) and its transamination product, 2-aminobutyric acid (α-amino-n-butyrate) in Lemna minor (common duckweed) plants (Rhodes et al., 1987). Although these compounds were highly toxic in ALS-inhibited Salmonella typhimurium (LaRossa et al., 1987), their contribution to phytotoxicity in higher plants is controversial. Shaner and Singh (1993) showed that neither compound accumulated in, nor were toxic to, maize plants. Further, antisense suppression of ALS activity in transgenic potato plants did not lead to an accumulation of 2-oxobutyrate (Hofgen et al., 1995). These apparently conflicting results indicate that the precise consequences of ALS inhibition may vary across plant species or families, and illustrate that our knowledge about the complex regulation of this pathway in plants is far from complete.

In addition to its normal condensation reactions, ALS can also catalyze an unusual oxygen-consuming reaction that results in enzymatic self-destruction. At low pyruvate concentrations, the enzyme converts pyruvate and oxygen to carbon dioxide and peracetic acid (Schloss et al., 1996), which leads to its rapid oxidative inactivation. Because ALS inhibitors bind the oxygen-sensitive form of the enzyme during catalysis, the rate of oxygen-dependent inactivation is enhanced. At higher pyruvate concentrations, peracetic acid reacts with pyruvate to form two molecules of acetate, effectively removing the acidic compound. This particular form of oxygen toxicity may be especially important under environmental or other stress conditions in which pyruvate concentrations are low. Another oxygen-related effect was recently reported in pea plants (Gaston et al., 2002). ALS inhibition by imazethapyr led to the induction of aerobic fermentation enzymes in response to a presumed accumulation of toxic acetaldehyde, ethanol and lactate products in roots. The change to anaerobic conditions may lead to the acidification of cytoplasm, as was shown in Escherichia coli where partial ALS inhibition caused intracellular acidification and induction of the σ^S-dependent stress response (Van Dyk et al., 1998). In pea roots, anaerobiosis was shown to reduce carbohydrate metabolism (Gaston et al., 2002), a result reminiscent of the plant stress response to hypoxia or other environmental stresses (Tagede et al., 1999). The similarity of these responses to seemingly disparate sources illustrates the crosstalk common among signal transduction pathways, and exemplifies the power of herbicides as specific tools to dissect such linkages.

Even though many ALS-inhibitor herbicides are applied at extremely low use rates (< 20 g/ha), some crop species like sugar beet and canola are extremely sensitive to their effects (Mangels, 1991). These characteristics, combined with the long soil half-lives of compounds like chlorsulfuron and triasulfuron (>120 days) lead to situations in which plants may be subjected to sublethal herbicide concentrations for extended periods of time. This situation has been addressed in regard to herbicide drift scenarios (Eberlein & Guttieri, 1994; Felsot et al., 1996) and the susceptibility of crops planted into soil containing herbicide residues (e.g. Wall, 1997). A small number of other studies investigated the short-term effects of low doses of ALS inhibitors. For example, pea plants maintained on sublethal doses of imazethapyr showed growth reductions and alterations in free amino acid pools, with reductions in levels of valine, but not the other branched-chain amino acids (Royuela et al., 2000). Low soil concentrations of chlorsulfuron inhibited root growth of peas and beans, likely by damaging root caps (Fayez & Kristen, 1996). Nitrogen metabolism was affected in soybean plants treated with low concentrations of metsulfuron-methyl, as indicated by an accumulation of asparagine (Jia et al., 2001).

In spite of these reports, the physiological and agronomic consequences of sublethal, long-term exposure to ALS-inhibitor herbicides have not been adequately addressed. Interestingly, one of the first documented effects of ALS inhibitors was a rapid cessation of cell division, with cells arresting in the G1 and G2 phases (Rost, 1984). This effect is due to the apparently absolute requirement for the branched-chain amino acids, particularly leucine, during cell division (Doering & Danner, 2000). Would reduced rates of cell division occur at sublethal doses, and if so, do they have measurable consequences? It seems intuitive that low levels of stress would affect crop competitiveness with weeds, particularly in fields supporting populations of herbicide-resistant weeds. This effect, although probably quite subtle, has not been quantified. Similarly, do such stresses cause detectable or commercially important reductions in the quality of harvested crops? Potatoes treated with simulated drift concentrations of imazamethabenz, imazethapyr or imazapyr were most sensitive at tuber initiation and tuber bulking stages, and tuber quality was affected more than tuber biomass accumulation (Eberlein & Guttieri, 1994). The doses used in this study were relatively high, but provide indications that subtle changes in crop quality may be overlooked if overall yield is not affected by very low doses. ALS inhibitors quickly inhibit DNA synthesis (Ray, 1982) by indirectly repressing genes encoding purine biosynthetic enzymes (Jia et al., 2000). Limiting levels of adenine and guanine could lead to base substitutions, or mutations, in replicating or newly synthesized DNA. To our knowledge, this effect has not been investigated.

The ALS inhibitor sulfometuron methyl was used as a specific metabolic probe to explore global changes in yeast gene expression (Jia et al., 2000). DNA microarrays showed that, of 1529 yeast genes analyzed, about 8% were induced greater than twofold within 15 min after a treatment causing 40% growth inhibition. Other short-term responses included a Gcn4p-mediated induction of genes involved in amino acid and cofactor biosynthesis, and the starvation response. Expression of genes for DNA repair enzymes was induced, which the authors propose was a result of α-ketoacid accumulation. Many genes involved in carbohydrate metabolism and nucleotide biosynthesis were significantly repressed. Extended exposure to the sublethal herbicide dose led to a relaxation of the initial response and the induction of sugar transporter and ergosterol biosynthetic genes. The results confirm that sublethal doses of a herbicide can lead to a multitude of changes in gene expression, metabolism and physiology, and underscore the utility of herbicides as metabolic probes.

7.3.1.1 Resistance

The first sulfonylurea and imidazolinone herbicides were marketed for cereal crops in the mid-1980s. These compounds quickly achieved a substantial market share because they were highly efficacious, controlled a broad spectrum of problem weeds, and had excellent crop safety. Many of the early commercialized herbicides have exceedingly long persistence in soil, a positive characteristic that improves their residual weed control, and in many cases extends weed suppression into the subsequent cropping seasons. However, the invincibility of the ALS inhibitors quickly faded, as weeds evolved resistance to these compounds faster than to any other family of herbicides (within 3 years of first use). There are several reasons for this rapid evolution, only one of which is the extremely strong selection pressure exerted by their exceptional efficacy and persistence. In contrast to most target-site inhibitors, these herbicides interact with the ALS enzyme at or near the allosteric regulatory domain (Schloss et al., 1988), a region that is typically much less conserved than the active site. For this reason, there was already a high incidence of semidominant mutations conferring resistance (estimates range from 1 per 10^5 to 10^6 plants, Gressel, 2002) before these herbicides were ever used. Further, there appears to be an extremely low fitness cost associated with these alleles (Bergelson et al., 1996), so that resistant weeds persist in fields for many years in the absence of ALS inhibitor use. In fact, some resistant biotypes of *Kochia scoparia* (kochia) may actually have a fitness advantage in most ecosystems, because they germinate faster at low temperatures than the wildtype (Dyer et al., 1993). Resistant biotypes in most cases contain amino acid substitutions in one of several conserved domains of ALS (reviewed in Saari et al., 1994; Guttieri et al., 1996), although some biotypes have enhanced rates of herbicide detoxification (Veldhuis et al., 2000). More than 50 ALS-inhibiting herbicides have been commercialized for use in many different cropping systems in recent years. The result has been an explosion of resistant weed populations, with more than 85 unique biotypes reported worldwide (Heap, 2004).

The phenomenon of ALS-inhibitor resistance has generated widespread interest in landscape- and population-scale responses to strong selection pressures. In addition, this type of resistance has necessitated significant changes in herbicide use patterns and agronomic weed management practices (Dyer, 1997). The labels of many ALS inhibitors have been changed to regulate how they are used and the amounts that can be applied in any one season. More than any other herbicide family, the ALS inhibitors have focused attention on the prevention and management of resistant weed biotypes. In particular, current weed management programs now take into account the cropping system, target weed species and characteristics of the relevant herbicides. The practice of rotating or mixing herbicides from different chemical families is a central tenet of such programs, in order to interrupt and reduce selection pressure for resistant biotypes. Ideally, modern weed management embraces an integrated approach in which herbicides are only one of many tools used.

7.3.2   Aromatic amino acid synthesis inhibitors

Probably the best-known inhibitor of amino acid biosynthesis is glyphosate, which inhibits the penultimate enzyme in the common shikimate pathway, EPSPS (EC 2.5.1.19). This pathway is responsible for the biosynthesis of the aromatic amino acids phenylalanine, tryptophan and tyrosine, critically important components of proteins, phenylpropanoids, phytoalexins, lignin and all other aromatic compounds. Glyphosate is considered by most as a nonselective herbicide, which indicates that it is lethal to all plant species (although see discussion later on resistant weeds). The herbicide is translocated very efficiently and accumulates in meristems, making it an excellent choice for perennial weed control. Glyphosate only has activity on foliage and green bark that it contacts, and since it is rapidly and tightly bound to soil colloids, it has no residual herbicidal activity in soil.

The shikimate pathway is unique to plants and microorganisms, so glyphosate has generally favorable toxicological properties for nontarget organisms. Such characteristics have led to several novel uses of this herbicide. Glyphosate resistance, conferred by an insensitive EPSPS or by glyphosate oxidoreductase (*gox*), has been used as a selectable marker in corn transformation protocols (Howe *et al.*, 2002). Somatic embryo production from sweetpotato leaf explants was enhanced by sublethal doses of glyphosate (Ramanjini-Gowda & Prakash, 1998), probably by inhibiting growth and promoting desiccation, a necessary step in embryo quiescence. Roberts *et al.* (2002) proposed that glyphosate might have utility as a broad-spectrum antimicrobial agent. In particular, the apicomplexan parasites *Toxoplasma gondii*, *Plasmodium falciparum* and *Cryptosporidium parvum* represent attractive targets, as they are the causative agents of the human diseases toxoplasmosis, malaria and cryptosporidiosis, respectively.

Like ALS inhibitors, the ultimate cause of plant death from glyphosate treatment is thought to be due to the depletion of amino acids, a resulting loss of protein biosynthesis (Franz *et al.*, 2000), and a global carbon drain due to deregulation of the shikimate pathway. In the presence of glyphosate, the activity of 3-deoxy-d-arabino-heptulosonate-7-phosphate synthase was induced, possibly due to decreased regulation by arogenate (Pinto *et al.*, 1988). Even without glyphosate, 20% of photosynthetically fixed carbon passes through the shikimate pathway, underscoring its fundamental importance in plant metabolism. In the presence of glyphosate, the loss of numerous compounds derived from aromatic amino acids can lead to other, sometimes unpredictable, consequences. For example, a number of compounds involved in inter- and intraspecies interactions are derived from the shikimate pathway, including phytoalexins, phenylpropanoids and other signaling compounds. Of particular interest is the effect of glyphosate on plant defense responses against pathogens (Levesque & Rahe, 1992). In the late 1990s, several outbreaks of sudden death syndrome (caused by *Fusarium solani*) and other diseases of soybeans coincided with the widespread adoption of glyphosate-resistant soybean cultivars, leading to speculation that the two events could be linked. A number of field and greenhouse studies followed (Harikrishnan & Yang, 2002; Nelson *et al.*, 2002; Njiti *et al.*, 2003), with most showing that susceptibility to pathogens was more related to the cultivar's host-resistance genotype than to glyphosate resistance. However, a few reports show that perturbation of the shikimate pathway with glyphosate can lead to increase in susceptibility. A glyphosate-resistant cotton cultivar was more sensitive to *Meloidogyne incognita* (root-knot nematode) than its nontransgenic parent cultivar (Colyer *et al.*, 2000). Sublethal glyphosate treatment of susceptible soybean plants suppressed production of glyceollin, a phytoalexin important in the resistance response to fungal infection (Keen *et al.*, 1982). Pretreatment of genetically resistant tomato plants with a sublethal dose of glyphosate caused them to become susceptible to *Fusarium oxysporum* colonization, most likely by interfering with lignification at penetration sites (Brammall & Higgins, 1988).

In a few cases, glyphosate has been used as a plant protective agent, by interfering with pathogen growth or virulence. Sharma *et al.* (1989) showed that glyphosate treatment of wheat straw infested with the root pathogen *Pyrenophora tritici-repentis* significantly reduced ascocarp production when the straw was incubated under simulated fallow conditions. Biological control agents for weeds have also been augmented by glyphosate treatment. Phytoalexin biosynthesis in the weedy species *Cassia obtusifolia* (sicklepod) was suppressed by glyphosate, making the weed more susceptible to infection by the mycoherbicide *Alternaria cassiae* (Sharon *et al.*, 1992). In a contrasting scenario, Kawate *et al.* (1997) examined the effect of glyphosate removal of weeds in fallow that are alternate hosts of crop pathogens. Soil populations of *F. solani* and *Pythium ultimum*, pathogens of field peas, were increased when henbit and downy bromegrass were killed with glyphosate in the preceding fallow season.

Tannins, polyphenolic compounds important for plant protection, are synthesized from gallic acid. However, the biosynthetic pathway for gallic acid was unknown until Ossipov *et al.* (2003) used glyphosate as a metabolic inhibitor in birch leaves to discover the new enzyme dehydroshikimate dehydrogenase, which catalyzes the direct conversion of 3-dehydroshikimic acid into gallic acid. In another example in which a herbicide was used as a metabolic probe, glyphosate treatment allowed the successful elucidation of the initial steps in the incompatible bean anthracnose host–parasite interaction (Johal & Rahe, 1988). Glyphosate pretreatment did not diminish the hypersensitive response when bean seedlings were exposed to *Colletotrichum lindemuthianum*. However, fungal hyphae remained viable and were able to outgrow hypersensitive cells in some cases. Further tests showed that phytoalexin accumulation was inhibited, leading to the conclusion that these compounds are a critical determinant of resistance in the bean–anthracnose system.

The commercialization of glyphosate-resistant soybean cultivars has led to concerns that glyphosate treatments may affect the health or population structure of rhizosphere microbial communities. In particular, the possible sensitivity of rhizobacteria may interfere with nodulation or nitrogen fixation. Initial studies showed that 11 *Rhizobium* species and strains, as well as *Agrobacterium rhizogenes* and *Agrobacterium tumefaciens*, possessed glyphosate degradation ability (Liu *et al.*, 1991). In contrast, Moorman *et al.* (1992) reported that USDA strains 110, 123 and 138 of *Bradyrhizobium japonicum* were sensitive to glyphosate and that protocatechuic acid, a shikimate pathway intermediate, accumulated in two of the strains. Later studies of soybeans nodulated with glyphosate-sensitive, intermediate and tolerant *B. japonicum* strains showed that nitrogenase activity was not inhibited, even though substantial concentrations of protocatechuic acid and other pathway intermediates accumulated (Hernandez *et al.*, 1999). Glyphosate inhibited growth of rhizobacteria that nodulate red clover, lucerne and birdsfoot trefoil (Mårtensson, 1992). In a comprehensive study, Reddy *et al.* (2000) did not detect significant effects of glyphosate treatment on nodulation, chlorophyll content or biomass accumulation of glyphosate-resistant soybean cultivars. Overall, it seems that various strains of rhizobacteria possess differential tolerance to glyphosate, and that glyphosate treatment of the legume host may cause a transient inhibition. Although these stresses are usually not reflected in nitrogen-fixing efficiency or crop yields, glyphosate could be used as a specific metabolic inhibitor to further dissect the signaling pathways and metabolite utilization patterns between host and symbiont.

A related possibility is that endophytic and rhizosphere microbial communities of glyphosate-resistant crop cultivars may have different population structures due to glyphosate treatment. In this regard, Biolog[TM] analysis was used to compare such communities associated with two conventional and one glyphosate-resistant canola cultivar under field conditions (Siciliano *et al.*, 1998). Microbial communities associated with roots of 'Quest', the glyphosate-resistant cultivar, had different Biolog[TM] and fatty acid methyl ester (FAME) profiles than the conventional cultivars. Glyphosate inhibited the growth of the vesicular–arbuscular mycorrhizal fungus *Glomus intraradices* (Wan *et al.*, 1998). The long-term ecological implications of these effects on soil microflora have not been investigated to our knowledge.

Sublethal exposure to glyphosate can affect overall plant architecture. Efficient translocation and a lack of metabolism in most plant species (Franz *et al.*, 2000) allows glyphosate to accumulate in and inhibit apical meristems. This inhibition encourages sprouting of axillary buds (Lee, 1984), which results in a 'witch's broom' appearance of affected plants. Glyphosate treatment of corn and cotton seedlings inhibited polar auxin transport (Baur, 1979) and was associated with enhanced rates of auxin metabolism in soybean and pea seed-lings (Lee, 1984). Although these alterations are generally associated with negative agronomic and economic effects of glyphosate drift on crop plants (e.g. Gilreath *et al.*, 2001), sublethal doses could also be used to alter landscap-ing or other horticultural plants to create aesthetically interesting growth habits.

The widespread use of glyphosate on resistant crop cultivars may affect the composition and structure of ecosystems that surround cropped fields. To test this idea, simulated glyphosate drift treatments were applied to 11 crop, weed and native species just before seed maturity to determine the effects on resulting seed and seedling vigor (Blackburn & Boutin, 2003). As might be expected for many herbicides, significant reductions in germination or seed-ling growth were observed for seven species, indicating that sublethal glypho-sate drift may have an effect on plant species composition and abundance.

7.3.2.1 Resistance

Weeds: Glyphosate was used for over 20 years without any indication of evolved resistance. This fact, in conjunction with the commercial difficulties in obtaining an enzymatically normal EPSPS for transgenic crop development, led some to claim that weed resistance would be very unlikely to develop in the field (Bradshaw *et al.*, 1997). However, the fact that glyphosate catabolic activities vary widely among plant species, coupled with the substantial selection pressure exerted by widespread transgenic crop use, provided strong indications that resistance would evolve (reviewed in Dyer, 1994). In 1996, a glyphosate-resistant rigid ryegrass was reported in Australia in an area where glyphosate had been applied continuously for 15 years (Sindel, 1996). Other reports include resistant rigid ryegrass in California (Simarmata *et al.*, 2001), *Elusine indica* (goosegrass) in Malaysia (Baerson *et al.*, 2002) and *Conyza canadensis* (horseweed) initially reported in Delaware (VanGessel, 2001) but now found throughout the Eastern and Midwestern USA. Of all these reports, only the Malaysian goosegrass has been shown to have an altered EPSPS, with a mutation of proline$_{106}$ to serine or threonine, resulting in a moderately resistant enzyme (Baerson *et al.*, 2002). While the mechanism of resistance in other biotypes has not been confirmed, there appear to be reductions in glyphosate translocation (Feng *et al.*, 1999; Lorraine-Colwill *et al.*, 1999). Westwood and Weller (1997) showed that different biotypes of field bindweed had variable tolerance to glyphosate due to multiple mechanisms including altered translocation, absorption, EPSPS induction and metabolism. The most tolerant biotype had an increase in shikimate pathway enzyme activity, and enhanced induction of EPSPS was observed after glyphosate exposure. Selec-tion for glyphosate resistance in tobacco cell cultures resulted in EPSPS gene amplification, which remained stable in the absence of herbicide (Jones *et al.*, 1996).

The widespread adoption of glyphosate-resistant crops will result in multiple applications of the herbicide during the cropping season and in consecutive years. Considering the naturally occurring inter- and intraspecific differential tolerance to glyphosate and the heavy reliance on this herbicide, we would expect a shift toward 'naturally' tolerant weed species. The exact mechanisms of this tolerance are not clear, although *Dicliptera chinensis*, a naturally tolerant grass, was shown to have multiple forms of EPSPS, as well as higher constitutive and inducible levels of EPSPS gene expression than related sensitive grass species (Yuan *et al.*, 2002). Similarly, cultured maize cell lines were reported to have two isozymes of EPSPS, only one of which was resistant to glyphosate (Forlani *et al.*, 1992). Regardless of the mechanisms involved, there is a need to provide farmers with the tools to avoid widespread resistance to glyphosate. Neve *et al.* (2003) developed a simulation model for the evolution of resistant rigid ryegrass biotypes under different cropping situations. In this model, the risks of evolving glyphosate resistance could be reduced by rotating between no-tillage and minimum tillage crop establishment systems or by rotating between glyphosate and paraquat for preseeding weed control. The strategy of using a 'double knockdown' strategy of sequential full-rate applications of glyphosate and paraquat reduced the risks of glyphosate and paraquat resistance development to < 2%. However, including glyphosate-resistant oilseed rape in the crop rotation significantly increased the predicted risk of resistance development in no-tillage systems, even when the double knockdown technique was used. Other practices such as high crop seeding rates and reducing weed seed production reduced resistance development. These types of simulation models are useful in combating the development and spread of resistant biotypes, especially in glyphosate-resistant cropping systems.

Crops: Resistance to glyphosate has been engineered into agronomic crops and is currently commercially available in soybean, corn, cotton and canola. Commercial release of Roundup Ready™ soybeans occurred in 1996 and Roundup Ready™ corn and cotton were released in 1998. Resistant cultivars of wheat, sugar beet, lettuce, turfgrass and several other species are in development. The most common method of conferring resistance has been to insert a resistant EPSPS gene from bacteria into crop plants (reviewed in Saroha *et al.*, 1998). The most effective gene used so far is from the CP4 strain of *Agrobacterium*, which exhibits a high level of glyphosate resistance and has similar kinetic efficiency to the native plant EPSPS. In addition to an insensitive EPSPS, most cultivars also contain the *gox* transgene encoding a glyphosate amine oxidase that metabolizes glyphosate to aminomethyl phosphonate and glyoxylate (Zhou *et al.*, 1995). Recently, Castle *et al.* (2004) isolated and characterized *N*-acetyltransferase (*gat*) enzymes from several microorganisms that can metabolize glyphosate. Although the native enzyme activity was insufficient to confer glyphosate tolerance to transgenic organisms, the researchers used DNA shuffling to improve enzyme efficiency by four orders of magnitude, and demonstrated glyphosate resistance in transgenic *E. coli*, Arabidopsis, tobacco and maize. These results indicate that *gat* may provide an additional strategy for obtaining glyphosate-resistant crop plants.

7.3.3   Fatty acid synthesis and elongation inhibitors

Two herbicide families with very different chemical structures, the aryloxy-phenoxy propionates and the cyclohexanediones, inhibit acetyl CoA-carboxylase (ACCase; EC 6.3.4.14), the first enzyme in fatty acid biosynthesis. ACCase catalyzes the ATP-dependent carboxylation of acetyl-CoA to form malonyl-CoA, which is the fatty acid moiety for fatty acid elongation and synthesis of secondary metabolites (Harwood, 1988a; Sasaki & Nagano, 2004). The enzyme is composed of two monomers with different functions: a biotin carboxylase complex and a carboxyltransferase complex. Higher plants have two forms of ACCase: a eukaryotic, cytosolic form that is mostly resistant to herbicides and a prokaryotic, plastidic form that is susceptible to herbicides (Konishi & Sasaki, 1994). The plastidic enzyme is further subdivided into two isoforms: a dimeric form found in most monocots and all dicots, and an unusual multidomain form found only in graminaceous monocots (Konishi *et al.*, 1996). This last form of the enzyme is the only one completely inhibited by the ACCase-inhibiting herbicides, which allows them to be used for specific control of weedy members of the Poacea in dicot crops.

A large part of our knowledge about plant ACCases derives from studies of this family of herbicides, their mechanism of toxicity and the mechanisms of evolved resistance (Harwood, 1996). For example, investigations on the family specificity mentioned earlier have provided detailed information on the different ACCase isoforms and their relative contributions to lipid metabolism (Harwood, 1988b; Gronwald, 1994; Price *et al.*, 2003). Herbicide specificity between the isoforms indicates that the plastidic form of the enzyme provides most of the lipids for membranes and cellular metabolism in the Poacea, and probably in other plant families. Early biochemical investigations on the mechanism of action for ACCase-inhibiting herbicides showed that they interacted with the carboxyltransferase domain of the enzyme (Rendina *et al.*, 1988). Subsequently, Nikolskaya *et al.* (1999) determined that herbicides interact with a 400 amino acid domain of the polypeptide. A comparison of ACCase protein sequences from naturally herbicide-sensitive and herbicide-resistant plant species showed that a leucine at position 1769 (numbering for wheat plastid ACCase ORF [AF029895]) conferred resistance while an iso-leucine conferred sensitivity (Zagnitko *et al.*, 2001). This amino acid substitution has been confirmed for resistant wild oat (Christoffers *et al.*, 2002) and green foxtail (Delye *et al.*, 2002), although there is some question of whether this substitution confers resistance to both aryloxyphenoxy propionate and cyclohexanedione herbicides (Delye *et al.*, 2003). Resistant and susceptible forms of ACCase have been used to screen analogs of known herbicides and novel compounds (Shukla *et al.*, 2004).

Metabolic engineering of fatty acid biosynthesis in plants has been of interest for some time (e.g. Roesler *et al.*, 1997; Sellwood *et al.*, 2000). Regulation of this pathway is complex, and ACCase-inhibitor herbicides have been used successfully to alter carbon flow. The slight differences in susceptibility of the cytosolic and plastidic ACCases to D-(+)-quizalofop-ethyl, an aryloxyphenoxy propionate herbicide, were exploited to enhance production of a secondary metabolite (Suzuki *et al.*, 2002). Transgenic tobacco plants expressing the biodegradable polymer poly-3-hydroxybutyrate in their plastids were treated with herbicide in order to partially inhibit the cytosolic enzyme and thus shunt more acetyl-CoA into the plastidial pathway. Production was enhanced about twofold, and the plants did not suffer the growth suppression common to transgenic Arabidopsis and other species engineered to overexpress this pathway. Further investigation of herbicide structure/function relationships and specific amino acid substitutions that confer differential resistance may provide additional strategies to manipulate carbon flow through this important pathway.

The carbamothioate and chloracetamide herbicides inhibit the synthesis of very long-chain fatty acids (VLCFAs) (Takahashi *et al.*, 2001). These soil-applied herbicides inhibit early seedling development, primarily the emergence and elongation of primary shoots. Recent work shows that these compounds inhibit several different elongases, each with different functions within the plant (Trenkamp *et al.*, 2004). Long-chain fatty acid:CoA ligase (EC 6.2.1.3) is inhibited by EPTC and triallate, herbicides in the carbamothioate family (Abulnaja & Harwood, 1991). All isozymes of this family convert free long-chain fatty acids into fatty acyl-CoA esters, and thereby play a key role in lipid biosynthesis and fatty acid degradation. In plants, VLCFAs are particularly important as membrane components and as waxes, primary constituents of the cuticle. One of the first documented injury symptoms of the carbamothioate EPTC was a reduction in cuticular wax deposition (Wilkinson & Hardcastle, 1970). Other severe injury symptoms include the failure of leaf emergence from the coleoptile and a general stunting of seedling growth (Gronwald, 1991), indicating that inhibition of VLCFA synthesis has multiple downstream repercussions. Indeed, there may be other functions for these enzymes in plants, as shown in mammalian systems, where long-chain fatty acid:CoA ligases are associated with the formation of xenobiotic acyl-coA thioesters and lipophilic xenobiotic conjugates (Knights, 2003). Interestingly, analysis of the Arabidopsis *cer* mutants shows that lesions in wax biosynthetic pathways also affect stomatal number and trichome development, suggesting that pathway intermediates or products could function as developmental signaling molecules for epidermal cell fate (Bird & Gray, 2003). It is thus conceivable that sublethal treatment with carbamothioate herbicides could phenocopy these mutations, although this has not been investigated to our knowledge. Schmutz *et al.* (1996) used EPTC as a specific inhibitor of fatty acid elongation in studies of the development and suberization of cotton fibers. Waxes have also been identified as key components of resistance mechanisms against fungal pathogens in a number of plant species (Niks & Rubiales, 2002). This role in plant defense could also be further dissected using carbamothioate herbicides as specific metabolic probes. Another group of herbicides, the chloroacetamides, inhibit plastidic VLCFA synthesis (Matthes & Böger, 2002) although the precise enzymatic step is not known. Chloroacetamides are soil applied and usually do not require any soil incorporation for activity. These compounds have widely used for over 50 years with little evolution of resistant biotypes. There appear to be genetic differences in susceptibility to this family, as the maize *se* and *sh2* endosperm mutants were more susceptible than 30 related commercial sweet corn cultivars (Bennett & Gorski, 1989).

Carbamothioate herbicides are applied to soil and incorporated prior to crop planting: crop selectivity depends upon seeding below the treated soil. Such manipulations are sometimes hard to achieve with precision, and so crop injury from these herbicides is a recurring problem. The potential for crop injury led to early research on compounds known as herbicide safeners, antidotes or protectants (Davies & Caseley, 1999). The first safeners thus identified were used to protect corn from EPTC injury, and include the compounds naphthalic anhydride, benoxacor, fenclorim and fluxofenim. Safeners are coated onto crop seeds or may be mixed with some commercial formulations of herbicides. They commonly induce one or more plant defense mechanisms, including glutathione content, glutathione transferase activity and cytochrome P450 activity, with a net result that crop plants are better able to detoxify the herbicide. Interestingly, these compounds protect monocot but not dicot crops from herbicide injury, for reasons that are not entirely clear (DeRidder *et al.*, 2002). Herbicide safeners are widely used commercially, but have not been exploited in physiological studies of plant response to other xenobiotic stresses. Similarly, the effects of such compounds on plant resistance to insects or pathogens have not been investigated.

Carbamothioate herbicides have been shown to alter the structure of soil microbial communities. In particular, consecutive usage in the same field selects for bacteria and actinomycetes that rapidly degrade the compounds, to the extent that they are essentially ineffective for weed control (Mueller *et al.*, 1989). Perhaps as an indirect result, these and other herbicides affect the kinetics of denitrification in field soils (Yeomans & Bremner, 1985). The possible effects of such population shifts and changes in nutrient cycling on subsequent crops or other plant species have not been investigated.

7.3.3.1 Resistance

As noted above, evolved resistance to ACCase inhibitors is most often conferred by a single amino acid substitution in the carboxyltransferase domain of the enzyme (Devine, 1997). However, some resistant biotypes have been reported with enhanced levels of herbicide metabolism (Maneechote et al., 1997). Only two studies have compared the relative fitness of resistant and susceptible biotypes, and in both cases the differences were not detectable in field or greenhouse experiments (Wiederholt & Stoltenberg, 1995, 1996).

Resistance to the carbamothioate herbicide triallate did not evolve until it had been used for more than 25 years. *Avena fatua* (wild oat) populations resistant to triallate were first documented in 1990 in Alberta (O'Donovan et al., 1994) and subsequently reported in Montana (Kern et al., 1996a). The resistant biotypes are unusual in this case, because they represent metabolic loss-of-function mutants. Carbamothioate herbicides are actually pro-herbicides, in that they require metabolic activation by sulfoxidase enzymes to become toxicologically active (Casida et al., 1974). In the Montana resistant biotype, the rate of metabolic activation was 10- to 15-fold slower than in the susceptible (Kern et al., 1996b). Both biotypes were equally susceptible to synthetic triallate sulfoxide, and the rates of metabolism of this toxic form were equivalent. Resistance was conferred by two recessive nuclear genes (Kern et al., 2002), which may encode the enzymes responsible for triallate sulfoxidation. The mechanism of resistance in the Canadian biotype appears to involve alterations in gibberellin biosynthesis (Rashid et al., 1998). Interestingly, both biotypes are cross-resistant to difenzoquat, an unrelated pyrazolium herbicide.

The exceedingly long, resistance-free usage period for triallate illustrates several important concepts about the plant response to herbicide selection pressure (Jasieniuk et al., 1996). First, even though triallate has a reasonably long soil half-life, its precise application requirements rarely allow it to achieve perfect weed control, and so the selection pressure that it exerts is reduced. Second, the target species (wild oat) has low seed production and poor seed dispersal, traits that tend to limit population sizes, and thus reduce the frequency of potentially resistant individuals. And third, the necessity of accumulating two separate recessive alleles to achieve resistance in self-pollinating wild oats would require a large number of individuals and generations. If carbamothioate herbicides in fact inhibit several elongases (Trenkamp et al., 2004), the accumulation of additional alleles would require even more time. It might be expected that the loss of two sulfoxidase-like activities would be associated with a fitness cost in the resistant biotype. Fitness comparisons have not been done for the Montana biotypes, but studies of the Canadian biotype did not support this idea, and in fact seed germination was higher in the resistant lines (O'Donovan et al., 1999).

7.3.4 Cellulose synthesis inhibitors

The herbicides dichlobenil and isoxaben are soil-applied herbicides with activity on a broad range of dicot and monocot weeds (Sabba & Vaughn, 1999). Early research on the mechanism of action of dichlobenil showed that it inhibited cellulose synthesis (Delmer *et al.*, 1987), and both herbicides probably inhibit cellulose synthase (EC 2.4.1.12). Isolation and cloning of members of this large gene family (Pear *et al.*, 1996) laid the groundwork for subsequent work that helped define the target of these herbicides. Recently, isoxaben-resistant Arabidopsis mutants were shown to contain point mutations in genes encoding one of the 10 isoforms of cellulose synthase (Desprez *et al.*, 2002). However, these and other mutants are not cross-resistant to all cellulose-inhibiting herbicides, indicating that the herbicides may act at different points in the cellulose biosynthetic pathway. The new herbicide triazofenamide also inhibits cellulose synthesis (Heim *et al.*, 1998).

The ectopic lignin (*eli*) mutants, which were isolated based on aberrant lignin deposition patterns, were recently shown to occur in a cellulose synthase gene (Cano-Delgado *et al.*, 2003). Interestingly, reduced rates of cellulose synthesis in the *eli* mutants apparently activate lignin synthesis and other defense responses through the jasmonate or ethylene signaling pathways. The possibility that sublethal doses of cellulose synthesis inhibitor herbicides might induce similar responses has not been investigated.

7.3.4.1 Resistance

Cell lines that can tolerate normally toxic concentrations of dichlobenil or isoxaben have been characterized. The basis for their resistance is that they have replaced the normal cell wall cellulose network with pectin and extensin (Wells *et al.*, 1994), a mechanism that has not been observed in whole plants. Other mutants have been obtained through mutagenesis, as noted earlier. In contrast, there are no reports of evolved resistance to these herbicides in weedy species. This lack of resistance is surprising because dichlobenil has been used for roadside weed control for over 30 years. There may be significant fitness costs associated with evolved resistance mechanisms that prevent their appearance.

7.3.5 Folic acid synthesis inhibitors

A single herbicide, asulam, inhibits 7,8-dihydropteroate synthase (EC 2.5.1.15), the penultimate enzyme in the biosynthesis of folic acid, an important component of purine nucleotides. This enzyme is also the target of sulfa drugs used against many microbial pathogens of animals and humans.

7.3.5.1 Resistance

Although asulam has a long history of usage for *Pteridium aquilinum* (bracken) control in many parts of the world (Burge & Kirkwood, 1992), there are no reports of evolved resistance in weedy species. In contrast, microbial resistance to sulfa and related drugs is widespread (Skold, 2001). Resistance is mediated by genes encoding insensitive forms of the target enzyme. Such a prokaryotic gene was used to create transgenic resistant tobacco (Guerineau *et al.*, 1990) and potato (Surov *et al.*, 1998) plants.

7.3.6 Nitrogen metabolism inhibitors

Plants convert inorganic nitrogen into organic compounds primarily through the addition of ammonia to glutamate to create glutamine, as catalyzed by glutamine synthetase (GS; E.C.6.3.1.2). The majority of GS activity occurs in chloroplasts, with a minor GS isoform in the cytosol. There are several related inhibitors of GS, including the natural product bialaphos (a fermentation product pro-herbicide from *Streptomyces* spp.), phosphinothricin (bialaphos that has been metabolically activated) and glufosinate (synthetically produced phosphinothricin) (Sherman *et al.*, 1996). Inhibition of GS leads to a rapid accumulation of ammonium to toxic levels. However, this is not the only mechanism of toxicity, as the loss of glutamine and glutamate indirectly decreases photosynthetic electron transport for the glycolate pathway. Glyoxylate accumulates to a level that reduces carbon fixation in the Calvin cycle, leading to an inhibition of the light reaction of photosynthesis (Lea & Ridley, 1989; Wild & Winder, 1993; Perez-Garcia *et al.*, 1998). The net result is an accumulation of toxic ROS that disrupts membranes and contributes to plant death.

Glufosinate is a nonresidual, mostly nonselective herbicide used in a number of noncrop and in-crop weed management situations. It is not translocated in plants as well as some other herbicides like glyphosate, and so its utility for perennial weed control is limited. However, the development of glufosinate-resistant crop plants has increased its usage.

7.3.6.1 Resistance

The *Streptomyces* spp. that synthesize bialaphos protect themselves through the action of an enzyme that acetylates and thereby inactivates the inhibitor (DeBlock *et al.*, 1987). The gene encoding phosphinothricin acetyltransferase from *Streptomyces viridochromogenes* is termed *pat* while its homologue from *Streptomyces hygroscopicus* is commonly known as *bar* (Wehrmann *et al.*, 1996). Both genes have been used widely and successfully as selectable markers in plant transformation research (Goodwin *et al.*, 2004).

Early attempts to obtain plant cell lines resistant to glufosinate resulted in alfalfa lines that overexpressed GS as a result of gene amplification (Donn *et al.*, 1984). Whole plants could not be regenerated from these lines, and other attempts to overproduce the enzyme have not produced commercially viable cultivars. Instead, the *bar* gene has been used to confer the ability to metabolize and inactivate glufosinate in transgenic crops. In fact, more plant species, including tobacco, canola, tomato and alfalfa, have been made resistant to glufosinate than to any other herbicide (Donn, 1997). These crops have not been as widely adopted as glyphosate-resistant crops for a number of reasons, including the relatively high cost of the herbicide. Another approach for introducing glufosinate resistance involved transferring the *E. coli gdhA* gene, encoding a NADPH-dependent glutamate dehydrogenase, to tobacco plants (Nolte *et al.*, 2004). Regenerated plants expressed about a sixfold increase in resistance, as compared to the 100-fold increase typically achieved from the *bar* transgene.

Since glufosinate inhibits GS from most eukaryotic and prokaryotic organisms, it has potential to be used in alternative approaches to plant pest management. For example, glufosinate was used to preferentially inhibit *Rhizoctonia solani*, the fungus causing leaf sheath blight, in transgenic rice expressing the *bar* gene (Uchimiya *et al.*, 1993). An attenuated zucchini yellow mosaic potyvirus-vector was used to introduce transient *bar* expression into several cucurbit crops that have troublesome weed management issues (Shiboleth *et al.*, 2001). Infected plants expressed the transgene and could therefore tolerate glufosinate treatment for several weeks after inoculation. *Xanthomonas campestris* is a plant pathogen with potential for use as a weed biological control agent. Charudattan *et al.* (1996) proposed that a bialaphos biosynthetic gene cassette could be introduced into this organism in order to improve its toxicity to weedy species.

7.3.7   Quinone synthesis inhibitors

Three relatively new herbicides, isoxaflutole, sulcotrione and mesotrione, inhibit ρ-hydroxyphenylpyruvate dioxygenase (HPDDase; EC 1.13.11.27), which catalyzes the conversion of ρ-hydroxyphenylpyruvate to homogentisic acid, the aromatic precursor for the biosynthesis of both α-tocopherol (vitamin E) and plastoquinone (Lee *et al.*, 1997; Pallett *et al.*, 2001). Plastoquinone in turn is an important recycled cofactor in the desaturation of phytoene as it is converted to carotene, lycopene and other carotenoids. HPDDase is a regulatory enzyme in this important pathway, since its overexpression in Arabidopsis led to increased levels of α-tocopherol in seeds and leaves (Tsegaye *et al.*, 2002). Similarly, overexpression of a barley HPDDase in tobacco plants caused seeds, but not leaves, to accumulate twofold higher levels of α-tocopherol (Falk *et al.*, 2003). In both cases, overexpression of this enzyme also conferred herbicide resistance. HPDDase is expressed during senescence and in response to oxidative stress (Falk *et al.*, 2002), reflecting the importance of α-tocopherol and quinones in reducing ROS and other damaging compounds.

The use of herbicides as metabolic probes for gene discovery is nicely illustrated by the role of sulcotrione in the identification and cloning of HPDDase (Zhen & Singh, 2001). As part of their studies on carotenoid biosynthesis, Norris *et al.* (1995) isolated and characterized albino Arabidopsis mutants. Two mutants, *pds1* and *pds2*, were originally assumed to map to phytoene desaturase based on their phenotypes and absence of chlorophyll and carotenoids. However, subsequent enzyme assays and genetic mapping did not support this idea. The *pds* albino phenotype did, however, resemble the symptoms caused by sulcotrione, which had recently been reported to be an inhibitor of quinone synthesis (Schultz *et al.*, 1993). Through a series of metabolite reversal experiments, cloning and genetic analyses, Norris *et al.* (1998) subsequently showed that *pds* mapped to HPDDase. Also, further analysis of the mutants showed that plastoquinone, and not tocopherol, is an essential component for phytoene desaturation. Thus, information on the herbicide site of action was exploited in functional genomic studies to gain additional knowledge about a key enzyme and discover important details of plant response to oxidative stress.

7.3.7.1 Resistance

As discussed above, overexpression of HPPDase confers resistance to the quinone synthesis-inhibiting herbicides. Mutant HPPDase enzymes have also been created and expressed in plants (Gressel, 2002). Consensus sequences of HPPDase from a number of organisms were identified and the *Pseudomonas* and *Synechocystis* genes were subjected to site-directed mutagenesis. The resulting mutated enzymes gave reasonable levels of resistance in transgenic tobacco plants. Although USDA-APHIS field applications have been submitted since 1999 for resistant corn and soybeans, the results of these tests have not been published. For weedy species, there have been no reports of evolved resistance as of this writing.

7.3.8 Carotenoid biosynthesis inhibitors

Herbicides in this group inhibit several enzymes in the carotenoid biosynthetic pathway and include: phytoene desaturase by norflurazon, fluridone and several experimental herbicides; zeta carotene desaturase by amitrol; and lycopene cyclase by amitrol. The precise site of action of clomazone is not known (Sandman & Böger, 1987a). Treatment results in bleaching of new foliage due to chlorophyll photooxidation in the absence of carotenoids.

Chlorophyll biosynthesis and the assembly of pigment–protein complexes in the thylakoids are also indirectly disrupted (Bramley & Pallett, 1993; Moskalenko & Karapetyen, 1996). Inhibition of phtyoene desaturase by fluridone is reversible and noncompetitive (Sandman *et al.*, 1996), although little else is known about herbicide binding. Treated plants accumulate high levels of phytoene and phytofluene. Amitrol, dichromate and several experimental herbicides inhibit zeta-carotene desaturation in the carotenoid biosynthetic pathway (Böger, 1996). The substituted triethylamines inhibit a cyclization step following lycopene synthesis, which results in an accumulation of lycopene. Several researchers have speculated that the site of action of clomazone is just prior to geranylgeranyl pyrophosphate (Duke & Kenyon, 1986) or at isopentenyl pyrophosphate isomerase (IPP isomerase; EC 5.3.3.2) (Sandman & Böger, 1986, 1987b). However, evidence by Lutzow *et al.* (1990), Croteau (1992) and Weimer *et al.* (1992) does not support these ideas, and so the search continues.

7.3.8.1   Resistance

The two reported cases of resistant weedy biotypes are *Hydrilla verticillata* (hydrilla) in Mississippi and *Raphanus raphanistrum* (wild radish) in Western Australia. No known mechanism for this resistance has been reported. However, there has been considerable research on the use of such resistance for crop improvement. Sandmann and coworkers purified, characterized and reconstituted the activity of phytoene desaturase from several sources after cloning the gene and overexpressing it in *E. coli* (e.g. Fraser *et al.*, 1992). They identified two different and unrelated types of phytoene desaturase. One is found in bacteria and fungi (*crtl* type), whereas the other is present in cyanobacteria, algae and higher plants (*pds* type) (Sandmann & Fraser, 1993). The bacterial and fungal forms were not inhibited by herbicides known to inhibit phytoene desaturase in higher plants, while the other form was sensitive. Different sources of phytoene desaturase yield different reaction products from phytoene. For example, phytoene desaturase from the cyanobacteria, *Synechococcus*, catalyzes two desaturation steps to yield zeta carotene, whereas the *Erwinia* enzyme catalyzes a four-step desaturation to yield lycopene (Linden *et al.*, 1991).

7.4   Induction of herbicide metabolism

Many plant species respond to herbicide stress by inducing detoxifying enzymes, and the systems and pathways are well characterized (see reviews by Cole, 1994; Hall *et al.*, 2001; Van Eerd *et al.*, 2003). In general, detoxification occurs via multistep biotransformation processes known as cometabolism, in which foreign organic compounds are subjected to reactions such as oxidation, reduction, hydrolysis and conjugation. The overall metabolism of herbicides in plants can be divided into four phases. Phase I is a direct change in the herbicide structure brought about by oxidation, reduction or hydrolysis reactions. Phase II is conjugation to cell constituents such as glucose, glutathione or amino acids. During Phase III, conjugates are transported across cell membranes into the vacuole or cell wall where further processing (Phase IV) can yield insoluble and bound residues. All of these processes result in a loss of herbicidal activity.

Phase I oxidation reactions that lead to herbicide detoxification are most often based on P_{450} enzymes (Cyt P450s) (reviewed in Mougin et al., 1991). These enzymes bind molecular oxygen, catalyze its activation and incorporate one of its atoms onto the herbicide molecule. They exist in large gene superfamilies, and individual members generally have very narrow substrate specificity. Almost half of the known Cyt P450 gene sequences are from plants (and Arabidopsis in particular), although relatively few have known functions. More is known about the wheat Cyt P450s than for any other species, primarily because they are responsible for wheat's tolerance to virtually all selective herbicides used in the crop. As discussed above, one of the mechanisms by which safeners protect crop plants from herbicide injury is by inducing Cyt P450s. Herbicides themselves can also induce Cyt P450s in some species. For example, several xenobiotics and herbicides induced the enzyme encoded by CYP76B1 from Jerusalem artichoke (Robineau et al., 1998). This gene was subsequently expressed in yeast where it enhanced metabolic rates of chlorotoluron and isoproturon. Similar approaches have been utilized to confer herbicide resistance to crop plants (reviewed in Ohkawa et al., 1999).

Primary oxidation of herbicides by Cyt P450s is often followed by conjugation to the tripeptide glutathione, glucose or other hydrophilic moieties. Glucosylated herbicides are then processed further, and become 'bound' residues in the extracellular matrix or are stored as water-soluble metabolites in the vacuole. The key enzymes of glutathione conjugation are glutathione transferases (usually abbreviated as GSTs), which catalyze nucleophilic attack of the thiol group onto electrolytic groups of oxidized herbicides. GST isozymes are found in large families that differ in number of members, herbicide specificity and inducibility across plant families. In many cases, these differences account for the range of susceptibility among plant species toward different herbicides. GST activity is inducible by safeners and some herbicides.

Metabolic pathways and specificities are highly diverse across the plant kingdom, and depend on the chemical structure of the xenobiotic, environmental conditions, metabolic factors and the ability of individual species to regulate these pathways. Interestingly, the so-called natural tolerance of crop species to selective herbicides is mostly due to their catabolic abilities, while the majority of cases of evolved resistance in weedy plants are due to mutations encoding altered target sites. Although this distinction may be due to the traditional methods of discovering herbicides (Gressel, 2002), it may also reflect the relative frequencies of mutations for these respective phenotypes in native populations.

7.4.1 Resistance

The multitude of Cyt P450, GST and other metabolic enzymes and systems in plants evolved because of plants' inability to escape potentially toxic concentrations of metals, salts and other 'natural' xenobiotics in soil. More recently, these same systems have been subjected to strong selection pressure through the widespread use of herbicides. Weedy species have responded in turn, and have developed resistant biotypes through overexpression or other elevation of metabolic activities. For example, velvetleaf, a common weed in corn fields, has evolved resistance to atrazine through enhanced GST activity (Gray *et al.*, 1995). In a more extreme example, biotypes of *Alopecurus myosuroides* (black-grass) have evolved resistance to at least four unrelated herbicides through a combination of enhanced Cyt P450 and GST activity, and elevated glutathione content (Cummins *et al.*, 1999). The latter case is particularly troubling from a weed management standpoint, and yet illustrates the highly flexible ability of some species to respond with multiple defenses to strong selection pressure.

7.5 Protoporphyrinogen oxidase inhibitors

Diphenyl ether (DPE), oxidiazon and N-phenylheterocycle herbicides inhibit the enzyme protoporphyrinogen oxidase (abbreviated PROTOX) (EC 1.3.3.4), a membrane-bound plastidic protein involved in chlorophyll and heme synthesis. This flavoprotein converts protoporphyrinogen IX to protoporphyrin IX in a six-electron aromatic oxidation. Interestingly, PROTOX inhibition leads to an accumulation of the product protoporphyrin IX through a convoluted series of steps. First, accumulated protoporphyrinogen IX molecules diffuse away from the reaction center in the chloroplast envelope membrane and into the cytoplasm (Jacobs & Jacobs, 1993). Protoporphyrinogen IX concentrations near the plasma membrane as high as 20 nmol/mg fresh weight of tissue have been reported (Dayan & Duke, 1996). Then enzymatic oxidation of protoporphyrinogen IX to protoporphyrin IX occurs in the cytoplasm (Yamato *et al.*, 1994, 1995) and at the plasma membrane (Jacobs *et al.*, 1991). Nonenzymatic oxidation (autooxidation) of protoporphyrinogen IX has also been implicated (Han *et al.*, 1995). The accumulating protoporphyrin IX now interacts with oxygen and light to form singlet oxygen and other ROS, which initiate lipid peroxidation, membrane disruption and plant death.

The relatively new PROTOX inhibitor flumioxazin is a soil-applied herbicide used in grape production. Its effects on grapevines varied by cultivar, inhibiting photosynthesis in field-grown plants, while stimulating photosynthesis in cuttings (Saladin *et al.*, 2003).

7.5.1 Resistance

There are now three reports of resistance to the PROTOX-inhibiting herbicides. Resistance has been reported for *Amaranthus rudis* (common waterhemp) in Kansas and Illinois, and *Euphorbia heterophylla* (wild poinsettia) in Brazil. The mechanism of resistance has not been elucidated in any of these cases. However, one hypothesis proposes that, since protoporphyrinogen IX diffusion from the chloroplast to the cytoplasm is central to herbicide mode of action, the regulation of protoporphyrinogen IX concentrations in the cytoplasm could be a mechanism for differential species sensitivity. Indeed, Jacobs *et al.* (1996) found an enzyme fraction in the cytoplasm of leaves, which can decompose protoporphyrinogen IX to nonporphyrin products. Rates of protoporphyrinogen IX destruction were low in sensitive cucumber leaves and high in more tolerant species such as mustard and radish. Several other tolerance mechanisms occur in plants, including:

(1) limited absorption across the cuticle (acifluorfen in soybean),
(2) metabolic degradation of the herbicide (acifluorfen in soybean),
(3) herbicide-resistant PROTOX enzyme (soybean cell line selected for tolerance to oxyfluorfen; Pornprom, *et al.*, 1994),
(4) increased specific activity of PROTOX (tobacco cell line selected for tolerance to DPEs; Ichinose *et al.*, 1995) and
(5) resistance to photooxidative damage by high endogenous levels of antioxidants (Knörzer *et al.*, 1996).

7.6 Mitotic disruptors

The dinitroaniline family of herbicides and pronamide are soil-applied herbicides that interfere with mitosis. These compounds bind to tubulin monomers and prevent their polymerization into microtubules, which prevents the normal movement of chromosomes during mitosis. Progression from prophase into metaphase is thus inhibited; the chromosomes coalesce in the center of the cell, and the nuclear envelope reforms, causing a polyploid nucleus. Cell division is halted during germination, and most seedlings are killed before they emerge from the soil. Injury symptoms include a cessation of root growth and a swelling of the root tip to create club-like projection.

Since tubulin and microtubules are essentially ubiquitous among eukaryotes, one may expect that dinitroaniline herbicides would show toxicity toward nontarget organisms. However, no effects have been observed in several mammalian systems, perhaps because of the substantial variability among the numerous known forms of tubulin (Fosket & Morejohn, 1992; Goddard *et al.*, 1994). In fact, selective inhibition of protozoan microtubule-related processes has been explored as a control measure for protozoan diseases in mammals (Chan & Fong, 1990; Chan *et al.*, 1993, 1995; Wang *et al.*, 1995).

Other herbicides that interfere with mitosis include: DCPA, propham and chloropropham (carbamates), dithiopyr and thiazopyr. The precise mechanism(s) of action for these herbicides are unknown. DCPA causes a significant disruption of cell wall formation during mitosis, in that walls form in random directions between the two daughter nuclei (Holmsen & Hess, 1984).

7.6.1   Resistance

Resistant biotypes of four weedy species have evolved after repeated use of dinitroaniline herbicides (reviewed in Smeda & Vaughn, 1994). In the southeastern USA, heavy use of these herbicides in cotton production set the stage for the appearance of resistant goosegrass. Similar use patterns in monoculture wheat production in the Canadian prairie provinces selected for resistant *Setaria viridis* (green foxtail) biotypes. Other resistant biotypes of *Sorghum halepense* (Johnsongrass) and *Amaranthus palmeri* (Palmer amaranth) have been reported. In addition, weed biotypes with cross-resistance to multiple herbicides, including rigid ryegrass in Australia and *A. myosuroides* (slender meadow foxtail) in the UK, are also resistant to dinitroaniline herbicides. The mechanism of resistance in the goosegrass biotype is due to a threonine$_{239}$ to isoleucine substitution in the α-tubulin component of the α/β-tubulin dimer (Anthony *et al.*, 1998). Other tubulin mutations have been reported in *Chlamydomonas* (Schibler & Huang, 1991) and tobacco cell cultures (Blume *et al.*, 1998) that have not evolved in weedy biotypes. Interestingly, tobacco T-DNA activation tagged mutants selected for resistance to aryl carbamate herbicides were also more tolerant to chilling injury (Ahad *et al.*, 2003). The authors proposed that adaptive microtubule reorganization patterns conferred both phenotypes, and that microtubule sensitivity was a limiting factor for chilling tolerance in tobacco.

7.7   Hormone disruptors

Several herbicide families interfere with phytohormone action or synthesis. The first family is the auxinic herbicides, so called because of their mechanistic resemblance to the phytohormone indole-3-acetic acid (IAA or auxin). The family is commercially important worldwide and is represented by several groups including phenoxys (2,4-D and MCPA), benzoics (dicamba and chloramben), pyridines (picloram and fluroxypyr) and quinoline carboxylic acids (quinclorac). Auxinic herbicides are mostly used to selectively control dicot weeds in turf or monocot crops. Most monocots are naturally tolerant to these herbicides due to metabolism and the location of meristematic tissues (Fedtke, 1982). Common symptoms after herbicide treatment of susceptible species include ethylene production, tissue epinasty, pronounced overexpression of auxin-inducible genes and a general induction of metabolism. Although the precise mechanism by which auxinic herbicides cause plant death is not known, available evidence supports the idea that they induce the same physiological and biochemical responses as IAA, only in an unregulated manner (Fedtke, 1982; Cobb, 1987; Coupland, 1994; Sterling & Hall, 1997; Grossmann & Kwiatkowski, 2000). The induction of 1--aminocyclopropane-1-carboxylic acid (ACC) synthase and resulting ethylene synthesis may also be an important component of their toxicity (Sterling & Hall, 1997; Grossmann & Kwiatkowski, 2000). It is generally assumed that auxinic herbicides are bound by the same receptors that bind auxin, and several auxin receptors have been identified in recent years (Napier *et al.*, 2002). In particular, the auxin-binding protein (ABP) in shoot tissues has been implicated in a number of auxin responses (Bauly *et al.*, 2000; Steffens *et al.*, 2001) and is thus a logical candidate as a herbicide receptor.

Quinclorac is an auxinic herbicide, but is unusual in that it has activity on a number of grass weeds including *Echinochloa*, *Digitaria* and *Setaria* species. Like other herbicides in this family, the precise molecular target for quinclorac is not known. There is some evidence that quinclorac inhibits cellulose synthesis (Koo *et al.*, 1996). However, the selective induction of ACC synthase activity, which ultimately leads to cyanide accumulation in the tissues of susceptible grasses, is more likely the primary mechanism of quinclorac action (Grossmann & Kwiatkowski, 2000).

7.7.1 Resistance

Weed resistance to auxinic herbicides has been slower to develop than for other herbicide families, perhaps because of moderate selection pressure, their use in mixtures with other herbicide families and their proposed multiple sites of physiological action (Jasieniuk *et al.*, 1996). Only 20 species have developed resistance to auxinic herbicides, which is surprising since members of this herbicide family have been in constant and widespread use for over 50 years. This phenomenon may indicate that, in addition to the reasons noted above, mutations that confer resistance but not a substantial fitness penalty are very rare.

Relatively little is known about mechanism(s) of auxinic herbicide resistance. Altered herbicide absorption, translocation and metabolism have been ruled out as mechanisms of resistance to mecoprop in *Stellaria media* (common chickweed) (Coupland, 1994), dicamba and picloram in *Sinapis arvensis* (syn *Brassica kaber*) (wild mustard) (Webb & Hall, 1995) and dicamba in *K. scoparia* (kochia) (Cranston *et al.*, 2001). In *S. arvensis*, ATP-dependent auxin- and auxinic herbicide-induced volume changes in resistant protoplasts were less than in susceptible protoplasts (Deshpande & Hall, 1996), and dicamba binding to ABPs was less competitive with IAA in resistant than in susceptible biotypes (Webb & Hall, 1995). Calcium flux after herbicide treatment was also different between biotypes, indicating the presence of alterations in auxin-mediated signal transduction (Wang *et al.*, 2001). Resistance in this species may be due to mutations in ABP (Zheng & Hall, 2001). Evolved resistance to quinclorac has been reported in smooth crabgrass and barnyardgrass, weedy species in rice cropping systems. The mechanism of resistance is not known.

The kochia HRd mutant is resistant to several auxinic herbicides and has several interesting auxin phenotypes (Goss & Dyer, 2003). The auxin-mediated growth processes of apical dominance and shoot gravitropism are impaired in the mutant, indicating that there may be a lesion in auxin binding or a signal transduction pathway. In addition, this and other kochia biotypes that are resistant to ALS-inhibiting herbicides present severe management issues for cereal producers in the Northern Great Plains, due to the lack of alternate control measures (Cranston *et al.*, 2001). This particular weedy species might be considered an archetypical 'canary in the coal mine' in regard to resistance. Kochia is an outcrosser with nonobligate allogamous mating behavior (Guttieri *et al.*, 1998), produces copious amounts of seed and has levels of genetic diversity among the highest measured for plant species (Mengistu & Messersmith, 2002). These characteristics, along with its highly efficient seed dispersal mechanism (tumbleweed), allow kochia to maintain huge amounts of diversity in many different ecological niches, and thus it can rapidly adapt to strong herbicide selection pressures. In this regard, it was one of the first species to evolve resistance to ALS inhibitors and auxinic herbicides.

In Arabidopsis, a number of auxin-insensitive and auxin-transport mutants have been obtained, including the *aux*, *tir*, *rcn*, *agr*, *axr*, *Atpin*, *ifl*, *mp*, *eir*, *sax*, *shy* and *iaa* mutants (reviewed in Friml & Palme, 2002; Swarup *et al.*, 2002). Many of these mutants have some degree of resistance to the synthetic auxins 2,4-D or NAA, although none seem to be allelic to the herbicide-resistant weedy biotypes. Further, characterization of the Arabidopsis mutants (obtained through mutagenesis) and the evolved weedy biotypes (selected in the field) will not only provide insight into the resistance mechanism(s), but will also add to our understanding of auxin's control of important plant growth processes.

7.8 Genome effects

As discussed above, there are many situations where herbicides can inflict stress without causing lethality. Under these conditions, there is increasing evidence that herbicides can indirectly cause genetic changes, in addition to their acute physiological and metabolic effects. Most herbicides are applied to seedlings or vegetative growth stages of plants, and so any genetic changes they might inflict would occur in somatic tissue only. However, sublethal levels of residual herbicides in soil or accumulated concentrations of herbicides that are metabolized slowly can certainly persist into reproductive growth stages, and thus could cause heritable changes.

It is important to point out that, as part of the registration process, herbicides are routinely screened for mutagenicity using the Ames test before commercialization, and so there is no evidence that herbicides directly cause mutations. However, the kinds of stress that herbicides induce are known to lead to short-term alterations in gene expression as well as long-term, heritable genetic changes. For example, auxinic herbicides and others can cause substantial modifications in gene expression patterns (Sterling & Hall, 1997). Some of these changes that differed between dicamba-resistant and susceptible kochia biotypes were identified using mRNA differential display and may be involved in the resistance mechanism (Kern *et al.*, 2004). Other herbicides that inhibit biosynthetic pathways can cause starvation for endproducts, which can in turn lead to a number of indirect genetic effects. For example, the microarray experiments discussed above showed that sublethal treatment with an ALS inhibitor caused global changes in yeast gene expression, many of which would have been unpredictable *a priori* (Jia *et al.*, 2000).

More severe impacts on nucleic acids can also be indirectly caused by herbicides. For example, herbicide-induced ROS that accumulate from photosynthetic inhibition can cause lesions in plastid and nuclear DNA. If these lesions form at faster rates it can be repaired by DNA repair enzymes, heritable mutations can result. ROS generated by paraquat led to increased levels of double-stranded DNA breaks and elevated the incidence of homologous recombination in Arabidopsis (Kovalchuk *et al.*, 2003). Pollen grains from a waxy (starch free) corn mutant were used to measure the frequency of mutations that restored starch biosynthesis and thus caused the pollen to be stained with iodine (Plewa, 1985). In atrazine-treated plants, the mutation frequency as measured in pollen was doubled over that of untreated plants. Other oxidative stresses can be caused by herbicides like glyphosate, by inhibiting the synthesis of flavonoids that normally protect DNA and other molecules against UV damage. Similar damage could result from HPDD inhibitors, which inhibit quinone synthesis and lead to ROS production.

Gene amplification, which often results from efforts to select herbicide-resistant lines from plant cell cultures, is another possible genetic response to herbicide stress. However, this reponse is likely an indirect result of somatic cell aberrations that occur during culture (Joyce *et al.*, 2003). There is little evidence that this mechanism also occurs in whole plants.

7.9 Summary and future prospects

Herbicides occupy a unique position among the abiotic stresses discussed in this book. First, they are applied deliberately to destroy vegetation, but imperfect applications and differential plant tolerance causes them to exert a wide variety of different stresses on plants. The direct responses caused by various herbicide chemistries and their biochemical targets are well characterized, but we are only beginning to learn about the numerous indirect responses and downstream effects. The significant variability among plant species and families in their abilities to respond to herbicide stress makes it difficult to generalize about these interactions. However, this same variation provides a rich spectrum of experimental systems in which to explore biochemical pathways and genetic mechanisms. The extremely strong selection pressure that herbicides impose has caused significant changes in species abundance and population complexity in agroecosystems and rangelands. Evolution of resistant biotypes in a number of cropping systems has led to significant changes in weed management strategies. Studies of these biotypes and their mechanisms of resistance can be exploited as paradigms of compressed evolution of plant response to strong stressors.

Herbicides' commercial success and commonplace usage often overshadow the fact that they are powerful and highly specific tools that can be used to dissect biochemical pathways and physiological responses. Much of our detailed knowledge about photosynthesis and plant metabolic pathways stems directly from herbicide research. Similar valuable contributions on plant catabolism have been realized from the use of herbicide safeners, although such studies are limited. In recent years, herbicides have been successfully used for gene discovery and in investigations of the transcriptome and proteome. Current work on stress-related genomic changes shows that herbicide stress can generate numerous and often unpredictable alterations in gene expression and signal transduction. We believe that such uses of herbicides are underexploited, and that many additional applications can be explored, including plant/pest interactions, secondary compound biosynthesis and metabolic engineering.

The continued commercial success of herbicides in the future will depend in large part on the willingness of producers to adopt proactive management strategies to avoid and combat resistance development. Recent examples of resistant biotypes with multiple mechanisms and enhanced metabolic activities do not bode well for the long-term future of many herbicides. Also, the heavy reliance on one herbicide, as in the case of Roundup Ready™ cultivars, will substantially increase selection pressure. It has already been clearly demonstrated (for paraquat-resistant biotypes) that strong selection caused by the persistence of herbicide application frequency can substitute for the herbicide's persistence of residual soil activity. Current resistance management recommendations of herbicide rotations and mixtures may delay the evolution of resistant biotypes, but its expansion across cropping systems and the involvement of new weed species seems inevitable.

References

Abulnaja, K. O. & Harwood, J. L. (1991) Thiocarbamate herbicides inhibit fatty acid elongation in a variety of monocotyledons. *Phytochemistry*, **30**, 1445–1448.

Ahad, A., Wolf, J. & Nick, P. (2003) Activation-tagged tobacco mutants that are tolerant to antimicrotubular herbicides are cross-resistant to chilling stress. *Transportation Research*, **12**, 615–629.

Anderson, M. P. & Gronwald, J. W. (1991) Atrazine resistance in a velvetleaf (*Abutilon theophrasti*) biotype due to enhanced glutathione *S*-transferase activity. *Plant Physiology*, **96**, 104–109.

Anthony, R. G., Waldin, T. R., Ray, J. A., Bright, S. W. J. & Hussey, P. J. (1998) Herbicide resistance caused by spontaneous mutation of the cytoskeletal protein tubulin. *Nature*, **393**, 260–262.

Bauly, J. M., Sealy, I. M., Macdonald, H., Brearley, J., Droge, S., Hillmer, S., Robinson, D. G., Venis, M. A., Blatt, M. R. & Lazarus, C. M. (2000) Overexpression of auxin-binding protein enhances the sensitivity of guard cells to auxin. *Plant Physiology*, **124**, 1229–1238.

Baur, J. R. (1979) Effect of glyphosate on auxin transport in corn and cotton tissues. *Plant Physiology*, **63**, 882–886.

Bearson, S. R., Rodriguez, D. J., Tran, M., Feng, Y., Biest, N. A. & Dill, G. M. (2002) Glyphosate-resistant goosegrass. Identification of a mutation in the target enzyme 5-enolpyruvylshikimate-3-phosphate synthase. *Plant Physiology*, **129**, 1265–1275.

Bennett, M. A. & Gorski, S. F. (1989) Response of sweet corn (*Zea mays*) endosperm mutants to chloracetamide and thiocarbamate herbicides. *Weed Technology*, **3**, 475–478.

Bergelson, J. & Purrington, C. B. (1996) Surveying patterns in the cost of resistance in plants. *American Naturalis*, **148**, 536–558.

Bergelson, J., Purrington, C. B,. Palm, C. J. & López-Guttiérrez, J.-C. (1996) Costs of resistance: a test using transgenic *Arabidopsis thaliana*. *Proceedings of the Royal Society* (*London*), **263**, 1659–1663.

Bird, S. M. & Gray, J. E. (2003) Signals from the cuticle affect epidermal cell differentiation. *New Phytologist*, **157**, 9–23.

Blackburn, L. G. & Boutin, C. (2003) Subtle effects of herbicide use in the context of genetically modified crops: a case study with glyphosate (Roundup®). *Ecotoxicology*, **12**, 271–285.

Blokhina, O., Virolainen, E. & Fagerstedt, K. V. (2003) Antioxidants, oxidative damage and oxygen deprivation stress: a review. *Annals of Botany*, **91**, 179–194.

Blume, Y. B., Strashnyuk, N. M., Smertenko, A. P., Solodushko, V. G., Sidorov, V. A. & Gleba, Y. Y. (1998) Alteration of beta-tubulin in *Nicotiana plumbagnifolia* confers resistance to amiprophos-methyl. *Theoretical and Applied Genetics*, **97**, 464–472.

Böger, P. (1996) Mode of action of herbicides affecting carotenogenesis. *Journal of Pesticide Science*, **21**, 473–478.

Bordas, B., Komives, T. & Lopata, A. (2003) Ligand-based computer-aided pesticide design. A review of applications of the CoMFA and CoMSIA methodologies. *Pest Management Science*, **59**, 393–400.

Bradshaw, L. D., Padgette, S. R., Kimball, S. L. & Wells, B. H. (1997) Perspectives on glyphosate resistance. *Weed Technology*, **11**, 189–198.

Bramley, P. M. & Pallet, K. E. (1993) Phytoene desaturase: a biochemical target of many bleaching herbicides. *Proceedings of the Brighton Crop Protection Conference – Weeds*, BCPC Publications, Hampshire, UK, pp. 713–722.

Brammall, R. A. & Higgins, V. J. (1988) The effect of glyphosate on resistance of tomato to fusarium crown and root rot disease and on the formation of host structural defensive barriers. *Canadian Journal of Botany*, **66**, 1547–1555.

Bridges, D. C. (1995) Ecology of weeds. In A. E. Smith (ed.) *Handbook of Weed Management Systems*, Chapter 2. Marcel Dekker, New York.

Bromilow, R. H. (2004) Paraquat and sustainable agriculture. *Pest Management Science*, **60**, 340–349.

Burge, M. N. & Kirkwood, R. C. (1992) The control of bracken. *Critical Reviews in Biotechnology*, **12**, 299–333.

Burnet, M. W. M., Loveys, B. R., Holtum, J. A. M. & Powles, S. B. (1993) Increased detoxification is a mechanism of simazine resistance in *Lolium rigidum. Pesticide Biochemistry and Physiology*, **46**, 207–218.

Cano-Delgado, A. S. Penfield, C., Smith, Catley, M. & Bevan, M. (2003) Reduced cellulose synthesis invokes lignification and defense responses in *Arabidopsis thaliana. Plant Journal*, **34**, 351–362.

Casida, J. E., Gray, R. A. & Tilles, H. (1974) Thiocarbamate sulfoxides: potent, selective, and biodegradable herbicides. *Science*, **184**, 573–574.

Castle, L. A., Shiehl, D. L., Gorton, R., Patten, P. A., Chen, Y. H., Bertain, S., Cho, H., Duck, N., Wong, J., Liu, D. & Lassner, M. W. (2004) Discovery and directed evolution of a glyphosate tolerance gene. *Science*, **304**, 1151– 1154.

Chan, M. M. & Fong, D. (1990) Inhibition of leishmanias but not host macrophages by the antitubulin herbicide trifluralin. *Science*, **249**, 924–926.

Chan, M. M., Grogl, M., Chen, C., Bienen, E. J. & Fong, D. (1993) Herbicides to curb human parasitic infections: *in vitro* and *in vivo* effects of trifluralin on the trypanosomatid protozoans. *Proceedings of the National Academy of Sciences of the United States of America*, **90**, 5657–5661.

Chan, M. M.-Y., Grogl, M., Callahan, H. & Fong, D. (1995) Efficacy of the herbicide trifluralin against four P-glycoprotein-expressing strains of *Leishmania. Antimicrobial Agents and Chemotherapy*, **39**, 1609–1611.

Charudattan, R., Prange, V. J. & Devalerio, J. T. (1996) Exploration of the use of the 'bialaphos genes' for improving bioherbicide efficacy. *Weed Technology*, **10**, 625–636.

Christoffers, M. J., Berg, M. L. & Messersmith, C. G. (2002) An isoleucine to leucine mutation in acetyl-CoA carboxylase confers herbicide resistance in wild oat. *Genome*, **45**, 1049–1056.

Cobb, A. H. (1987) Herbicides action and interaction at the plant cell membrane. *Pesticide Science*, **19**, 329–330.

Cole, D. J. (1994) Detoxification and activation of agrochemicals in plants. *Pesticide Science*, **42**, 209–222.

Colyer, P. D., Vernon, P. R., Caldwell, W. D. & Kirkpatrick, T. L. (2000) Root-knot nematode reproduction and root galling severity on related conventional and transgenic cotton cultivars. *Journal of Cotton Science (Online)*, **4**, 232–236.

Conard, S. G. & Radosevich, S. R. (1979) Ecological fitness of *Senecio vulgaris* and *Amaranthus retroflexus* biotypes susceptible or resistant to atrazine. *Journal of Applied Ecology*, **16**, 171–177.

Coupland, D. (1994) Resistance to auxinic analog herbicides. In S. B. Powles & J. A. M. Holtum (eds) *Herbicide Resistance in Plants*, Chapter 6. Lewis Publishers, London.

Cranston, H. J., Kern, A. J., Hackett, J. L., Miller, E. K., Maxwell, B. D. & Dyer, W. E. (2001) Dicamba resistance in kochia. *Weed Science*, **49**, 164–170.

Croteau, R. (1992) Clomazone does not inhibit the conversion of isopentenyl pyrophosphate to geranyl, farnesyl, or geranylgeranyl pyrophosphate in vitro. *Plant Physiology*, **98**, 1515–1517.

Cummins, I., Cole, D. J. & Edwards, R. (1999) A role for glutathione transferases functioning as glutathione peroxidases in resistance to multiple herbicides in black-grass. *Plant Journal: Cell and Molecular Biology*, **18**, 285–292.

Davies, J. & Caseley, J. C. (1999) Herbicide safeners: a review. *Pesticide Science*, **55**, 1043–1058.

Dayan, F. E. & Duke, S. O. (1996) Porphyrin-generating herbicides. *Pesticide Outlook*, **7**, 22–27.

DeBlock, M., Botterman, J., Vandewiele, M., Dockx, J., Thoen, C., Gossele, V., Movva, N. R., Thompson, C., Van Montagu, M. & Leemans, J. (1987) Engineering herbicide resistance in plants by expression of a detoxifying enzyme. *EMBO Journal*, **6**, 2513–2518.

Delmer, D. P., Read, S. M. & Copper, G. (1987) Identification of a receptor protein in cotton fibers for the herbicide 2,6-dichlorobenzonitrile. *Plant Physiology*, **84**, 415–420.

Delye, C., Wang, T. & Darmency, H. (2002) An isoleucine–leucine substitution in chloroplastic acetyl-CoA carboxylase from green foxtail (*Setaria viridis* L. Beauv.) is responsible for resistance to the cyclohexanedione herbicide sethoxydim. *Planta*, **214**, 421–427.

Delye, C., Zhang, X. Q., Chalopin, C., Michel, S. & Powles, S. B. (2003) An isoleucine residue within the carboxyl-transferase domain of multidomain acetyl-coenzyme A carboxylase is a major determinant of sensitivity to aryloxyphenoxypropionate but not to cyclohexanedione inhibitors. *Plant Physiology*, **132**, 1716–1723.

DeRidder, B. P., Edwards, R. Goldsbrough, P. B. Dixon, D. P. & Beussman, D. J. (2002) Induction of glutathione S-transferases in Arabidopsis by herbicide safeners. *Plant Physiology*, **130**, 1497–1505.

Deshpande, S. & Hall, J. C. (1996) ATP-dependent auxin- and auxinic herbicide-induced volume changes in isolated protoplast suspensions from *Sinapis arvensis* L. *Pesticide Biochemistry and Physiology*, **56**, 26–43.

Desprez, T., Vernhettes, S., Fagard, M., Refregier, G., Desnos, T., Aletti, E., Py, N., Pelletier, S. & Hofte, H. (2002) Resistance against herbicide isoxaben and cellulose deficiency caused by distinct mutations in same cellulose synthase isoform CESA6. *Plant Physiology*, **128**, 482–490.

Devine, M. D. (1997) Mechanisms of resistance to acetyl-coenzyme A carboxylase inhibitors: a review. *Pesticide Science*, **51**, 259–264.

Doering, C. B. & Danner, D. J. (2000) Amino acid deprivation induces translation of branched-chain alpha-ketoacid dehydrogenase kinase. *American Journal of Physiology – Cell Physiology*, **279**, 1587–1594.

Donn, G. (1997) Herbicide resistant crops generated by biotechnology. In R. DePrado (ed.) *Weed and Crop Resistance to Herbicides*. Kluwer Academic Publishers, Dordrecht, The Netherlands, pp. 217–227.

Donn, G., Tischer, E., Smith, J. A. & Goodman, H. M. (1984) Herbicide-resistant alfalfa cells: an example of gene amplification in plants. *Journal of Molecular and Applied Genetics*, **2**, 621–636.

Duggleby, R. G. & Pang, S. S. (2000) Acetohydroxyacid synthase. *Journal of Biochemistry and Molecular Biology*, **33**, 1–36.

Duke, S. O. & Kenyon, W. H. (1986) Effects of dimethazone (FMC 57020) on chloroplast development. II. Pigment synthesis and photosynthetic function in cowpea (*Vigna unguiculata* L.) primary leaves. *Pesticide Biochemistry and Physiology*, **25**, 11–18.

Dyer, W. E. (1994) Mechanisms of plant resistance to glyphosate. In S. Powles & J. Holtum (eds) *Herbicide Resistance in Plants: Biology and Biochemistry*. CRC Press, Boca Raton, FL, pp. 229–241.

Dyer, W. E. (1997) Technological, ecological and social aspects of herbicide-resistant crops. In R. DePrado, J. Jorrín & L. García-Torres (eds) *Weed and Crop Resistance to Herbicides*. Kluwer Academic Publishers, Dordrecht, The Netherlands, pp. 249–258.

Dyer, W. E., Chee, P. W. & Fay, P. K. (1993) Rapid germination of sulfonylurea-resistant *Kochia scoparia* accessions is associated with elevated seed levels of branched chain amino acids. *Weed Science*, **41**, 18–22.

Eberlein, C. V. & Guttieri, M. J. (1994) Potato (*Solanum tuberosum*) response to simulated drift of imidazolinone herbicides. *Weed Science*, **42**, 70–75.

Falk, J., Andersen, G., Kernebeck, B. & Krupinska, K. (2003) Constitutive overexpression of barley 4-hydroxyphenylpyruvate dioxygenase in tobacco results in elevation of the vitamin E content in seeds but not in leaves. *FEBS Letters*, **540**, 35–40.

Falk, J., Krauss, N., Daehnhardt, D. & Krupinska, K. (2002) The senescence associated gene of barley encoding 4-hydroxyphenyl-pyruvate dioxygenase is expressed during oxidative stress. *Journal of Plant Physiology*, **159**, 1245–1253.

Fayez, K. A. & Kristen, Y. (1996) The influence of herbicides on the growth and proline content of primary roots and on the ultrastructure of root caps. *Environmental and Experimental Botany*, **36**, 71–81.

Fedtke, C. (1982) *Biochemistry and Physiology of Herbicide Action*. Springer-Verlag, New York, 202 pp.

Felsot, A. S., Reisenauer, G., Mink, G. I. & Bhatti, M. A. (1996) Biomonitoring with sentinel plants to assess exposure of nontarget crops to atmospheric deposition of herbicide residues. *Environmental Toxicology and Chemistry*, **15**, 452–459.

Feng, P. C. C., Pratley J. E. & Bohn, J. A. (1999) Resistance to glyphosate in *Lolium rigidum*. II. Uptake, translocation, and metabolism. *Weed Science*, **47**, 412–415.

Forlani, G., Nielsen E. & Bacchi, M. L. (1992) A glyphosate resistant 5-enol-pyruvyl-shikimate-3-pho synthase confers tolerance to a maize cell line. *Plant Science*, **85**, 9–15.

Fosket, D. E. & Morejohn, L. C. (1992) Structural and functional organization of tubulin. *Annual Review of Plant Physiology and Plant Molecular Biology*, **43**, 201–240.

Franz, J. F., Mao, M. K. & Sikorski, J. A. (2000) *Glyphosate – A Unique, Global Herbicide*. American Chemical Society Monograph Series. American Chemical Society, Washington, DC, 688 pp.

Fraser, P. D., Misawa, N., Linden, H., Yamano, S., Kobayashi, K. & Sandmann, G. (1992) Expression in *Escherichia-coli* purification and reactivation of the recombinant *Erwinia uredovora* phytoene desaturase. *Journal of Biological Chemistry*, **267**, 19891–19895.

Friml, J. & K. Palme. (2002) Polar auxin transport – old questions and new concepts? *Plant Molecular Biology*, **49**, 273–284.

Fuerst, E. P. & Vaughn, K. C. (1990) Mechanisms of paraquat resistance. *Weed Technology*, **4**, 150–156.

Garg, A. K., Kim, J. K., Owens, T. G., Ranwala, A. P., Do Choi, Y., Kochian, L. V. & Wu, R. J. (2002) Trehalose accumulation in rice plants confers high tolerance levels to different abiotic stresses. *Proceedings of the National Academy of Sciences of the United States of America*, **99**, 15898–15903.

Gaston, S., Zabalza, A., González, E. M., Arrese-Igor, C., Aparicio-Tejo, P. M. & Royuela, M. (2002) Imazethapyr, an inhibitor of the branched-chain amino acid biosynthesis, induces aerobic fermentation in pea plants. *Physiologia Plantarum*, **114**, 524–532.

Gilreath, J. P., Chase, C. A. & Locascio, S. J. (2001) Crop injury from sublethal rates of herbicide. I. Tomato. *HortScience*, **36**, 669–673.

Goddard, R. H., Wick, S. M., Silflow, C. D. & Snustad, P. D. (1994) Microtubule components of the plant cell cytoskeleton. *Plant Physiology*, **104**, 1–6.

Goodwin, J. L., Pastori, G. M., Davey, M. R. & Jones, H. D. (2004) Selectable markers: antibiotic and herbicide resistance. *Methods in Molecular Biology*, **286**, 191–202.

Goss, G. A. & Dyer, W. E. (2003) Physiological characterization of auxinic herbicide-resistant biotypes of *Kochia scoparia*. *Weed Science*, **51**, 839–844.

Gray, J. A., Stoltenberg, D. E. & Balke, N. E. (1995) Absence of herbicide cross-resistance in two atrazine-resistant velvetleaf (*Abutilon theophrasti*) biotypes. *Weed Science*, **43**, 352–357.

Gressel, J. (2002) *Molecular Biology of Weed Control*. Taylor & Francis, London, 504 pp.

Gronwald, J. W. (1991) Lipid biosynthesis inhibitors. *Weed Science*, **39**, 435–449.

Gronwald, J. W. (1994) Herbicides inhibiting acetyl-CoA carboxylase. *Biochemical Society Transactions*, **22**, 616–621.

Grossmann, K. & Kwiatkowski, J. (2000) The mechanism of quinclorac selectivity in grasses. *Pesticide Biochemistry and Physiology*, **66**, 83–91.

Guerineau, F., Broks, L., Meadows, J., Lucy, A., Robinson, C. & Mullineaux, P. (1990) Sulfonamide resistance gene for plant transformation. *Plant Molecular Biology*, **15**, 127–136.

Guttieri, M. J., Eberlein, C. V., Mallory-Smith, C. A. & Thill, D. C. (1996) Molecular genetics of target-site resistance to acetolactate synthase inhibiting herbicides. In T. M. Brown (ed.) *Molecular Genetics and Evolution of Pesticide Resistance*. American Chemical Society, Washington, DC, pp. 10–16.

Guttieri, M. J., Eberlein, C. V. & Souza, E. J. (1998) Inbreeding coefficients of field populations of *Kochia scoparia* using chlorsulfuron resistance as a phenotypic marker. *Weed Science*, **46**, 521–525.

Hall, J. C., Hoagland, R. E. & Zablotowicz, R. M. (eds) (2001) *Pesticide Biotransformation in Plants and Microorganisms: Similarities and Divergences*. Oxford University Press, Oxford, UK.

Han, O., Kim, O., Kim, C., Park, R.-D. & Guh, J.-O. (1995) Role of autooxidation of protoporphyrinogen IX in the action mechanism of diphenyl ether herbicides. *Bulletin of the Korean Chemical Society*, **16**, 1013–1014.

Harikrishnan, R. & Yang, X. B. (2002) Effects of herbicides on root rot and damping-off caused by *Rhizoctonia solani* in glyphosate-tolerant soybean. *Plant Disease*, **86**, 1369–1373.

Harwood, J. L. (1988a) Fatty acid metabolism. *Annual Review of Plant Physiology and Plant Molecular Biology*, **39**, 110–121.

Harwood, J. L. (1988b) The site of action of some selective graminaceous herbicides is identified as acetyl-CoA carboxylase. *Trends in Biochemical Sciences*, **13**, 330–331.

Harwood, J. L. (1996) Recent advances in the biosynthesis of plant fatty acids. *Biochimica et Biophysica Acta*, **1301**, 7–56.

Heap, I. (2004) *The International Survey of Herbicide Resistant Weeds*. Online. Internet. August 13, Available www.weedscience.com.

Heim, D. R., Larrinua, I. M., Murdoch, M. G. & Roberts, J. L. (1998) Triazofenamide is a cellulose biosynthesis inhibitor. *Pesticide Biochemistry and Physiology*, **59**, 163–168.

Hernandez, A., Garcia-Plazaola, J. I. & Becerril, J. M. (1999) Glyphosate effects on phenolic metabolism of nodulated soybean (*Glycine max* L. Merr.). *Journal of Agricultural and Food Chemistry*, **47**, 2920–2925.

Hofgen, R., Laber, B., Schuttke, I., Klonus, A.-K., Streber, W. & Pohlenz, H.-D. (1995) Repression of acelolactate synthase activity through antisense inhibition: molecular and biochemical analysis of transgenic potato (*Solanum tuberosum* L. cv. Desiree) plants. *Plant Physiology*, **107**, 469–477.

Hole, S. J. W., Howe, P. W. A., Stanley, P. D. & Hadfield, S. T. (2000) Pattern recognition analysis of endogenous cell metabolites for high throughput mode of action identification: removing the postscreening dilemma associated with whole-organism high throughput screening. *Journal of Biomolecular Screening*, **5**, 335–342.

Holmsen, J. D. & Hess, F. D. (1984) Growth inhibition and disruption of mitosis by DCPA in oat (*Avena sativa*) roots. *Weed Science*, **32**, 732–738.

Howe, A. R., Gasser, C. S., Brown, S. M., Padgette, S. R., Hart, J., Parker, G. B., Fromm, M. E. & Armstrong, C. L. (2002) Glyphosate as a selective agent for the production of fertile transgenic maize (*Zea mays* L.) plants. *Molecular Breeding*, **10**, 153–164.

Jacobs, J. M. & Jacobs, N. J. (1993) Porphyrin accumulation and export by isolated barley (*Hordeum vulgare*) plastids. *Plant Physiology*, **101**, 1181–1187.

Jacobs, J. M., Jacobs, N. J. & Duke, S. O. (1996) Protoporphyrinogen destruction by plant extracts and correlation with tolerance to protoporphyrinogen oxidase-inhibiting herbicides. *Pesticide Biochemistry and Physiology*, **55**, 77–83.

Jacobs, J. M., Jacobs, N. J., Sherman, T. D. & Duke, S. O. (1991) Effect of diphenyl ether herbicides on oxidation of protoporphyrinogen to protoporphyrin in organellar and plasma membrane enriched fractions of barley. *Plant Physiology*, **97**, 197–203.

Jasieniuk, M., Brûlé-Babel, A. L. & Morrison, I. N. (1996) The evolution and genetics of herbicide resistance in weeds. *Weed Science*, **44**, 176–193.

Jia, M. H., LaRossa, R. A., Lee, J.-M., Rafalski, A., DeRose, E., Gonye, G. & Xue, Z. (2000) Global expression profiling of yeast treated with an inhibitor of amino acid biosynthesis, sulfometuron methyl. *Genomics*, **3**, 83–92.

Jia, M., Keutgen, N., Matsuhashi, S., Mitzuniwa, C., Ito, T., Fujimura, T. & Hashimoto, S. (2001) Ion chromatographic analysis of selected free amino acids and cations to investigate the change of nitrogen metabolism by herbicide stress in soybean (*Glycine max*). *Journal of Agricultural and Food Chemistry*, **49**, 276–280.

Johal, G. S. & Rahe, J. E. (1988) Glyphosate hypersensitivity and phytoalexin accumulation in the incompatible bean anthracnose host-parasite interaction. *Physiological and Molecular Plant Pathology*, **32**, 267–282.

Jones, J. D., Goldsbrough P. B. & Weller, S. C. (1996) Stability and expression of amplified EPSPS genes in glyphosate resistant tobacco cells and plantlets. *Plant Cell Reports*, **15**, 431–436.

Joyce, S. M., Cassells, A. C. & Jain, S. M. (2003) Stress and aberrant phenotypes in *in vitro* culture. *Plant Cell Tissue and Organ Culture*, **74**, 103–121.

Kawate, M. K., Kraft, J. M., Ogg, A. G. & Colwell, S. G. (1997) Effect of glyphosate-treated henbit (*Lamium amplexicaule*) and downy brome (*Bromus tectorum*) on *Fusarium solani* f.sp. pisi and *Pythium ultimum*. *Weed Science*, **45**, 739–743.

Keen, N. T., Holliday, M. J. & Yoshikawa, M. (1982) Effects of glyphosate on glyceollin production and expression of resistance to *Phytophthora megasperma* f sp glycinea in soybean (*Glycine max*) cultivar Harosoy 63. *Phytopathology*, **72**, 1467–1469.

Kern, A. J., Chaverra, M. E., Cranston, H. J. & Dyer, W. E. (2004) Dicamba responsive genes in herbicide-resistant and susceptible biotypes of *Kochia scoparia*. *Weed Science*, **54**, (in press).

Kern A. J., Colliver, C. T., Maxwell, B. D., Fay, P. K. & Dyer, W. E. (1996a) Characterization of wild oat (*Avena fatua* L.) populations and an inbred line with multiple herbicide resistance. *Weed Science*, **44**, 847–852.

Kern A. J., Peterson D. M., Miller E. K., Colliver C. T. & Dyer, W. E. (1996b) Triallate resistance in *Avena fatua* L. is due to reduced herbicide activation. *Pesticide Biochemistry and Physiology*, **56**, 163–173.

Kern, A. J., Murray, B. E., Jasieniuk, M., Maxwell, B. D. & Dyer, W. E. (2002) Triallate resistance in *Avena fatua* L. is conferred by two recessive nuclear genes. *Journal of Heredity*, **93**, 48–50.

Knights, K. M. (2003) Long-chain-fatty-acid coA ligases: the key to fatty acid activation, formation of xenobiotic acyl-coA thioesters and lipophilic xenobiotic conjugates. *Current Medicinal Chemistry – Immunology, Endocrine and Metabolic Agents*, **3**, 235–244.

Knörzer, O. C., Durner, J. & Böger, P. (1996) Alterations in the antioxidative system of supension-cultured soybean cells (*Glycine max*) induced by oxidative stress. *Physiologia Plantarum*, **97**, 388–396.

Konishi, T. & Sasaki, Y. (1994) Compartmentalization of two forms of acetyl-CoA carboxylase in plants and the origin of their tolerance toward herbicides. *Proceedings of the National Academy of Sciences of the United States of America*, **91**, 3598–3601.

Konishi, T., Shinohara, K., Yamada, K. & Sasaki, Y. (1996) Acetyl-CoA carboxylase in higher plants: most plants other than gramineae have both the prokaryotic and the eukaryotic forms of this enzyme. *Plant and Cell Physiology*, **37**, 117–122.

Koo, S. J., Neal, J. C. & DiTomaso, J. M. (1996) 3,7-Dichloroquinolinecarboxylic acid inhibits cell-wall biosynthesis in maize roots. *Plant Physiology*, **112**, 1383–1389.

Kovalchuk, I., Filkowski, J., Smith, K. & Kovalchuk, O. (2003) Reactive oxygen species stimulate homologous recombination in plants. *Plant Cell and Environment*, **26**, 1531–1539.

LaRossa, R. A. & Schloss, J. V. (1984) The sulfonyl urea herbicide sulfometuron-methyl is an extremely potent and selective inhibitor of acetolactate synthase (EC 4.1.3.18) in *Salmonella typhimurium*. *Journal of Biological Chemistry*, **259**, 8753–8757.

LaRossa, R. A., Van Dyk, T. K. & Smulski, D. R. (1987) Toxic accumulation of alpha-ketobutyrate caused by inhibition of the branched-chain amino acid biosynthetic enzyme acetolactate synthase in *Salmonella typhimurium*. *Journal of Bacteriology*, **169**, 1372–1378.

Lea, P. J. & Ridley, S. M. (1989) Glutamine synthase and its inhibition. *Seminar Series – Society for Experimental Biology*, **38**, 137–170.

Lee, D. L., Prisbylla, M. P., Cromartie, T. H., Dagarin, D. P., Howard, S. W., Provan, W. M., Ellis, M. K., Fraser, T. & Mutter, L. C. (1997) The discovery and structural requirements of inhibitors of p-hydroxyphenylpyruvate dioxygenase. *Weed Science*, **45**, 601–609.

Lee, T. T. (1984) Release of lateral buds from apical dominance by glyphosate in soybean (*Glycine max*) cultivar Evans and pea (*Pisum sativum*) cultivar Alaska seedlings. *Journal of Plant Growth Regulation*, **3**, 227–236.

Levesque, C. A. & Rahe, J. E. (1992) Herbicide interactions with fungal root pathogens, with special reference to glyphosate. *Annual Review of Phytopathology*, **30**, 579–602.

Linden, H., Misawa, N., Chamovitz, D., Pecker, I., Hirschberg, J. & Sandmann, G. (1991) Functional complementation in Escherichia coli of different phytoene desaturase genes and analysis of accumulated carotenes. *Zeitschrift fur Naturforschung*, **46c**, 1045–1051.

Liu, C. M., Mclean, P. A., Sookdeo, C. C. & Cannon, F. C. (1991) Degradation of the herbicide glyphosate by members of the family rhizobiaceae. *Applied and Environmental Microbiology*, **57**, 1799–1804.

Loosli, R. (1995) Epidemiology of atrazine. *Reviews of Environmental Contamination and Toxicology*, **143**, 47–57.

Lorraine-Colwill, D. F., Hawkes, T. R., Williams, P. H., Warner, S. A. J., Sutton, P. B., Powles, S. B. & Preston, C. (1999) Resistance to glyphosate in *Lolium rigidum*. *Pesticide Science*, **55**, 489–491.

Lutzow, M., Beyer, P. & Kleinig, H. (1990) The herbicide Command does not inhibit the prenyl diphosphate-forming enzymes in plastids. *Zeitschrift fur Naturforschung*, **45c**, 856–858.

Mack, R. N., Simberloff, D., Lonsdale, W. M., Evans, H., Clout, M. & Bazzaz, F. (2000) Biotic invasions: causes, epidemiology, global consequences and control. *Ecological Applications*, **10**(3), 689–710.

Malik, J., Barry, G. & Kishore, G. (1989) The herbicide glyphosate. *BioFactors*, **2**, 17–25.

Maneechote, C., Preston, C. & Powles, S. B. (1997) A diclofop-methyl-resistant *Avena sterilis* biotype with a herbicide-resistant acety-coenzye A carboxylase and enhanced metabolism of diclofop-methyl. *Pesticide Science*, **49**, 105–114.

Mangels, G. (1991) Behavior of the imidazolinone herbicides in soil – a review of the literature. In D. L. Shaner, S. L. O'Connor (eds) *The Imidazolinone Herbicides*. CRC Press, Boca Raton, FL, pp. 191–210.

Mårtensson, A. (1992) Effects of agrochemicals and heavy metals on fast-growing rhizobia and their symbiosis with small-seeded legumes. *Soil Biology and Biochemistry*, **24**, 435–445.

Masabni, J. G. & Zandstra, B. H. (1999) A serine to threonine mutation in linuron-resistant *Portulaca oleracea*. *Weed Science*, **47**, 393–400.

Matthes, B. & Böger, P. (2002) Chloroacetamides affect the plasma membrane. *Zeitschrift fur Naturforschung Section C – Journal of Biosciences*, **57**, 843–852.

Mengistu, L. W. & Messersmith, C. G. (2002) Genetic diversity of kochia. *Weed Science*, **50**,498–503.

Moorman, T. B., Becerril, J. M., Lydon, J. & Duke, S. O. (1992) Production of hydroxybenzoic acids by *Bradyrhizobium japonicum* strains after treatment with glyphosate. *Journal of Agricultural and Food Chemistry*, **40**, 289–293.

Mori, I. C. & Schroeder, J. I. (2004) Reactive oxygen species activation of plant Ca^{2+} channels. A signaling mechanism in polar growth, hormone transduction, stress signaling, and hypothetically mechanotransduction. *Plant Physiology*, **135**, 702–708.

Moskalenko, A. A. & Karapetyan, N. V. (1996) Structural role of carotenoids in photosynthetic membranes. *Zeitschrift fur Naturforschung*, **51c**, 763–771.

Mougin, C., Polge, N., Scalla, R. & Cabanne, F. (1991) Interactions of various agrochemicals with cytochrome P-450-dependent monooxygenases of wheat cells. *Pesticide Biochemistry and Physiology*, **40**, 1–11.

Mueller, J. G., Skipper, H. D., Lawrence, E. G. & Kline, E. L. (1989) Bacterial stimulation by carbamothioate herbicides. *Weed Science*, **37**, 424–427.

Naber, J. D. & van Rensen, J. J. S. (1991) Activity of photosystem II herbicides is related with the residence times at the D1 protein. *Zeitschrift fur Naturforschung*, **46c**, 575–578.

Napier, R. M., David, K. M. & Perrot-Rechenmann, C. (2002) A short history of auxin-binding proteins. *Plant Molecular Biology*, **49**, 339–348.

Navas, M. L. (1991) Using plant populations biology in weed research: a strategy to improve weed management. *Weed Research*, **31**, 171–179.

Nelson, K. A., Hammerschmidt, R. & Renner, K. A. (2002) Cultivar and herbicide selection affects soybean development and the incidence of Sclerotinia stem rot. *Agronomy Journal*, **94**, 1270–1281.

Neve, P., Diggle, A. J., Smith, F. P. & Powles, S. B. (2003) Simulating evolution of glyphosate resistance in *Lolium rigidum*. II. Past, present and future glyphosate use in Australian cropping. *Weed Research*, **43**, 418–427.

Nikolskaya, T., Zagnitko, O., Tevzadze, G., Haselkorn, R. & Gornicki, P. (1999) Herbicide sensitivity determinant of wheat plastid acetyl-CoA carboxylase is located in a 400-amino acid fragment of the carboxyltransferase domain. *Proceedings of the National Academy of Sciences of the United States of America*, **96**, 14647–14651.

Niks, R. E. & Rubiales, D. (2002) Potentially durable resistance mechanisms in plants to specialised fungal pathogens. *Euphytica*, **124**, 201–216.

Niyogi, K. K. (1999) Photoprotection revisited: genetic and molecular approaches. *Plant Physiology and Plant Molecular Biology*, **50**, 333–359.

Njiti, V. N., Lighfoot, D. A., Schroeder, D. & Myers, O. (2003) Roundup ready soybean: glyphosate effects on *Fusarium solani* root colonization and sudden death syndrome. *Agronomy Journal*, **95**, 1140–1145.

Nolte, S. A., Young, B. G., Mungur, R. & Lightfoot, D. A. (2004) The glutamate dehydrogenase gene *gdhA* increased the resistance of tobacco to glufosinate. *Weed Research*, **44**, 335–339.

Norris, S. R., Barrette, R. R. & DellaPenna, D. (1995) Genetic dissection of carotenoid synthesis in Arabidopsis defines plastoquinone as an essential component of phytoene desaturation. *Plant Cell*, **7**, 2139–2149.

Norris, S. R., Shen, X. & DellaPenna, D. (1998) Complementation of the Arabidopsis pds1 mutation with the gene encoding p-hydroxyphenylpyruvate dioxygenase. *Plant Physiology*, **117**, 1317–1323.

O'Donovan, J. T., Newman, J. C., Blackshaw, R. E., Harker, N. E., Derksen, D. A. & Thomas, A. G. (1999) Growth, competitiveness, and seed germination of triallate/difenzoquat-susceptible and -resistant wild oat populations. *Canadian Journal of Plant Science*, **79**, 303–312.

O'Donovan, J. T., Sharma, M. P., Harker, K. N., Maurice, D., Baig, M. N. & Blackshaw, R. E. (1994) Wild oat (*Avena fatua*) populations resistant to triallate are also resistant to difenzoquat. *Weed Science*, **42**, 195–199.

Ohkawa, H., Tsujii, H. & Ohkawa, Y. (1999) The use of cytochrome P450 genes to introduce herbicide tolerance in crops: a review. *Pesticide Science*, **55**, 867–874.

Ossipov, V., Salminen, J. P., Ossipova, S., Haukioja, E. & Pihlaja, K. (2003) Gallic acid and hydrolysable tannins are formed in birch leaves from an intermediate compound of the shikimate pathway. *Biochemical Systemetics and Ecology*, **31**, 3–16.

Ott, K. H., Aranibar, N., Singh, B. & Stockton, G. W. (2003) Metabonomics classifies pathways affected by bioactive compounds. Artificial neural network classification of NMR spectra of plant extracts. *Phytochemistry*, **62**, 971–985.

Pallett, K. E., Cramp, S. M., Little, J. P., Veerasekaran, P., Crudace, A. J. & Slater, A. E. (2001) Isoxaflutole: the background to its discovery and the basis of its herbicidal properties. *Pest Management Science*, **57**, 133–142.

Pear, J. R., Kawagoe, Y., Schreckengost, W. E., Delmer, D. P. & Stalker, D. M. (1996) Higher plants contain homologs of the bacterial *celA* genes encoding the catalytic subunit of cellulose synthase. *Proceedings of the National Academy of Sciences of the United States of America*, **93**, 12637–12642.

Perez-Garcia, A., Pereira, S., Pissarra, J., Garcia-Gutierrez, A., Cazorla, F. M., Salema, R., DeVicente, A. & Canovas, F. M. (1998) Cytosolic localization in tomato mesophyll cells of a novel glutamine synthetase induced in response to bacterial infection or phosphinothricin treatment. *Planta*, **206**, 426–434.

Pinto, J. E. B. P., Dyer, W. E., Weller, S. C. & Herrman, K. M. (1988) Glyphosate induces 3-deoxy-D-arabino-heptulosonate-synthase in potato (*Solanum tuberosum* L.) cells grown in suspension culture. *Plant Physiology*, **87**, 891–893.

Plewa, M. J. (1985) Mutation testing with maize. *Basic Life Science*, **34**,323–328.

Pornprom, T., Matsumoto, H., Usui, K. & Ishizuka, K. (1994) Characterization of oxyfluorfen tolerance in selected soybean cell line. *Pesticide Biochemistry and Physiology*, **50**, 107–114.

Powles, S. B. & Holtum, J. A. M. (eds.) (1994) *Herbicide Resistance in Plants – Biology and Biochemistry*. CRC Press, Boca Raton, FL.

Price, L. J., Herbert, D., Cole, D. J. & Harwood, J. L. (2003) Use of plant cell cultures to study graminicide effects on lipid metabolism. *Phytochemistry*, **63**, 533–541.

Ramanjini-Gowda, P. H. & Prakash, C. S. 1998. Herbicide glyphosate at sublethal concentrations enhances somatic embryo development in sweetpotato (*Ipomoea batatas* L.). *Current Science Bangalore*, **75**, 508–510.

Rashid, A., O'Donovan, J. T., Khan, A. A., Blackshaw, R. E., Harker, K. N. & Pharis, R. P. (1998) A possible involvement of gibberellin in the mechanism of *Avena fatua* resistance to triallate and cross-resistance to difenzoquat. *Weed Research*, **3**, 461–466.

Ray, T. B. (1982) The mode of action of chlorsulfuron: a new herbicide for cereals. *Pesticide Biochemistry and Physiology*, **17**, 10–17.

Reddy, K. N., Hoagland, R. E. & Zablotowicz, R. M. (2000) Effect of glyphosate on growth, chlorophyll, and nodulation in glyphosate-resistant and susceptible soybean (*Glycine max*) varieties. *Journal of New Seeds*, **2**, 37–52.

Rendina, A. R., Felts, J. M., Beaudoin, J. D., Craig-Kennard, A. C., Look, L. L., Paraskos, S. L. & Hagenah, J. A. (1988) Kinetic characterization stereoselectivity and species selectivity of the inhibition of plant acetyl-coenzyme A carboxylase by the aryloxyphenoxypropionic acid grass herbicides. *Archives of Biochemistry and Biophysics*, **265**, 219–225.

Rhodes, D., Hogan, A. L., Deal, L., Jamison, G. C. & Haworth, P. (1987) Amino acid metabolism of *Lemna minor* L. III. Responses to chlorsulfuron. *Plant Physiology*, **84**, 775–780.

Roberts, C. W., Roberts, F., Lyons, R. E., Kirisits, M. J., Mui, E. J., Finnerty, J., Johnson, J. J., Ferguson, D. J. P., Coggins, J. R., Krell, T., Coombs, G. H., Milhous, W. K., Kyle, D. E., Tzipori, S., Barnwell, J., Dame, J. B., Carlton, J. & McLeod, R. (2002) The shikimate pathway and its branches in apicomplexan parasites. *Journal of Infectious Diseases*, **185**, S25–S36.

Robineau, T., Batard, Y., Nedelkina, S., Cabello-Hurtado, F., LeRet, M., Sorokine, O., Didierjean, L. & Werck-Reichhart, D. (1998) The chemically inducible plant cytochrome P450 CYP76B1 actively metabolizes phenylureas and other xenobiotics. *Plant Physiology*, **118**, 1049–1056.

Roesler, K., Shintani, D., Savage, L., Boddupalli, S. & Ohlrogge, J. (1997) Targeting of the Arabidopsis homomeric acetyl-coenzyme A carboxylase to plastids of rapeseeds. *Plant Physiology*, **113**, 75–81.

Rost, T. L. (1984) The comparative cell cycle and metabolic effects of chemical treatments on root tip meristems. III. Chlorsulfuron. *Journal of Plant Growth Regulation*, **3**, 51–63.

Royuela, M., González, A., González, E. M., Arrese-Igor, C., Aparicio-Tejo, P. M. & González-Murua, C. (2000) Physiological consequences of continuous, sublethal imazethapyr supply to pea plants. *Journal of Plant Physiology*, **157**, 345–354.

Ryan, G. F. (1970) Resistance of common groundsel to Simazine and Atrazine. *Weed Science*, **18**, 614–616.

Saari, L. L., Cotterman, J. C. & Thill, D. C. (1994) Resistance to acetolactate synthase-inhibitor herbicides. In S. B. Powles & J. A. M. Holtum (eds) *Herbicide Resistance in Plants: Biology and Biochemistry*. Lewis Publishers, Boca Raton, FL, pp. 80–139.

Sabba, R. P. & Vaughn, K. C. (1999) Herbicides that inhibit cellulose biosynthesis. *Weed Science*, **47**, 757–763.

Saladin, G., Magne, C. & Clement, C. (2003) Effects of flumioxazin herbicide on carbon nutrition of *Vitis vinifera* L. *Journal of Agricultural and Food Chemistry*, **51**, 4017–4022.

Sandmann, G. & Böger, P. (1986) Interference of dimethazone with formation of terpenoid compounds. *Zeitschrift fur Naturforschung*, **41c**, 729–732.

Sandmann, G. & Böger, P. (1987a) Herbicides affecting plant pigments. *Proceedings of the British Crop Protection Conference – Weeds*. BCPC Publications, Hampshire, UK, pp. 139–148.

Sandmann, G. & Böger, P. (1987b) Interconversion of prenyl pyrophosphates and subsequent reactions in the presence of FMC 57020. *Zeitschrift fur Naturforschung*, **42c**, 803–807.

Sandmann, G. & Fraser, P. D. (1993) Differential inhibition of phytoene desaturases from diverse origins and analysis of resistant cyanobacterial mutants. *Zeitschrift fur Naturforschung*, **48c**, 307–311.

Sandmann, G., Misawa, N. & Böger, P. (1996) Steps towards genetic engineering of crops resistant against bleaching herbicides. In S. O. Duke (ed.) *Herbicide Resistant Crops. Agricultural, Environmental, Economic, Regulatory, and Technical Aspects*. Lewis Publishers, Boca Raton, FL, pp. 189–200.

Saroha, M. K., Sridhar, P. & Malik, V. S. (1998) Glyphosate-tolerant crops: genes and enzymes. *Journal of Plant Biochemistry and Biotechnology*, **7**, 65–72.

Sasaki, Y. & Nagano, Y. (2004) Plant acetyl-CoA carboxylase: structure, biosynthesis, regulation, and gene manipulation for plant breeding. *Bioscience, Biotechnology and Biochemistry*, **68**, 1175–1184.

Schibler, M. J. & Huang, B. (1991) The *colR4* and *colR15* beta tubulin mutations in *Chlamydomonas reinhardtii* confer altered sensitivities to microtubule inhibitors and herbicides by enhancing microtubule stability. *Journal of Cell Biology*, **113**, 605–614.

Schloss, J. V., Ciskanik, L. M. & Van Dyk, D. E. (1988) Origin of the herbicide binding site of acetolactate synthase. *Nature*, **331**, 360–362.

Schloss, J. V., Hixon, M. S., Chu, F., Chang, S. & Duggleby, R. G. (1996) Products formed in the oxygen-consuming reactions of acetolactate synthase and pyruvate decarboxylase. In H. Bisswanger & A. Schellenberger (eds) *Biochemistry and Physiology of Thiamin Diphosphate Enzymes*. Intemann, Prien, Germany, pp. 580–585

Schmutz, A., Buchala, A. J. & Ryser, U. (1996) Changing the dimensions of suberin lamellae of green cotton fibers with a specific inhibitor of the endoplasmic reticulum-associated fatty acid elongases. *Plant Physiology*, **110**, 403–411.

Schultz, A., Ort, O., Beyer, P. & Kleinig, H. (1993) SC-0051, a 2-benzoylcyclohexane-1,3-dione bleaching herbicide, is a potent inhibitor of the enzyme ρ-hydroxyphenylpyruvate dioxygenase. *FEBS Letters*, **318**, 162–166.

Sellwood, C., Slabas, A. R. & Rawsthorne, S. (2000) Effects of manipulating expression of acetyl-CoA carboxylase I in *Brassica napus* L. embryos. *Biochemical Society Transactions*, **28**, 598–600.

Shaaltiel, Y. & Gressel, J. (1986) Multienzyme oxygen radical detoxifying system correlated with paraquat resistance in *Conyza bonbariensis*. *Pesticide Biochemistry and Physiology*, **26**, 22–28.

Shaner, D. L. & Singh, B. K. (1993) Phytotoxicity of acetohydroxyacid synthase inhibitors is not due to the accumulation of 2-ketobutyrate and/or 2-aminobutyrate. *Plant Physiology*, **103**, 1221–1226.

Sharma, U., Pfender, W. F. & Adee, E. A. (1989) Effect of glyphosate herbicide on pseudothecia formation by *Pyrenophora tritici-repentis* in infested wheat straw. *Plant Disease*, **73**, 647–650.

Sharon, A., Amsellem, Z. & Gressel, J. (1992) Glyphosate suppression of an elicited defense response increased susceptibility of *Cassia obtusifolia* to a mycoherbicide. *Plant Physiology*, **98**, 654–659.

Sherman, T. D., Vaughn, K. C. & Duke, S. O. (1996) Mechanisms of action and resistance to herbicides. In S. O. Duke (ed.) *Herbicide-resistant Crops: Agricultural, Environmental, Economic, Regulatory, an Technical Aspects*. CRC Press, Boca Raton, FL, pp. 13–36.

Shiboleth, Y. M., Arazi, T., Wang, Y. & Gal-On, A. (2001) A new approach for weed control in a cucurbit field employing an attenuated potyvirus-vector for herbicide resistance. *Journal of Biotechnology*, **92**, 37–46.

Shukla, A., Nycholat, C., Subramanian, M. V., Anderson, R. J. & Devine, M. D. (2004) Use of resistant ACCase mutants to screen for novel inhibitors against resistant and susceptible forms of ACCase from grass weeds. *Journal of Agricultural and Food Chemistry*, **11**, 5144–5150.

Siciliano, S. D., Theoret, C. M., DeFreitas, J. R., Hucl, P. J. & Germida, J. J. (1998) Differences in the microbial communities associated with the roots of different cultivars of canola and wheat. *Canadian Journal of Microbiology*, **44**, 844–851.

Simarmata, M., Kaufmann, J. E. & Penner, D. (2001) Progress in determining the origin of the glyphosate-resistant ryegrass in California. *WSSA Abstracts*, **41**, 95.

Sindel, B. (1996) Glyphosate resistance discovered in annual ryegrass. *Resistant Pest Management*, **8**, 5–6.

Singh, B. K. (1999) Biosynthesis of valine, leucine, and isoleucine. In B. K. Singh (ed.) *Plant Amino Acids: Biochemistry and Biotechnology*. Marcel Dekker, New York, NY, pp. 227–247.

Skold, O. (2001) Resistance to trimethoprim and sulfonamides. *Veterinary Research*, **32**, 261–273.

Smeda, R. J. & Vaughn, K. C. (1994) Resistance to dinitroaniline herbicides. In S. B. Powles & J. A. M. Holtum (eds) *Herbicide Resistance in Plants – Biology and Biochemistry*. CRC Press, Boca Raton, FL, pp. 215–228.

Steffens, B., Feckler, C., Palme, K., Christian, M., Bottger, M. & Luthen, H. (2001) The auxin signal for protoplast swelling is perceived by extracellular ABP1. *Plant Journal: Cell and Molecular Biology*, **27**, 591–599.

Sterling, T. M. & Hall, J. C. (1997) Mechanism of action of natural auxins and the auxinic herbicides. In R. M. Roe (ed.) *Herbicide Activity: Toxicology, Biochemistry and Molecular Biology*. IOS Press, Amsterdam, The Netherlands.

Sulmon, C., Gouesbet, G., Couée, I. & El Amrani, A. (2004) Sugar-induced tolerance to atrazine in *Arabidopsis* seedlings: interacting effects of atrazine and soluble sugars on *psbA* mRNA and D1 protein levels. *Plant Science*, **167**, 913–923.

Surov, T., Aviv, D., Aly, R., Joel, D. M., Goldman-Guez, T. & Gressel, J. (1998) Generation of transgenic asulam-resistant potatoes to facilitate eradication of parasitic broomrapes (*Orobanche* spp.), with the *sul* gene as the selectable marker. *Theoretical and Applied Genetics*, **96**, 132–137.

Suzuki, Y., Kurano, M., Arai, Y., Nakashita, H., Doi, Y., Usami, R., Horikoshi, K. & Yamaguchi, I. (2002) Enzyme inhibitors to increase poly-3-hydroxybutyrate production by transgenic tobacco. *Bioscience, Biotechnology and Biochemistry*, **66**, 2537–2542.

Swarup, R., Parry, G., Graham, N., Allen, T. & Bennett, M. (2002) Auxin cross-talk: integration of signaling pathways to control plant development. *Plant Molecular Biology*, **49**, 411–426.

Tagede, M., Dupuis, I. & Kuhlemeier, C. (1999) Ethanolic fermentation: new functions for an old pathway. *Trends in Plant Science*, **4**, 320–325.

Takahashi, H. Ohki, A., Kato, S., Tanaka, A., Sato, Y., Matthes, B., Böger, P. & Wakabayashi, K. (2001) Inhibition of very-long-chain acid biosynthesis by 2-chloro-N-(3-methoxy-2-146.

Trenkamp, S., Martin, W. & Tietjen, K. (2004) Specific and differential inhibition of very-long-chain fatty acid elongases from *Arabidopsis thaliana* by different herbicides. *Proceedings of the National Academy of Sciences of the United States of America*, **101**, 11903–11908.

Tsegaye, Y., Shintani, D. K. & DellaPenna, D. (2002) Overexpression of the enzyme p-hydroxyphenolpyruvate dioxygenase in Arabidopsis and its relation to tocopherol biosynthesis. *Plant Physiology and Biochemistry*, **40**, 913–920.

Turcsanyi, E., Suranyi, G., Lehoczki, E. & Borbely, G. (1994) Superoxide dismutase activity in response to paraquat resistance in *Conyza canadensis* L. *Journal of Plant Physiology*, **144**, 599–606.

Uchimiya, H., Christensen, A. H., Anzai, H., Toki, S., Quail, P. H., Ooba, S., Nojirl, C., Iwata, M., Takamatsu, S. & Samarajeewa, P. K. (1993) Bialaphos treatment of transgenic rice plants expressing a *bar* gene prevents infection by the sheath blight pathogen (*Rhizoctonia solani*). *Biotechnology*, **11**, 835–836.

Van Dyk, T. K., Ayers, B. L., Morgan, R. W. & Larossa, R. A. (1998) Constricted flux through the branched-chain amino acid biosynthetic enzyme acetolactate synthase triggers elevated expression of genes regulated by rpoS and internal acidification. *Journal of Bacteriology*, **180**, 785–792.

Van Eerd, L. L., Hoagland, R. E., Zablotowicz, R. M. & Hall, C. J. (2003) Pesticide metabolism in plants and microorganisms. *Weed Science*, **51**, 472–495.

VanGessel, M. J. (2001) Glyphosate-resistant horseweed from Delaware. *Weed Science*, **49**, 703–705.

Veldhuis, L. J., Hall, L. M., O'Donovan, J. T., Dyer, W. & Hall, J. C. (2000) Metabolism-based resistance of a wild mustard (*Sinapis arvensis* L.) biotype to ethametsulfuron-methyl. *Journal of Agricultural and Food Chemistry*, **48**, 2986–2990.

Wall, D. A. (1997) Effect of crop growth stage on tolerance to low doses of thifensulfuron: tribenuron. *Weed Science*, **45**, 538–545.

Wan, M. T., Rahe, J. E. & Watts, R. G. (1998) A new technique for determining the sublethal toxicity of pesticides to the vesicular–arbuscular mycorrhizal fungus *Glomus intraradices*. *Environmental Toxicology and Chemistry*, **17**, 1421–1428.

Wang, A., Band, R. N. & Kopachik, W. (1995) Effects of trifluralin on growth and differentiation of the amoebo-flagellate *Naegleria*. *FEMS Microbiology Letters*, **127**, 99–103.

Wang, Y., Deshpande, S. & Hall, J. C. (2001) Calcium may mediate auxinic herbicide resistance in wild mustard. *Weed Science*, **49**, 2–7.

Warwick, S. I. & Black, L. D. (1994) Relatives fitness of herbicide-resistant and susceptible biotypes of weeds. *Phytoprotection*, **75**, 37–49.

Webb, S. R. & Hall, J. C. (1995) Auxinic herbicide-resistant and -susceptible wild mustard (*Sinapis arvensis* L.) biotypes: effect of auxinic herbicides on seedling growth and auxin-binding activity. *Pesticide Biochemistry and Physiology*, **52**, 137–148.

Wehrmann, A., Van Vliet, A., Opsomer, C., Botterman, J. & Schulz, A. (1996) The similarities of *bar* and *pat* gene products make them equally applicable for plant engineers. *Nature Biotechnology*, **14**, 1274–1278.

Weimer, M. R., Balke, N. E. & Buhler, D. D. (1992) Herbicide clomazone does not inhibit *in vitro* geranylgeranyl synthesis from mevalonate. *Plant Physiology*, **98**, 427–432.

Wells, B., McCann, M. C., Shedletzky, E., Delmer, D. & Roberts, K. (1994) Structural features of cell walls from tomato cells adapted to grow on the herbicide 2,6-dichlorobenzonitrile. *Journal of Microscopy*, **173**, 155–164.

Westwood, J. H. & Weller, S. C. (1997) Cellular mechanisms influence differential glyphosate sensitivity in field bindweed. *Weed Science*, **45**, 2–11.

Wiederholt, R. J. & Stoltenberg, D. E. (1995) Cross-resistance of a large crabgrass (*Digitaria sanguinalis*) accession to aryloxyphenoxypropionate and cyclohexanedione herbicides. *Weed Technology*, **9**, 518–524.

Wiederholt, R. J. & Stoltenberg, D. E. (1996) Absence of differential fitness between giant foxtail (*Setaria faberi*) accessions resistant and susceptible to acetyl-coenzyme A carboxylase inhibitors. *Weed Science*, **44**, 18–24.

Wild, A. & Wendler, C. (1993) Effect of glufosinate on amino acid content. Photorespiration and photosynthesis. *Pesticide Science*, **30**, 422–424.

Wilkinson, R. E. & Hardcastle, W. S. (1970) EPTC effects on total leaflet fatty acids and hydrocarbons. *Weed Science*, **18**, 125–128.

Yamato, S, Katagiri, M. & Ohkawa, H. (1994) Purification and characterization of a protoporphyrinogen-oxidizing enzyme with peroxidase activity and light-dependent herbicide resistance in tobacco cultured cells. *Pesticide Biochemistry and Physiology*, **50**, 72–82.

Yamato, S., Suzuki, Y., Katagiri, M. & Ohkawa, H. (1995) Protoporphyrinogen-oxidizing enzymes of tobacco cells with respect to light-dependent herbicide mode of action. *Pesticide Science*, **43**, 357–358.

Yeomans, J. C. & Bremner, J. M. (1985) Denitrification in soil effects of herbicides. *Soil Biology and Biochemistry*, **17**, 447–452.

Yuan, C. I., Chaing, M. Y. & Chen, Y. M. (2002) Triple mechanisms of glyphosate-resistance in a naturally occurring glyphosate-resistant plant *Dicliptera chinensis*. *Plant Science*, **16**, 543–554.

Zagnitko, O., Jelenska, J., Tevzadze, G., Haselkorn, R. & Gornicki, P. (2001) An isoleucine/leucine residue in the carboxyltransferase domain of acetyl-CoA carboxylase is critical for interaction with aryloxyphenoxypropionate and cyclohexanedione inhibitors. *Proceedings of the National Academy of Sciences of the United States of America*, **98**, 6617–6622.

Zheng, H.-G. & Hall, J. C. (2001) Understanding auxinic herbicide resistance in wild mustard: physiological, biochemical, and molecular genetic approaches. *Weed Science*, **49**, 276–281.

Zhen, R. G. & Singh, B. K. (2001) From inhibitors to target site genes and beyond: herbicidal inhibitors as powerful tools for functional genomics. *Weed Science*, **49**, 266–272.

Zhou, H., Arrowsmith, J. W., Fromm, M. E. & Hironika, C. M. (1995) Glyphosate tolerant CP4 and GOX genes as a selectable marker in wheat transformation. *Plant Cell Reports*, **15**, 159–163.

8 Integration of abiotic stress signaling pathways

Manu Agarwal and Jian-Kang Zhu

8.1 Introduction

Abiotic stresses are major impediments to plant growth and crop production. Tapping the genetic potential depends on both genotype (g) and environment (e), and therefore, the interaction g × e is a critical determinant for plant productivity. Stress perception and accompanying physiological, molecular and biochemical changes in plants largely depend on factors such as genotype, developmental stage and severity of stress. Often, at particular times, a multitude of stresses can change the ability of a plant to perceive stress.

Usually one kind of stress is accompanied or followed by another stress. For example, heat stress is accompanied by drought stress due to physical loss of water, and cold stress is followed by drought stress due to physiological unavailability of water. Because of the multiplicity of environmental insults a plant faces at a particular time, abiotic stress signaling is a very complex phenomenon. Plants have means to avoid and encounter these stresses. On one hand they are capable of producing inducible and appropriate responses that result in a specific change suited for that particular stress condition (Knight & Knight, 2001), while on the other, considerable overlap between abiotic stress signaling components exists, with nodal points at which stress signaling pathways integrate.

The first step in a signal transduction pathway is the perception of a signal, which is almost immediately followed by the generation of secondary signals. Secondary signals are generally nonprotein molecules such as Ca^{2+}, inositol phosphates (IPs) and reactive oxygen species (ROS), and the change in their levels provides a means to generate further signals (Xiong et al., 2002a). Each of the secondary signals is capable of initiating a protein phosphorylation cascade, which in turn can regulate the activity of specific transcription factors or target genes. Additionally, these signals give rise to further signaling molecules that can either modulate the levels of secondary signals, and thereby provide an additional checkpoint for a signal to flow in a specific manner, or initiate a protein phosphorylation cascade like their predecessors. Apart from these components that directly participate in stress signaling are components that do not relay signals themselves but are equally important. These components participate in modulating signaling components and include protein modifiers, scaffolds and adaptors (Xiong & Zhu, 2001).

A quick way to investigate components involved in a particular stress and components that overlap between different stresses is to analyze the whole genome of the transcriptome, then group the components according to function. Microarray technology has become a powerful tool for the systematic analysis of expression profiles of large numbers of genes (Richmond & Somerville, 2000). Transcript expression profiles at the magnitude of the near-whole genome have been reported in response to drought, cold and high salinity (Kawasaki et al., 2001; Seki et al., 2001, 2002a; Chen et al., 2002; Fowler & Thomashow, 2002; Kreps et al., 2002). Use of 7000 full-length cDNAs revealed 299 drought-inducible, 54 cold-inducible, 213 high-salinity and 245 abscisic acid (ABA)–inducible genes (Seki et al., 2002b). This analysis revealed considerable crosstalk between the signaling components of these stresses. Overlap of gene expression is greatest in response to drought, high salinity and ABA, and partial in response to cold and osmotic stress. In a separate study, 2409 unique stress-regulated genes were identified in Arabidopsis plants subjected to salt, mannitol and cold stress (Kreps et al., 2002). Most of the genes were found to be stimulus specific. However, at 3 hours of stress, nearly 5% of the genes analyzed were induced by all the stresses.

Microarray technology has been applied to study the expression levels of 402 transcription factor genes (Chen et al., 2002). This study revealed that 28 transcription factor genes were induced by stress. Many genes responsive to cold stress had elements similar to both ABA-responsive elements (ABRE) and dehydration-responsive elements (DRE) in their promoters (Chen et al., 2002). This observation gives further clues about crosstalk in abiotic stress signaling.

Studies to date have revealed many linear pathways. However, these pathways are actually part of a more complex signaling network, and it is imperative to connect these linear pathways for a holistic view of the signaling. In this chapter, we review individual signaling components that act as nodal points in this network. Important components that have been shown to interact with each other and constitute a branch of signaling pathway in abiotic stress are also discussed.

8.1.1 Sensors

Because of a plethora of stresses and their individual effects at the cellular level, it is easy to imagine the presence of sensors that specifically promote transduction of a particular stress signal to cellular targets. This implies the presence of different sensors for cold, drought or salinity. In spite of recent accelerations in gene discovery, only a few studies of function implicate a particular protein as a potential sensor. Most of the information on sensors of abiotic stress has been derived from the work on transmembrane two-component systems in simpler organisms such as Synechocystis and Bacilus *subtilis*. A typical two-component system consists of a membrane-localized histidine kinase that senses the input signal and a response regulator (Rres) that mediates the output. Transmembrane two-component histidine kinases HIK33 (*Synechocystis*) and DesK (*B. subtilis*) have been suggested as thermosensors (Suzuki *et al.*, 2000; Aguilar *et al.*, 2001).

Supplementary insights for the function of HIK33 were recently provided by Inaba *et al.* (2003) who assessed the expression of cold-inducible genes by engineering membrane rigidity. The authors used a double mutant of *Synechocystis desA$^-$/desD$^-$* with only monounsaturated lipid molecules, which therefore increased membrane rigidity. Microarray analysis of double mutants revealed the expression changes of cold-inducible genes, which were grouped on the basis of their induction levels into three categories: high, moderate or unaffected. Surprisingly, a triple mutant of *desA$^-$/desD$^-$/hik33$^-$* inhibited the cold inducibility of some genes in the moderate category (Inaba *et al.*, 2003). The results have two important implications: HIK33 senses the membrane rigidity to regulate the expression of cold-inducible genes, and other cold sensors may coexist. HIK33 was also proposed to function as an osmosensor (Mikami *et al.*, 2002). Thus, HIK33 represents a nodal point at which cold and osmotic stresses are integrated in *Synechocystis*. Other histidine kinases from cyanobacteria HIK16 and HIK34 could be putative salt sensors, and HIK41 could be a signal transducer (Marin *et al.*, 2003).

Budding yeast (*Saccharomyces cerevisiae*) undergoes rapid phosphorylation and activation of a mitogen-activated protein kinase (MAPK)-high-osmolarity glycerol response 1 (HOG1) in response to high osmolarity. HOG1 is activated by a MAPK kinase, PBS2, which in turn is activated by further upstream kinases. In-depth mutational analysis has revealed the presence of at least two osmosensors (SLN1 and SHO1) that independently activate the HOG1 pathway through a multistep phosphotransfer reaction (Maeda *et al.*, 1994, 1995; Posas *et al.*, 1996; Raitt *et al.*, 2000; O'Rourke *et al.*, 2002). *Arabidopsis thaliana* histidine kinase1 (AtHK1) has been considered as a candidate osmosensor because of its increased expression under high salt (250 mM NaCl) and low temperature (4°C) conditions (Urao *et al.*, 1999). AtHK1 confers a high degree of tolerance to osmotic stress in the double mutant *sln1Δsho1Δ*, which indicates that it can activate the HOG1 MAPK pathway in yeast (Urao *et al.*, 1999). Recently, the *Arabidopsis* histidine kinase Cre1, which functions as a cytokinin receptor (Inoue *et al.*, 2001; Yamada *et al.*, 2001), was observed to function as a sensor for changes in turgor pressure when expressed in yeast (Reiser *et al.*, 2003).

In addition to two-component histidine kinases, receptor-like protein kinases have been implicated in abiotic stress response. A transcript of *Nicotiana tabacum* clone, NtC7, a transmembrane protein, was found to accumulate rapidly and transiently with 1 h of salt and osmotic stress treatment. Transgenic plants overexpressing NtC7 exhibited a marked tolerance with 12-h mannitol treatment, which suggests that NtC7 functions to maintain osmotic adjustment (Tamura *et al.*, 2003). Since a receptor-like protein kinase from *Arabidopsis*, AtRLK3, has high accumulation of its transcript under oxidative stress, it might have a role in oxidative stress response (Czernic *et al.*, 1999). Cold, drought or salinity are known to induce Ca^{2+} influx in cell cytoplasm (Sanders *et al.*, 1999; Knight, 2000), and channels responsible for Ca^{2+} influx might be potential sensors for these stress signals (Xiong *et al.*, 2002a). The *A. thaliana* genome has 11 genes for histidine kinases (Hiks) and 22 genes for Rres (Arabidopsis Genome Initiative, 2000). Undoubtedly, other Hiks might be part of the two-component system in sensing other abiotic stresses. However, the mere presence of these systems in genomes does not guarantee their function as sensors. Future studies should aim at determining their exact roles, intermediates and targets in a signaling cascade.

8.1.2   ROS

Oxygen is a double-edged sword for plants. It is required for normal growth but unavoidably leads to the formation of ROS as a by-product of aerobic metabolic processes such as respiration and photosynthesis (Apel & Hirt, 2004). Molecular O_2 can be converted to superoxide anion by a single electron transfer, which subsequently undergoes a series of reductions to form H_2O_2 and hydroxyl radicals. These three molecules act as ROS and can cause oxidative damage to proteins, DNA and lipids. The generation of ROS is also mediated by an increase in activity of certain oxidases and peroxidases because of environmental perturbations. This process, also known as oxidative burst, has generally been associated with elicitor responses (Lamb & Dixon, 1997), although instances of oxidative bursts during abiotic stress exist (Rao *et al.*, 1996; Wohlgemuth *et al.*, 2002).

The cellular levels of ROS increase because of an imbalance between the production and removal of ROS. Stress conditions such as drought and low and high temperatures can disrupt this equilibrium and lead to oxidative stress (Prasad *et al.*, 1994; Foyer *et al.*, 2001). Recent evidence suggests that ROS can also act as signaling molecules. H_2O_2 is more likely to be a signaling molecule: because of its ability to diffuse through aqueous and lipid phases, it is uncharged and has a longer half-life than superoxide and hydroxyl species. It is not clear whether ROS act as signaling molecules or whether they create an intracellular signal via the oxidation of molecular substrates. The possibility of protein conformational changes in the presence of ROS is high because of the oxidation potential of thiol residues in proteins. Both *Escherichia coli* transcription factor (TF) OxyR and yeast TF YAP-1 are of paramount importance in oxidative stress signaling. ROS regulate the activity of these proteins through cysteine thiol modifications (Zheng *et al.*, 1998; Delaunay *et al.*, 2000, 2002). Plant genes important for oxidative stress tolerance have been identified by complementing these mutants. *Arabidopsis* annexin-like protein and CEO1 protein rescued the *E. coli oxyR* and yeast *yap1* mutants, respectively, from oxidative damage (Gidrol *et al.*, 1996; Belles-Boix *et al.*, 2000).

In addition to regulating protein activity by covalent modifications, ROS are also known to activate specific MAPKs (Kovtun *et al.*, 2000; Samuel *et al.*, 2000; Desikan *et al.*, 2001). H_2O_2 has been shown to activate MAPK3 and MAPK6 in *Arabidopsis* via ANP1, a mitogen-activated protein kinase kinase kinase (MAPKKK). Tobacco plants constitutively overexpressing active tobacco NPK-1 (orthologue of ANP1) have enhanced tolerance to cold, salt and high temperature (Kovtun *et al.*, 2000). Another important protein associated with MAPK-mediated H_2O_2 signaling in plants is nucleotide diphosphate kinase 2 (NDPK2) from *Arabidopsis*. Plants overexpressing AtNDPK2 had lower levels of ROS than wild-type plants and were tolerant to multiple environmental stresses (Moon *et al.*, 2003). Recently, a ser/thr kinase, OXI1, from *Arabidopsis* was shown to be required for full activation of MAPK3 and MAPK6. *OXI1* expression and activity are induced by H_2O_2. *OXI1* gene expression is also induced by cold, osmotic stress and heat, thus raising the possibility of OXI1 being a universal mediator of oxidative bursts and stress in *Arabidopsis* (Rentel *et al.*, 2004). Since the disruption of cellular redox balance is a key phenomenon associated with abiotic and biotic stresses, oxidative stress response may be a central component for cross-tolerance. Consistent with this notion, salicylic acid (SA) and osmotic stress activate the same MAPK (Hoyos & Zhang, 2000; Mikolajczyk *et al.*, 2000). These results suggest that various stresses induce ROS production, which in turn activates MAPK pathways. Although the intermediates and targets of the MAPK activation are unknown, ROS-induced activation of MAPKs appears to be a generalized pathway for cellular responses to multiple stresses.

The plant hormone ABA accumulates in response to abiotic stress and induces a range of stress adaptation responses including stomatal closure. ABA induces H_2O_2 production (Guan *et al.*, 2000; Pei *et al.*, 2000), and therefore many processes controlled by ABA may also be mediated by ROS production that follows ABA signaling. The interlocking of ABA and H_2O_2 signaling has been unknotted by genetic analysis of *Arabidopsis* mutants. In the *gca2* mutant, ABA-induced H_2O_2 synthesis was unaffected, although the H_2O_2-mediated calcium channel activation and stomatal closure were lacking (Pei *et al.*, 2000), which suggests that GCA2 protein transduces signals generated by H_2O_2. A similar genetic analysis of ABA-insensitive mutants *abi1* and *abi2* revealed that ABA-induced ROS production is lacking in *abi1* but not in *abi2* mutant plants, thus indicating that ABI1 is upstream and ABI2 is downstream of ROS signaling (Murata *et al.*, 2001). Another upstream gene to the ROS production, *OST1*, was recently identified by Mustilli *et al.* (2002). *OST1* codes for a protein kinase activated by ABA in roots and guard-cell protoplasts. ABA-induced ROS production was defective in *ost1* mutants, and stomata from *ost1* plants could still close in response to H_2O_2 (Mustilli *et al.*, 2002).

8.1.3 Calcium

Ca^{2+} is another nonprotein molecule that undoubtedly is a nodal player in plant responses to abiotic stress. Increased Ca^{2+} levels have been noticed with many abiotic stresses, including cold, drought and salinity (Bowler & Fluhr, 2000; Knight & Knight, 2001; Sanders *et al.*, 2002). Both temporal and spatial Ca^{2+} changes seem to provide some specificity in abiotic stress signaling. Furthermore, Ca^{2+}, by itself, is important for the activity of many different proteins, thus providing additional specificity for cellular responses during a particular stress (Sanders *et al.*, 2002). One aspect of Ca^{2+} signaling is coding and decoding the signal brought about by different Ca^{2+}-permeable channels. Decoding the signal is mediated by calcium sensors, which, either on their own or by protein interaction, modulate gene expression or protein activity changes. In this section, we discuss Ca^{2+} signal encoders and mechanisms governing their action during abiotic stress. In the following sections, we discuss the decoders of the Ca^{2+} signal.

The precise mechanisms for Ca^{2+} signal generation (change in levels of Ca^{2+}) are just beginning to be unraveled. In the cytosol, a Ca^{2+} spike is generated and maintained (until the signal is decoded) by the interplay between influx and efflux. Although previous research has focused on the components controlling Ca^{2+} influx, reports from nonplant systems show that efflux systems can also play an active role in signal transduction (Camacho & Lechleiter, 1993; Lechleiter *et al.*, 1998; Roderick *et al.*, 2000). Even though reports indicating a similar scenario in plants are lacking, efflux systems may nevertheless be important for attenuating the Ca^{2+} signal or compartmentalizing it to the endoplasmic reticulum and vacuole, which can be sources for later regulated Ca^{2+} release. The efflux channels may also be important for Ca^{2+} homeostasis, and their modified expression or deregulation results in disturbance of ionic balance (Hirschi, 1999). Disturbed Ca^{2+} homeostasis by overexpression of a Ca^{2+}/H^+ antiporter calcium exchanger 1 (*CAX1*), which loads Ca^{2+} into vacuoles, renders the transgenic plants hypersensitive to ionic and cold stress. The hypersensitive response is abrogated by increasing Ca^{2+} in the medium (Hirschi, 1999), which suggests that Ca^{2+} transient is important during ionic and cold stress. Vacuolar release of Ca^{2+} under cold stress has been observed in tobacco and *Arabidopsis* (Knight *et al.*, 1996); therefore, conceivably, the constitutive expression of *CAX1* in transgenic plants may perturb the Ca^{2+} concentrations in the cytosol during cold stress.

The elevation of cytosolic Ca^{2+} in response to signals can be due to influx of Ca^{2+} from apoplasts or its release from intracellular stores including the endoplasmic reticulum, vacuoles, mitochondria, chloroplasts and nucleus (Reddy, 2001). Ostensibly, channels giving rise to Ca^{2+} transient exist in both plasma membranes and endomembranes. However, few reports on the regulatory mechanism governing Ca^{2+} transport by these channels exist. The magnitude and kinetics of Ca^{2+} accumulation during stress appears to be dependent on the type of the stress imposed. ABA and mechanical stress signals primarily mobilize Ca^{2+} from intracellular stores, whereas the influx of extracellular Ca^{2+} is a key component in the transduction of low-temperature signals (Wood et al., 2000). Ca^{2+} influx and the accompanying expression of a cold acclimation-specific (CAS) gene in Medicago sativa has been shown to depend on membrane rigidity (Orvar et al., 2000). Cytoskeletal reorganization and membrane rigidity have also been shown to be important for Ca^{2+} influx and cold activation of the BN115 promoter in Brassica napus (Sangwan et al., 2001). These studies suggest that Ca^{2+} channels functioning in cold stress are mechanosensitive in nature. Cyclic ADP-ribose (cADPR) has also been shown to facilitate Ca^{2+} release from internal stores (Navazio et al., 2000; Sangwan et al., 2001). cADPR-gated channels were also suggested to be involved in ABA signal transduction (Wu et al., 1997; McAinsh & Hetherington, 1998). In addition, hyperpolarization-activated channels are known to be involved in increasing cytosolic Ca^{2+} that follows ABA application (Grabov & Blatt, 1998; Hamilton et al., 2000).

High-affinity inositol 1,4,5-triphosphate (IP_3) binding sites on the endoplasmic reticulum suggest the presence of IP_3-gated Ca^{2+} channels (Martinec et al., 2000). Though a direct role of these channels in Ca^{2+} influx has yet to be demonstrated, a distinct role of IP_3 in the transduction of cold, salt and hyperosmotic-stress signal dehydration indicates that the stress response is mediated through these channels (Drobak & Watkins, 2000; DeWald et al., 2001; Takahashi et al., 2001; Xiong et al., 2001a). In addition to these channels, other selective and nonselective Ca^{2+} channels are present in plants (Sanders et al., 2002). Their precise role in Ca^{2+} influx during abiotic stress signaling remains to be determined.

8.1.4   Phospholipids

Phospholipids, an integral part of the plasma membrane, can generate a multitude of signal molecules. Because of their close proximity to membrane receptors, it is easy to envisage the production of these messengers as a predominant means of transducing a signal. However, because of the complexity of intracellular communication, phospholipid-based signaling represents only the tip of the iceberg. Nevertheless, signaling molecules generated by phospholipids play a major role in stress signaling (Munnik & Meijer, 2001).

Different kinds of phospholipids are classified on the basis of a head group, which is attached to the phosphatidyl moiety (for details refer to review by

Meijer & Munnik, 2003). Phospholipids with inositol and its phosphorylated forms as head groups constitute an important but complex group of signal or signal precursors (Meijer & Munnik, 2003). Phosphorylated forms of phosphatidyl inositol (PI) arise by phosphorylation reactions mediated by specific kinases. In plants, isomers of the phosphorylated forms – phosphatidyl inositol monophosphate (PIP) and phosphatidyl inositol bisphosphate (PIP$_2$) – provide an additional complexity during signaling. The degradation of phospholipids is mediated by phospholipases, and perhaps the most well studied is phospholipase C (PLC), which degrades phosphatidyl inositol 4,5-bisphosphate [PI(4,5)P$_2$] to produce inositol 1,4,5-triphosphate (IP$_3$) and diacylglcerol (DAG), which individually act as secondary messengers.

Genes involved in the synthesis and degradation of PI(4,5)P$_2$ are modulated by certain environmental cues (Figure 8.1). This observation was evident by the upregulation of the PI(4,5)P$_2$ biosynthesis gene, phosphatidyl inositol 4-phosphate 5-kinase (*PIP5K*), during water and ABA stress in *Arabidopsis* (Mikami *et al.*, 1998) and by the upregulation of PI(4,5)P$_2$-degrading enzyme, *PLC*, by drought or salt stress (Hirayama *et al.*, 1995; Kopka *et al.*, 1998). Such a regulation may be important for both the generation and attenuation or transference (forming further downstream messengers by degrading PI[4,5]P$_2$) of the signal. An increase in [PI(4,5)P$_2$] levels was also observed with hyperosmotic stress (Pical *et al.*, 1999; DeWald *et al.*, 2001).

The degradation of PI(4,5)P$_2$ produces IP$_3$ and DAG, which can trigger Ca^{2+} release or activate protein kinase C (PKC), respectively. The absence of PKC from plant genomes (Meijer & Munnik, 2003) and rapid conversion of DAG to phosphatidic acid (PA) suggests that PA is more likely to be a signaling molecule than DAG (Munnik, 2001). Therefore, PLC activity can be seen to produce IP$_3$ and PA rather than IP$_3$ and DAG. The role of exogenous IP$_3$, in releasing Ca^{2+} from intracellular stores, is well documented (Alexandra *et al.*, 1990; Sanders *et al.*, 2002). IP$_3$ levels increase transiently in response to salt stress (Srivastava *et al.*, 1989; DeWald *et al.*, 2001; Takahashi *et al.*, 2001) and ABA (Lee *et al.*, 1996; Sanchez & Chua, 2001; Xiong *et al.*, 2001a).

Levels of IP$_3$ show an increase in guard cells of *Solanum tuberosum* and *Vicia faba* in response to ABA (Lee *et al.*, 1996). Cellular levels of IP$_3$ have a direct effect on ABA-induced gene expression. ABA induction of *KIN1* (Brunette *et al.*, 2003) and *RD29A*, *KIN1* and *RD22* (Sanchez & Chua, 2001) is delayed or decreased in plants overexpressing inositol polyphosphate 5-phosphatase (In5Pase), an enzyme involved in the catabolism of IP$_3$. The transduction of the signal can be modulated by both the synthesis and degradation of messengers, and therefore degradation of phosphoinositide messengers appears to be an important step for signal transduction. IP$_3$ breakdown occurs either through the addition or removal of phosphate from IP$_3$. In animal cells, the predominant pathway for IP$_3$ phosphorylation is through a 3-kinase

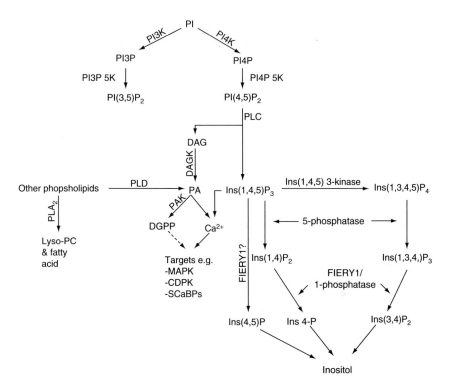

Figure 8.1 Biosynthesis and degradation of phospholipids. Phosphatidylinositol (PI) is converted into phosphatidylinositol-3-phosphate (PI3P) by phosphoinositide-3-kinase (PI3K) or to phosphatidylinositol-4-phosphate (PI4P) by phosphoinositide-4-kinase (PI4K). PI3P and PI4P are converted into phosphatidylinositol-3,5-bisphosphate [PI(3,5)P$_2$] and phosphatidylinositol-4,5-bisphosphate [PI(4,5)P$_2$] by PI3P-5-kinase and PI4P-5-kinase, respectively. PI(4,5)P$_2$ is degraded into inositol-1,4,5-triphosphate (IP$_3$) and diacylglycerol (DAG), which act as signaling molecules. Pathways for further degradation of IP$_3$ and other phopholipids shown in the figure are described in the text.

pathway resulting in accumulated inositol 1,3,4,5-tetraphosphate [Ins(1,3,4,5)P$_4$] or through IP$_3$ dephosphorylation mediated by a 5-phosphatase pathway giving rise to accumulated inositol 1,4-bisphosphate [Ins(1,4)P$_2$] (Berridge & Irvine, 1989).

IP$_3$ degradation can also occur by inositol polyphosphate 1-phosphatase (In1Pase), as was revealed by genetic and biochemical analysis of the *fiery1* (*fry1*) *Arabidopsis* mutant (Xiong *et al.*, 2001a). *fry1* was isolated in a genetic screen using a firefly luciferase under control of the *RD29A* promoter. The mutant exhibits early and increased induction of stress-responsive genes under cold, drought, salt and ABA treatments. *FRY1* codes for a bifunctional enzyme with 3'(2')5'-bisphosphate nucleotidase and In1Pase activity and is identical to *SAL1* (Quintero *et al.*, 1996). The *fry1* mutants accumulated substantially higher levels of IP_3 than the wild type, which indicates that loss/reduction of In1Pase activity was responsible for gene expression changes in *fry1* (Xiong *et al.*, 2001a). SAL1 has been previously shown to hydrolyze $Ins(1,4)P_2$ and ionisitol 1,3,4-trisphosphate $[Ins(1,3,4)P_3]$ (Quintero *et al.*, 1996). In addition to hydrolysis of $Ins(1,4)P_2$ and $Ins(1,3,4)P_3$, FRY1/SAL1 also hydrolyzed IP_3, albeit its activity for $Ins(1,4)P_2$ and $Ins(1,3,4)P_3$ was higher. Since IP_3 can be degraded into $Ins(1,4)P_2$ and $Ins(1,3,4)P_3$ by Ins5Pase, a mutation in FRY1 may indirectly cause IP_3 accumulation because of reduced hydrolysis of $Ins(1,4)P_2$ and $Ins(1,3,4)P_3$ (Xiong *et al.*, 2001a, 2002a).

In addition to PLC, phospholipase D (PLD) is an important component of stress signal transduction. PLD hydrolyzes phospholipids at the terminal phosphodiester bond, generating a free head group and PA. PLD activity is stimulated by high osmoticum in *Chlamydomonas* and *Arabidopsis* (Frank *et al.*, 2000; Katagiri *et al.*, 2001). ABA also increases PLD activity and gene expression (Ritchie & Gilroy, 1998; Jacob *et al.*, 1999). Despite its induction by osmotic stress and ABA, excess PLD may have a negative impact on membrane stabilization, because surplus PA levels do not favor the formation of bilayers (Wang, 1999). PA is a very dynamic molecule, and increased PA levels have been observed under osmotic, ABA and oxidative stress (Munnik, 2001). PA is further phosphorylated to diacylglycerol pyrophosphate (DGPP) by phosphatidic acid kinase (PAK). Since excess PA can have a negative impact on plant stress tolerance, attenuation of the PA signal may also be important. Consistent with this notion, increased PAK activity has been reported under hyperosmotic stress and drought (Carman & Zeimetz, 1996; Pical *et al.*, 1999; Munnik *et al.*, 2000).

Membrane lipids can also be hydrolyzed by phospholipase A2 (PLA2) in addition to PLC and PLD. PLA2 specifically hydrolyzes the acyl ester bond of glycerophospholipids at the sn-2 position, resulting in release of free fatty acids and a lyso-phospholipid (L-PA). PLA2 activity is stimulated by hyperosmotic stress in plants (Munnik & Meijer, 2001) and algae (Einspahr *et al.*, 1988; Meijer *et al.*, 1999). The dynamic nature of phospholipids (i.e. their property of interconversion in different forms by phosphorylation/dephosphorylation, proximity to receptors and ability to generate second messengers) clearly indicates that they are an integral point in stress signaling.

8.1.5   SOS pathway

Salinity stress is a major impediment in plant agriculture and influences the geographical distribution of crop plants. Plant growth is affected by high salinity in multiple ways (Chinnusamy & Zhu, 2004), and plants respond to salinity stress by maintaining cellular homeostasis, controlling stress damage and repair and regulating growth (Zhu, 2002). Increased Ca^{2+} levels have been associated with salt stress in plants (Knight *et al.*, 1997). How this Ca^{2+}

signature generated in response to salt stress is transduced to regulate ion homeostasis has been answered by the identification of the salt-overly-sensitive (SOS) pathway in *Arabidopsis*. Systematic molecular, genetic and biochemical analyses have led to the identification of three major components, namely, SOS3, SOS2 and SOS1, which constitute a signaling pathway to transduce the Ca^{2+} signal, thereby maintaining ion homeostasis during salt stress (Zhu, 2002; Chinnusamy & Zhu 2004). The salt stress-induced Ca^{2+} signal is perceived by the myristoylated and membrane-associated Ca^{2+} sensor SOS3, which recruits a ser/thr protein kinase, SOS2, to the plasma membrane. The SOS3–SOS2 complex phosphorylates and activates SOS1, a plasma membrane-localized Na^+/H^+ antiporter, which results in efflux of Na^+ ions, thereby restoring cellular ionic balance (Figure 8.2).

The *sos3* mutant of *Arabidopsis* is hypersensitive to salt stress. Molecular cloning revealed that *SOS3* encodes a Ca^{2+} sensor and shares significant sequence similarity with calcineurin B subunit from yeast and neuronal calcium sensors from animals (Liu & Zhu, 1998; Ishitani *et al.*, 2000). The SOS3 protein is predicted to contain three Ca^{2+}-binding sites and an N-terminal myristoylation motif. Both Ca^{2+}-binding sites and myristoylation are required for SOS3 function (Ishitani *et al.*, 2000).

Genetic screening for salt tolerance genes led to the identification of the *SOS2* locus in *Arabidopsis*. Mutations in SOS2 cause Na^+ and K^+ imbalance and render increased sensitivity toward growth inhibition by Na^+ (Liu *et al.*, 2000). SOS2 is a ser/thr-type protein kinase capable of autophosphorylation and has N-terminal catalytic and C-terminal regulatory domains. Both domains are essential for salt tolerance (Liu *et al.*, 2000). Two-hybrid analysis revealed that the N- and C-terminal regions of SOS2 interact with each other, which suggests that the regulatory domain may inhibit kinase activity by blocking substrate access to the catalytic site. SOS3 binds to the regulatory region of SOS2 and activates its protein kinase activity in a Ca^{2+}-dependent manner (Halfter *et al.*, 2000). A 21-amino acid region (309–329) in SOS2, known as the FISL motif, is necessary and sufficient for SOS2 binding to SOS3, and deletion of this motif constitutively activates SOS2, which indicates that the SOS3-binding motif can also serve as a kinase autoinhibitory loop of the SOS2 kinase domain (Guo *et al.*, 2001). A mutation in the activation domain of the SOS2 kinase (Thr168 to Asp) also renders SOS2 constitutively active. Overexpression of constitutively active SOS2 partially rescues salt hypersensitivity in *sos2* and *sos3* mutant plants (Guo *et al.*, 2004), which suggests that SOS3 and SOS2 function in the same signaling pathway during salt stress.

The isolation of the *SOS1* locus revealed that the SOS pathway regulates cellular ion homeostasis under salt stress. *SOS1* encodes a 127-kDa protein with 12 transmembrane domains (Shi *et al.*, 2000). The upregulation of *SOS1* gene expression in response to NaCl stress was abrogated in *sos3* or *sos2*

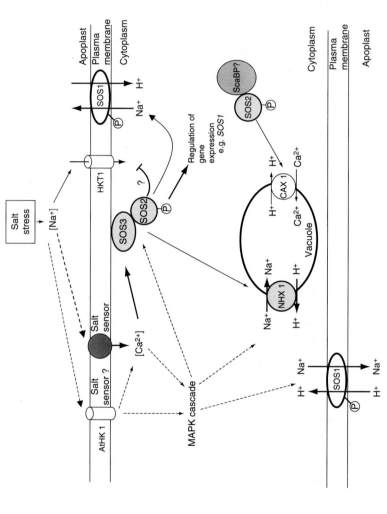

Figure 8.2 The SOS pathway for regulation of ion homeostasis during salt stress. Salt stress induced Ca^{2+} signals are perceived by SOS3, which activates and recruits SOS2 kinase to plasma membrane. Activated SOS2 phosphorylates SOS1 Na^+/H^+ antiporter, which then pumps excess Na^+ out of cytosol. SOS3/SOS2 complex also regulates cytosolic Na^+ levels by activating vacuolar transporter NHX1 to compartmentalize these ions. This complex also restricts entry of Na^+ ions, by inhibiting plasma membrane transporter HKT1 activity. SOS2 along with yet unknown ScaBP also regulate the cytosolic Ca^{2+} levels by modulating activity of vacuolar Ca^{2+} channel, CAX1. AtHK1 is a Na^+ sensor and initiates MAPK cascade, which possibly can play a role in controlling the activity of SOS components.

mutant plants, which indicates that *SOS1* expression is controlled by the SOS2–SOS3 pathway. SOS1 demonstrates Na^+/H^+ exchange activity in highly purified plasma membrane vesicles (Qiu *et al.*, 2002). Vesicles isolated from *sos2* and *sos3* plants have less Na^+/H^+ antiport activity than do wild-type plants, and the activity was restored when constitutively active SOS2 protein kinase was added *in vitro* (Qiu *et al.*, 2002). Coexpression of all three functional components of the SOS pathway in a salt-sensitive yeast mutant led to a great increase in SOS1-dependent Na^+ tolerance. In addition, the SOS3–SOS2 complex promoted the phosphorylation of SOS1 (Quintero *et al.*, 2002). All these results indicate that a Ca^{2+} signature, developed due to salt stress, is transduced via the SOS3–SOS2 pathway to activate SOS1, which then regulates ion homeostasis.

To maintain ion homeostasis, the SOS pathway may also turn off Na^+ influx in addition to activating Na^+ efflux. High-affinity K^+ transporters can also cause the conditional influx of Na^+. Plants showing reduced expression of wheat *HKT1* (achieved with plants expressing *HKT1* in the antisense orientation) had significantly less uptake of ^{22}Na and survived better under salinity conditions (Laurie *et al.*, 2002). In agreement with these results, a mutation in *AtHKT1* can suppress the salt-sensitive phenotype of *sos3* (Rus *et al.*, 2001). Thus, the SOS3–SOS2 complex might also negatively regulate AtHKT1 function. Ionic balance can be maintained by regulating the levels of Na^+ in the cytoplasm, either by controlling the influx/efflux or by compartmentalization. A vacuolar Na^+/H^+ transporter, AtNHX1, showed significantly higher, greatly reduced and unchanged activity in *sos1*, *sos2* and *sos3* mutants, respectively, which suggests that AtNHX1 is a target of SOS2 (Qiu *et al.*, 2004). Although the role of the SOS pathway or its components in other stresses is not currently clear, dissection of the SOS pathway has illuminated previously unknown aspects of salt-stress signaling.

8.1.6 SOS3-like Ca²⁺-binding proteins and SOS2-like protein kinases

The *Arabidopsis* genome encodes at least eight SOS3-like Ca^{2+}-binding proteins (ScaBPs) and 23 SOS2-like protein kinases (PKS; Guo *et al.*, 2001; Gong *et al.*, 2004). RNAi lines were generated for each of the *SCaBPs* and *PKSs*, to study whether the interaction between a Ca^{2+} sensor and a protein kinase coregulate (Guo *et al.*, 2002a) plant responses to ABA, sugars and cellular pH homeostasis. RNAi lines of *SCaBP5* (also known as *CBL1*; Kudla *et al.*, 1999) and *PKS3* were found to be hypersensitive to ABA. The mutant plants exhibited substantially reduced transpiration water loss. In yeast two-hybrid and *in vitro* and *in vivo* protein interaction assays, SCaBP5 interacted with PKS3, which in turn interacted with ABI2 and to a lesser extent ABI1 (Guo *et al.*, 2002a). The study demonstrated that SCaBP5 and PKS3 may transduce the Ca^{2+} signal in response to ABA and may function together with

ABI1 and ABI2. A higher expression of ABA-induced genes such as *COR47*, *COR15A* and *RD29A* was observed in mutant plants compared with wild-type plants. In *scabp5/pks3*, double-mutant ABA hypersensitivity was not additive, which suggests the action of these proteins in the same pathway. SCaBP5 and PKS3 were found to transduce Ca^{2+} signals in response to ABA specifically, and transgenic plants overexpressing constitutively active PKS3 (PKS3 deleted for FISL motif) were ABA insensitive.

In contrast to results found by Guo *et al.* (2002a, 2002b), CBL1 has been shown to regulate plant responses to drought and salt stress but not ABA (Albrecht *et al.*, 2003). Albrecht and coworkers suggested that CBL1 might constitute an integration node in the plant response to abiotic stimuli. Recently, a CBL-interacting protein kinase 3 (CIPK3, similar to PKS12) was shown to be involved in the regulation of ABA and cold signal transduction in *Arabidopsis* (Kim *et al.*, 2003). CIPK3 represents a crosstalk node between ABA-dependent and -independent pathways in stress responses (Kim *et al.*, 2003). Transgenic plants overexpressing a constitutively active form of PKS18 (PKS18 T/D) were hypersensitive to ABA during seed germination and seedling growth, whereas RNAi lines of the same kinase conferred ABA insensitivity (Gong *et al.*, 2002a). Plants overexpressing the constitutive form of PKS11 (PKS11 T161D) were more resistant to glucose-containing growth medium (Gong *et al.*, 2002b). Significant advances have been made in understanding the functions of SCaBPs and PKSs since the discovery of SOS3 and its interacting partner SOS2; however, only a few reports on their combinatory interaction affecting a particular plant process exist. In addition, the targets of most of the PKS kinases are unknown at present.

8.1.7   CDPKs

Calcium-dependent protein kinases (CDPKs) are one of the largest subfamilies of plant protein kinases. CDPKs have a hybrid structure of a Ca^{2+} sensor and a kinase, thus both a calmodulin-like domain containing the EF hand and an N-terminal ser/thr protein kinase domain (Ludwig *et al.*, 2004). CDPKs have been implicated to play an important role in signaling of abiotic stresses. Transcript induction of CDPKs has been observed in response to diverse stimuli. During cold stress, CDPKs from alfalfa show differential regulation: *M. sativa* calcium kinase 1 (*MsCK1*) is upregulated, whereas *MsCK2* expression is downregulated (Monroy & Dhindsa, 1995). The cold induction of maize CDPK-*Zea mays CDPK1* (*ZmCDPK1*; Berberich & Kusano, 1997) has also been observed. Certain CDPKs are induced during osmotic/dehydration stress, including a CDPK from rice (*OsCPK7*; Saijo *et al.*, 2000), *Arabidopsis* (*AtCPK10* and *AtCPK11*; Urao *et al.*, 1994), mung bean (*VrCPK1*; Botella *et al.*, 1996), common ice plant (*MsCDPK1*; Patharkar & Cushman, 2001) and tobacco (*NtCDPK2* and *NtCDPK3*; Romeis *et al.*, 2000).

The use of reverse genetics or overexpression approaches has thrown light on the function of some of these CDPKs. Transgenic rice plants over- and underexpressing *OsCDPK7* exhibited increased and decreased tolerance, respectively, against cold, drought and salt (Saijo *et al.*, 2000). Constitutively active *Arabidopsis* CDPK (AtCDPK1 and AtCDPK1a) activated a stress-inducible promoter *HVA1* in maize protoplasts (Sheen, 1996). Results from these studies showed that CDPKs can activate a G-box type ABRE-containing gene (*RAB16* in OsCDPK7 and *HVA1* in AtCDPK1). Little is known about the downstream targets of CDPKs. In common ice plant (*Mesembryanthemum crystallinum*), CDPK substrate protein1 (CSP1) has been identified as a substrate for MsCDPK1 (Patharkar & Cushman, 2000). MsCDPK1 phosphorylates CSP1 in a Ca^{2+}-dependent manner. Salt stress induced the colocalization of both MsCDPK1 and CSP1 in the nucleus (Patharkar & Cushman, 2000). Future studies will further reveal the integration of abiotic stresses at the level of CDPKs. The functional significance of CDPKs is just beginning to emerge, and some CDPKs may be important integral points in abiotic stress signaling, because they can confer tolerance to more than one stress.

8.1.8   MAPKs

Like CDPKs, MAPKs are in large numbers in plants. MAPK cascades provide one of the strongest evidence of crosstalk during abiotic stress signaling in plants. The activation of MAPK is brought about by at least two or more kinases (MAPKK and MAPKKK), which present a further complexity in the MAPK cascade. MAPKs are implicated in abiotic, biotic, developmental and hormone signaling (Ligterink & Hirt, 2001). The *Arabidopsis* genome has 60 MAPKKKs, 10 MAPKKs and 20 MAPKs. Signals perceived by 60 MAPKKKs have to be transduced through 10 MAPKKs to 20 MAPKs. This observation indirectly implies that for six MAPKKKs, there is one MAPKK, and for each MAPKK, there are two MAPKs; hence MAPKKs represent an important converging point in the MAPK cascade. Consistent with this observation, several studies have demonstrated that Arabidopsis AtMEKK1 (a MAPKKK) can interact with and activate four different MAPKKs (Ichimura *et al.*, 1998; Asai *et al.*, 2002). In addition, a plant MAPKK can interact with and activate more than one MAPK-stress-induced MAPKK (SIMKK), and pathogen-responsive MAPKK (PRKK) can activate up to three different types of MAPKs (Cardinale *et al.*, 2002).

Salt stress activates AtMEKK1 and upregulates its gene expression (Ichimura *et al.*, 1998). In a yeast two-hybrid assay, AtMEKK1 interacted with a MAPKK, AtMKK2 (has high sequence homology to AtMEK1), and a MAPK, AtMPK4. Coexpression of AtMEKK1 either with AtMKK2 or with AtMEK1 complemented the growth defect (under high osmotic conditions) of the *pbs2* mutant in yeast. The coexpression of AtMPK4 with AtMEK1 complemented a growth defect of yeast *bck1* and *mpk1* mutants (Ichimura *et al.*, 1998). These results indicate that the AtMEKK1, AtMKK2/AtMEK1 and AtMPK4 may constitute a MAP kinase cascade.

ROS and salt stress both lead to activation of common MAPKs such as AtMPK3, AtMPK4 and AtMPK6 (Mizoguchi *et al.*, 1996; Ichimura *et al.*, 2000). An H_2O_2-inducible *Arabidopsis* MAPKKK, ANP1, can phosphorylate AtMPK3 and AtMPK6, and constitutive expression of a tobacco ANP1 ortho-logue, NPK1, provides enhanced tolerance to multiple environmental stresses (Kovtun *et al.*, 2000). Similarly, the overexpression of H_2O_2-inducible NDPK2 led to enhanced tolerance against multiple stresses (Moon *et al.*, 2003). NDPK2 interacts with AtMPK3 and AtMPK6, and the H_2O_2-mediated induction of these kinases was highly reduced in *atndpk2* mutants. Both ANP1/NPK1 and NDPK2 may be involved in abiotic stress signaling during oxidative stress, and activation of AtMPK3/AtMPK6 may be a central phe-nomenon for ROS-mediated cross-tolerance. AtMEKK1 can activate AtMKK4/AtMKK5, which in turn can activate AtMPK3/AtMPK6 to elicit a defense response (Asai *et al.*, 2002). These studies reveal that certain MAPKs are important for both abiotic and biotic stresses and might represent a nodal point where both stress pathways converge.

A number of reports indicate that MAPKs are deactivated by protein phosphatases. AtMPK4 was shown to be a dephosphorylation target of a tyrosine-specific phosphatase AtPTP1 (Huang *et al.*, 2000). The *AtPTP1* gene is upregulated by salt stress and downregulated by cold stress (Xu *et al.*, 1998), which indicates that AtPTP1 can control the activity of AtMPK4 during these stresses. A MAP kinase phosphatase (MKP1) interacts with AtMPK6, AtMPK4 and AtMPK3, the strongest interaction being with AtMPK6, which suggests the control of AtMPK6 activity. As well, *in vivo* activity of AtMPK6 was controlled by MKP1 (Ulm *et al.*, 2002). This observation suggests the involvement of MKP1 in salt stress, because AtMPK6 (and AtMPK3 and 4) are induced during salt stress in Arabidopsis. *mkp1* mutant plants showed increased resistance to salt stress, which suggests that MKP1 is a negative regulator of salt stress signaling through MAPK6. The mRNA levels of a putative Na^+/H^+ exchanger (At4g23700) were increased in *mkp1* mutants, which indicates that this trans-porter may be positively regulated by the MAPK cascade during salt stress and negatively regulated by MKP1 (Ulm *et al.*, 2002).

8.1.9   *ICE1 pathway for cold regulation*

Many plant species native to temperate regions regularly encounter subzero temperatures. To survive these freezing temperatures, plants have evolved a mechanism known as cold acclimation, which enables the plants to acquire tolerance against freezing temperatures by a prior exposure to nonfreezing temperatures. The hallmark of cold acclimation is induction of a set of genes known as *COR* genes (cold responsive) (Thomashow, 1999; Viswanathan & Zhu, 2002). *COR* genes often are responsive to other stresses such as drought, salt and ABA (Zhu, 2001). One of the *COR* genes, *COR78* (also known as low-temperature-inducible 78 [*LTI78*] or responsive-to-dehydration 29A [*RD29A*]), has been an excellent paradigm for studying gene regulation during cold and osmotic stress.

Analyses of the promoter region of *RD29A* identified a 9-bp sequence, dehydration responsive element/C-repeat (*DRE/CRT*), that confers responsiveness to low temperature, drought and high salinity but not ABA (Yamaguchi-Shinozaki & Shinozaki, 1994). The *DRE/CRT* sequence is a hallmark of some of the *COR* genes. Some *COR* genes also contain *ABRE*. Stockinger *et al.* (1997) employed the yeast one-hybrid analysis to isolate a transcription factor that binds to *DRE/CRT* elements. This transcription factor, C-repeat binding factor 1 (CBF1), belongs to the AP2/EREBP family of transcription factors. Additional homologues of CBF1 (also known as DRE-binding proteins, DREBs) have been isolated from *Arabidopsis* (Gilmour *et al.*, 1998; Liu *et al.*, 1998). *DREB* genes can be grouped according to their responsiveness to low temperature and drought. *DREB1A* (*CBF3*), *DREB1B* (*CBF1*) and *DREB1C* (*CBF2*) are grouped together because of their responsiveness to cold but not drought, whereas *DREB2A* and *DREB2B* are responsive to drought but not low temperature (Liu *et al.*, 1998) (Figure 8.3).

DREB1/CBF genes are also transiently regulated by cold stress (Gilmour *et al.*, 1998; Liu *et al.*, 1998). The expression of *CBF* genes is induced within 15 min of cold stress, which led Thomashow's group (Gilmour *et al.*, 1998) to propose the constitutive presence of a hypothetical activator of *CBF* genes that activates their expression upon cold stress. This activator, the inducer of CBF expression (ICE1), is proposed to be modified posttranslationally during cold stress, leading to the activation of *CBF* genes. In a genetic screen of an ethyl methanesulphonate (EMS)-mutagenized population of *CBF3:LUC* plants, *ICE1* was recently cloned (Chinnusamy *et al.*, 2003). *ice1* mutants are sensitive to both chilling and freezing stress and have reduced expression of *CBF3* and its target *COR* genes. *ICE1* encodes a myc-like bHLH transcription factor localized in the nucleus. Results of gel shift assays showed that ICE1 specifically binds MYC recognition sequences of the *CBF3* promoter. Overexpression of ICE1 in *CBF3:LUC* plants increased the luminescence of these plants only under cold stress, which further suggests the requirement of cold-induced posttranslational modification of ICE1 for *CBF3* expression. Further analysis with ICE1 overexpression lines showed enhanced expression of *CBF3*, *RD29A* and *COR15A* genes during cold stress in these plants. Overexpression lines survived better than did wild-type plants under freezing stress. All these studies demonstrate that ICE1 is a major determinant of cold-regulated gene expression and cold acclimation in *Arabidopsis* (Chinnusamy *et al.*, 2003). How a cold signal may modify ICE1 activity is still unknown.

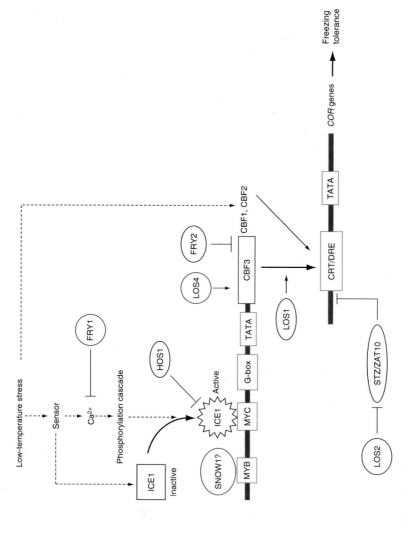

Figure 8.3 Regulation of a cold responsive transcriptome and freezing tolerance in Arabidopsis. ICE1 is a constitutive transcriptional factor, which is activated upon imposition of cold stress and binds to the myc-elements in the CBF3 promoter, thereby inducing CBF3 expression. CBFs bind to the CRT/DRE *cis*-element on COR genes and induce their expression leading to acquired freezing tolerance. ICE1 can interact with a hypothetical protein SNOW1, which binds to myb-elements in CBF3 promoter, and regulate the CBF3 expression. HOS1, a putative E3 ubiquitin-conjugating enzyme, appears to target ICE1 for degradation. Other proteins shown in the figure are described in the text.

Another important gene that regulates cold response is *HOS1*, which was cloned by Lee *et al.* (2001). *hos1* mutant plants show enhanced induction of *CBF* transcription factors and their downstream target genes by cold stress. *HOS1* encodes a novel protein that has a RING finger motif (Lee *et al.*, 2001). A similar RING motif is found in a group of inhibitor-of-apoptosis (IAP) proteins in animals. These proteins are known to act as E3 ubiquitin ligases, which can target specific proteins for degradation (Yang *et al.*, 2000). HOS1 is localized in the cytoplasm at ambient temperatures but shows nuclear localization under cold stress. On the basis of the downregulation of *CBF* transcripts by HOS1 and its putative role as an E3 ubiquitin ligase, it is tempting to speculate that HOS1 might target the ICE1 protein for degradation and regulate *CBF* expression by controlling the turnover of the ICE1 protein.

In addition to ICE1 and HOS1, several other proteins related to the CBF pathways have been identified by mutational analysis. A negative regulator of *CRT/DRE* gene expression was identified by screening either T-DNA mutants (Koiwa *et al.*, 2002) or an EMS-mutagenized population (Xiong *et al.*, 2002b) of *RD29A:LUC*-harboring C24 plants. Koiwa *et al.* (2002) isolated two non-allelic mutants, *cpl1* and *cpl3*, which showed enhanced luminescence and transcript accumulation in response to cold, ABA and NaCl (*cpl1*), or ABA (*cpl3*). *CPL1* and *CPL3* code for C-terminal domain phosphatase-like proteins. Similar genes from nonplant systems have been shown to dephosphorylate a conserved heptapeptide repeat in the C-terminal domain of RNA polymerase II (RNAPol II). CPL1 seems to regulate transcription in the elongation stage rather than at the initiation phase, because the luciferase transcript did not have an increased synthesis rate.

fiery2, a mutant allelic to *cpl1*, also exhibited high *RD29A-LUC* expression relative to wild-type plants when treated with cold, salt or ABA. *RD29A*, *COR15A*, *KIN1* and *COR47* were highly induced in *fry2* mutant plants (Xiong *et al.*, 2002b). Expression of the genes *CBF1*, *CBF2* and *CBF3* upstream of the *COR* genes was also higher in the *fry2* mutant, which indicates that FRY2 acts as a negative regulator of *CBF/DREB* transcription, which in turn downregulates the expression of their target genes.

The importance of proteins involved in RNA processing in cold-regulated gene expression was observed by Gong *et al.* (2002). *los4* mutant plants were isolated because of their reduced luminescence in a genetic screen of an EMS-mutagenized population of *RD29A-LUC* plants. The expression of *CBF* and their downstream genes was reduced or delayed in *los4* mutants. The mutant plants were sensitive to chilling stress, which was overcome by ectopic expression of CBF3. The *LOS4* gene encodes for a DEAD-box RNA helicase and directly or indirectly controls the stability or other aspects of *CBF* transcripts under cold stress (Gong *et al.*, 2002).

Proteins involved in translation also appear to affect cold-induced genes. The *los1* mutant was identified by using a mutational screen similar to that used for *los4* (Guo *et al.*, 2002b). The *los1* mutants had reduced expression of *RD29A*, *COR15A* and *COR47* genes under cold stress, but the expression of *CBF* genes was superinduced. The mutant plants were defective in acquired tolerance to freezing stress. Using map-based cloning, the *LOS1* gene was cloned and found to encode a translation elongation factor-2-like protein. The *los1* mutant plants were also blocked in *de novo* protein synthesis specifically in cold. These results suggest that the synthesis of CBF proteins is critical for cold responses.

RD29A can be a target of additional transcription factors. Homologues of STZ/ZAT10, a zinc finger protein, recognize similar sequences containing the core sequence 'AGT' (Takatsuji & Matsumoto, 1996). *RD29A* promoter has a tandem repeat of two AGT sequences (Lee *et al.*, 2002), thereby suggesting that *RD29A* may be a target of the STZ/ZAT10 transcription factor. Consistent with this observation, overexpression of ZAT10 led to repression of *RD29A* promoter activity in transient assays (Lee *et al.*, 2002).

STZ/ZAT10 itself was regulated by LOS2, which codes for a bifunctional enolase. The *los2* mutant was isolated on the basis of reduced luminescence of *RD29A:LUC* plants under cold stress. A mutation in *LOS2* reduced the ability of plants to accumulate transcripts of several *COR* genes but had no effect on transcript accumulation of the *CBF2* gene. The LOS2 protein was able to bind the human enolase-binding site in a c-myc promoter. In addition, LOS2 retarded the *ZAT10* promoter fragment (which has sequence similarity to the human enolase-binding sequence of the c-myc promoter) in electromobility shift assays. Increased expression of *ZAT10* in *los2* mutant plants in response to cold stress is indicative of LOS2 negatively regulating *STZ/ZAT10* expression. Thus, LOS2 is a positive regulator of *RD29A* and increases its expression by negatively regulating the expression of *STZ/ZAT10*, a transcriptional repressor of *RD29A*.

Studies by various groups have demonstrated the existence of specific pathways for cold-regulated gene expression, but perhaps the most remarkable pathway is the regulation of *CBF* and *COR* genes by ICE1. The interaction between ICE1 and other proteins involved in cold regulation might exist and may provide further fine-tuning of cold-induced gene expression.

8.2   Regulation of gene expression by ABA

ABA is an important plant hormone that regulates various aspects of plant development and most importantly it plays a critical role in the regulation of water balance and osmotic stress tolerance. Consistent with its importance for stress tolerance, both the synthesis and degradation of ABA is regulated by osmotic stress. ABA synthesis is a multistep procedure and involves C_{40} carotenoid cleavage by a specific dioxygenase, which results in the formation of a C_{25} by-product and a C_{15} compound, xanthonin. Xanthonin is further converted into abscisic aldehyde and, ultimately, ABA (Taylor *et al.*, 2000).

The important enzymes involved in ABA biosynthesis are zeaxanthin epoxidase (ZEP, also known as LOS6/ABA1), which converts zeaxanthin to viloxanthin, and 9-*cis* epoxycarotenoid dioxygenase (NCED), which cleaves neoxanthin into xanthonin. In addition to these enzymes, ABA3 and AAO3, which help convert ABA aldehyde to ABA, are important for ABA biosynthesis. The *NCED* gene (which was first cloned from the maize *vp14* mutant) seems to be rate limiting in ABA biosynthesis (Qin & Zeevart, 1999) and is upregulated by osmotic stress (Qin & Zeevart, 1999; Thompson *et al.*, 2000; Iuchi *et al.*, 2001). Transcripts of *ABA1* (Liotenberg *et al.*, 1999), *AAO3* (Seo *et al.*, 2000) and *ABA3* (Xiong *et al.*, 2001b) are also upregulated by osmotic stress.

The synthesis of ABA is under a feedback regulation, as was shown by the increased expression of *ABA1*, *AAO3* and *ABA3* with application of ABA (Xiong *et al.*, 2001b; Cheng *et al.*, 2002; Xiong *et al.*, 2002c). Consistent with these observations, reduced induction of *ABA3* and *AAO3* in the ABA-insensitive mutant *abi* was ABA dependent (Xiong *et al.*, 2001b). Notably, the effect of ABA on the feedback regulation of ABA biosynthetic genes has not been confirmed at the protein level in most cases. In addition to cellular levels of ABA being regulated by ABA, an RNA-binding protein has been demonstrated to regulate ABA biosynthesis. SAD1, a polypeptide similar to Sm-like U6 small nuclear ribonucleoproteins (SnRNPs) required for mRNA splicing, export and degradation was found to modulate ABA signal transduction and biosynthesis (Xiong *et al.*, 2001c).

In the *sad1* mutant plants, the degree of induction of *AAO3* and *ABA3* was substantially lower than that in wild-type plants, which indicates that SAD1 controls the feedback loop of ABA biosynthesis (Xiong *et al.*, 2001c). Given the importance of ABA in regulating key processes in the cell normally and under stress, genes involved in the catabolism of ABA may further throw light on processes governing ABA balance in a cell. ABA degradation seems to be controlled by a cytochrome P450 monooxygenase (Krochko *et al.*, 1998), which catalyzes C-8' hydroxylation to produce 8'-hydroxy ABA. Genome sequencing revealed the presence of nearly 272 genes for cytochrome P450 in *Arabidopsis* (Schuler & Werck-Reichhart, 2003). Recently, members of the CYP707A family of cytochrome P450 were identified to encode ABA 8' hydroxylases (Kushiro *et al.*, 2004). During drought stress, all *CYP707A* genes were upregulated. *cyp707a2* mutants exhibited hyperdormancy in seeds and had very high ABA levels compared to wild-type plants, which indicates a key role of CYP707A2 in degrading ABA (Kushiro *et al.*, 2004). Little information exists on the signaling components between signal perception and initiation of ABA biosynthesis and degradation. ROS have been shown to mediate ABA biosynthesis (Zhao *et al.*, 2001), which suggests that

ABA biosynthesis may involve ROS and Ca^{2+}-dependent protein phosphorylation cascades (Xiong *et al.*, 2002a).

Several stress-responsive genes have been proposed to be regulated in both an ABA-dependent and an ABA-independent manner (Shinozaki & Yamaguchi-Shinozaki, 2000; Zhu, 2002). The availability of ABA-deficient and -insensitive mutants led to the proposed existence of both ABA-dependent and -independent pathways. In agreement with this, the accumulation of the *RD29A* transcript by osmotic stress was not completely abolished in these mutants (Yamaguchi-Shinozaki & Shinozaki, 1993a). The *DRE cis*-element in the *RD29A* promoter was found to be necessary and sufficient for the osmotic stress-mediated induction of *RD29A*, which confirmed the presence of an ABA-independent pathway (Yamaguchi-Shinozaki & Shinozaki, 1994). AP2-type transcription factors DREB2A and DREB2B bind to the *DRE* element and transactivate stress-responsive gene expression in an ABA-independent manner. However, cloning of *CBF4*, a DREB1-related transcription factor, suggested that *DRE* elements may also be regulated in an ABA-dependent manner. The *CBF4* transcript was upregulated by ABA and drought. The overexpression of CBF4 resulted in the constitutive expression of *DRE/CRT*-containing genes and enhanced drought and freezing tolerance (Haake *et al.*, 2002).

Although activation of *DRE* by osmotic stress is independent of ABA, ABA might still be required for full activation of *DRE*-containing stress-responsive genes. This suggestion was shown by analysis of gene expression in the ABA-biosynthetic mutant *los5/aba3*. The expression of the stress-responsive genes *RD29A*, *RD22*, *COR15A*, *COR47* and *P5CS* was severely diminished or completely blocked during osmotic stress in *los5*. *LOS5* codes for a molybdenum cofactor sulphurase, required in the last step of ABA biosynthesis (Xiong *et al.*, 2001b).

The activation of basic leucine zipper *trans*-acting proteins is mediated by ABA. These factors bind to *ABRE* to regulate cold- and osmotic stress-responsive genes. Four *ABRE*-binding factors (ABF1–4) have been cloned from *Arabidopsis*. These factors induce *ABRE*-driven reporter gene expression in yeast (Choi *et al.*, 2000). The interdependence of *ABRE* and *DRE* in activating the expression of the *RD29A* gene was recently shown by Narusaka *et al.* (2003). Results of transactivation experiments involving *Arabidopsis* leaf protoplasts showed that DREBs and ABREs cumulatively transactivate the expression of a GUS reporter gene fused to a 120-bp promoter region (containing both *DRE* and *ABRE*) of *RD29A* (Narusaka *et al.*, 2003).

C_2H_2-type zinc finger protein from soybean *SCOF1* is induced by cold and ABA. The constitutive overexpression of SCOF1 results in enhanced *COR* gene expression and enhanced cold tolerance of nonacclimated transgenic *Arabidopsis* and tobacco plants. SCOF1 does not bind directly to either *CRT* or *ABRE* elements but enhances DNA binding of SGBF-1 to *ABRE* (Kim *et al.*,

2001). SCOF1 also enhances *ABRE*-dependent gene expression mediated by SGBF-1. Thus, SCOF-1 functions as a positive regulator of *COR* gene expression mediated by *ABRE* via protein–protein interaction.

ABA-mediated expression in some genes also occurs through non-ABRE *cis*-elements. A classic example of this is the *RD22* gene whose expression is induced by dehydration and ABA (Yamaguchi-Shinozaki & Shinozaki, 1993b). The promoter of the *RD22* gene does not contain the *ABRE* consensus sequence. However, a 67-bp promoter region of *RD22* can regulate drought-inducible gene expression (Iwasaki *et al.*, 1995). This 67-bp region contains MYC and MYB recognition sites. Proteins that bind to this region include AtMYC2 (a basic helix–loop–helix transcription factor) and AtMYB2. Both these factors cooperate and activate reporter gene expression fused to the 67-bp region (Abe *et al.*, 1997). Recently, the importance of AtMYC2 and AtMYB2 was shown when plants overexpressing either of these showed markedly increased ABA-induced expression of *RD22* and *AtADH1* (Abe *et al.*, 2003).

8.3 Conclusions and perspectives

It is evident that abiotic stress signaling is a complex phenomenon. Although we have tried to cover aspects of molecules that seem to integrate abiotic stress signaling, the overall picture is still obscure. Understanding abiotic stress signaling during 1990s has seen significant progress and has proceeded from the identification of relevant genes and proteins to revealing their functional necessity *in planta*. Conventional genetic screening based on stress injury or tolerance, together with reporter-gene-based molecular screens, has generated a wealth of information about the components of stress signaling. However, many links are missing, and many signaling components and their interacting partners need to be revealed. Forward and reverse genetic approaches, together with cell biology and biochemistry, will be required to establish the role of a molecule in the integration of signaling.

8.4 Summary

The perception of abiotic stress and the signaling events that follow lead to global changes in plants. These changes, which occur with the ultimate aim of protecting the plant, provide clues to engineering-enhanced tolerance against abiotic stresses. Certain changes were large enough to catch the attention of researchers years ago, whereas certain smaller but unequivocally important changes had to wait for the development of adequate technology. Because of the severity of one stress over another, in perceiving multiple stresses at a given time, plants have balanced the need (based on the available resources) to counteract these stresses by either modulating specific components or involving additional components that can crosstalk and lead to a more apt response. Recent progress has led to identification of the components that help plants in mounting these responses. In *Arabidopsis*, genetic analysis has defined the SOS pathways, for ion homeostasis, and inducer of C-repeat binding protein expression 1 (ICE1), for cold regulation. Multiple convergent points exist in stress signaling, and at some of these points, like the MAPK cascade, both abiotic and biotic stresses appear to converge.

References

Abe, H., Urao, T., Ito, T., Seki, M., Shinozaki, K. & Yamaguchi-Shinozaki, K. (2003) Arabidopsis AtMYC2 (bHLH) and AtMYB2 (MYB) function as transcriptional activators in abscisic acid signaling. *Plant Cell*, **15**(1), 63–78.

Abe, H., Yamaguchi-Shinozaki, K., Urao, T., Iwasaki, T., Hosokawa, D. & Shinozaki, K. (1997) Role of arabidopsis MYC and MYB homologs in drought- and abscisic acid-regulated gene expression. *Plant Cell*, **9**(10), 1859–1868.

Aguilar, P. S., Hernandez-Arriaga, A. M., Cybulski, L. E., Erazo, A. C. & de Mendoza, D. (2001) Molecular basis of thermosensing: a two-component signal transduction thermometer in *Bacillus subtilis*. *EMBO Journal*, **20**(7), 1681–1691.

Albrecht, V., Weinl, S., Blazevic, D., D'Angelo, C., Batistic, O., Kolukisaoglu, U., Bock, R., Schulz, B., Harter, K. & Kudla, J. (2003) The calcium sensor CBL1 integrates plant responses to abiotic stresses. *Plant Journal*, **36**(4), 457–470.

Alexandra, J., Lassalles, J. P. & Kado, R. T. (1990) Opening of Ca^{2+} stores in isolated beet root vacuole membrane by inositol 1,4,5-triphosphate. *Nature*, **343**(6258), 567–570.

Apel, K. & Hirt, H. (2004) Reactive oxygen species: metabolism, oxidative stress, and signal transduction. *Annual Review on Plant Physiology and Plant Molecular Biology*, **55**, 373–399.

Arabidopsis Genome Initiative. (2000) Analysis of the genome sequence of the flowering plant *Arabidopsis thaliana*. *Nature*, **408**(6814), 796–815.

Asai, T., Tena, G., Plotnikova, J., Willmann, M. R., Chiu, W. L., Gomez-Gomez, L., Boller, T., Ausubel, F. M. & Sheen, J. (2002) MAP kinase signalling cascade in Arabidopsis innate immunity. *Nature*, **415**(6875), 977–983.

Belles-Boix, E., Babiychuk, E., Van Montagu, M., Inze, D. & Kushnir, S. (2000) CEO1, a new protein from *Arabidopsis thaliana*, protects yeast against oxidative damage. *FEBS Letters*, **482**(1–2), 19–24.

Berberich, T. & Kusano, T. (1997) Cycloheximide induces a subset of low temperature-inducible genes in maize. *Molecular and General Genetics*, **254**(3), 275–283.

Berridge, M. J. & Irvine, R. F. (1989) Inositol phosphate and cell signaling. *Nature*, **341**(6239), 197–205.

Botella, J. R., Arteca, J. M., Somodevilla, M. & Arteca, R. N. (1996) Calcium-dependent protein kinase gene expression in response to physical and chemical stimuli in mungbean (*Vigna radiata*). *Plant Molecular Biology*, **30**(6), 1129–1137.

Bowler, C. & Fluhr, R. (2000) The role of calcium and activated oxygens as signals for controlling cross-tolerance. *Trends in Plant Science*, **5**(6), 241–246.

Brunette, R. N., Gunesekara, B. M. & Gillaspy, G. E. (2003) An *Arabidopsis* inositol 5-phosphatase gain-of-function alters abscisic acid signaling. *Plant Physiology*, **132**(2), 1011–1019.

Camacho, P. & Lechleiter, J. D. (1993) Increased frequency of calcium waves in Xenopus laevis oocytes that express a calcium-ATPase. *Science*, **260**(5105), 226–229.

Cardinale, F., Meskiene, I., Ouaked, F. & Hirt, H. (2002) Convergence and divergence of stress-induced mitogen-activated protein kinase signaling pathways at the level of two distinct mitogen-activated protein kinase kinases. *Plant Cell*, **14**(3), 703–711.

Carman, G. M. & Zeimetz, G. M. (1996) Regulation of phospholipid biosynthesis in the yeast *Saccharomyces cerevisiae. Journal of Biological Chemistry*, **271**(23), 13293–13296.

Chen, W., Provart, N. J., Glazebrook, J., Katagiri, F., Chang, H. S., Eulgem, T., Mauch, F., Luan, S., Zou, G., Whitham, S. A., Budworth, P. R., Tao, Y., Xie, Z., Chen, X., Lam, S., Kreps, J. A., Harper, J. F., Si-Ammour, A., Mauch-Mani, B., Heinlein, M., Kobayashi, K., Hohn, T., Dangl, J. L., Wang, X. & Zhu, T (2002) Expression profile matrix of Arabidopsis transcription factor genes suggests their putative functions in response to environmental stresses. *Plant Cell*, 14(3), 559–574.

Chinnusamy, V. & Zhu, J. K. (2004) Plant salt tolerance. In H. Hirt & K. Shinozaki (eds) *Plant Responses to Abiotic Stress*. Springer-Verlag, Berlin, pp. 241–270.

Chinnusamy, V., Ohta, M., Kanrar, S., Lee, B. H., Hong X., Agarwal, M. & Zhu J. K. (2003) ICE1: a regulator of cold-induced transcriptome and freezing tolerance in Arabidopsis, *Genes & Development*, **17**(8), 1043–1054.

Chinnusamy, V., Schumaker, K. & Zhu, J. K. (2004) Molecular genetic perspectives on cross-talk and specificity in abiotic stress signalling in plants. *Journal of Experimental Botany*, **55**(395), 225–236.

Choi, H., Hong, J., Ha, J., Kang, J. & Kim, S. Y. (2000) ABFs, a family of ABA-responsive element binding factors. *Journal of Biological Chemistry*, **275**(3), 1723–1730.

Czernic, P., Visser, B., Sun, W., Savoure, A., Deslandes, L., Marco, Y., Van Montagu, M. & Verbruggen, N. (1999) Characterization of an *Arabidopsis thaliana* receptor-like protein kinase gene activated by oxidative stress and pathogen attack. *Plant Journal*, **18**(3), 321–327.

Delaunay, A., Isnard, A. D. & Toledano, M. B. (2000) H2O2 sensing through oxidation of the Yap1 transcription factor. *EMBO Journal*, **19**(19), 5157–5166.

Delaunay, A., Pflieger, D., Barrault, M. B., Vinh, J. & Toledano, M. B. (2002) A thiol peroxidase is an H2O2 receptor and redox-transducer in gene activation. *Cell*, **111**(4), 471–481.

Desikan, R., Hancock, J. T., Ichimura, K., Shinozaki, K. & Neill, S. J. (2001) Harpin induces activation of the Arabidopsis mitogen-activated protein kinases AtMPK4 and AtMPK6. *Plant Physiology*, **126**(4), 1579–1587.

DeWald, D. B., Torabinejad, J., Jones, C. A., Shope, J. C., Cangelosi, A. R., Thompson, J. E., Prestwich, G. D. & Hama, H. (2001) Rapid accumulation of phosphatidylinositol 4,5-bisphosphate and inositol 1,4,5-trisphosphate correlates with calcium mobilization in salt-stressed Arabidopsis. *Plant Physiology*, **126**(2), 759–769.

Drobak, B. K. & Watkins, P. A. (2000) Inositol (1,4,5) trisphosphate production in plant cells: an early response to salinity and hyperosmotic stress. *FEBS Letters*, **481**(3), 240–244.

Einspahr, K. J., Maeda, M. & Thompson, G. A. Jr. (1988) Concurrent changes in *Dunaliella salina* ultrastructure and membrane phospholipid metabolism after hyperosmotic shock. *Journal of Cell Biology*, **107**(2), 529–538.

Fowler, S. & Thomashow, M. F. (2002) Arabidopsis transcriptome profiling indicates that multiple regulatory pathways are activated during cold acclimation in addition to the CBF cold response pathway. *Plant Cell*, **14**(8), 1675–1690.

Foyer, C. H., Theodoulou, F. L. & Delrot, S. (2001) The functions of inter- and intracellular glutathione transport systems in plants. *Trends in Plant Science*, **6**(10), 486–492.

Frank, W., Munnik, T., Kerkmann, K., Salamini, F. & Bartels, D. (2000) Water deficit triggers phospholipase D activity in the resurrection plant *Craterostigma plantagineum. Plant Cell*, **12**(2), 111–123.

Gidrol, X., Sabelli, P. A., Fern, Y. S. & Kush, A. K. (1996) Annexin-like protein from *Arabidopsis thaliana* rescues delta oxyR mutant of *Escherichia coli* from H2O2 stress. *Proceedings of the National Academy of Sciences of the United States of America*, **93**(20), 11268–11273.

Gilmour, S. J., Zarka, D. G., Stockinger, E. J., Salazar, M. P., Houghton, J. M. & Thomashow, M. F. (1998) Low temperature regulation of the Arabidopsis CBF family of AP2 transcriptional activators as an early step in cold-induced COR gene expression. *Plant Journal*, **16**(4), 433–442.

Gong, D., Gong, Z., Guo, Y., Chen, X. & Zhu, J. K. (2002a) Biochemical and functional characterization of PKS11, a novel Arabidopsis protein kinase. *Journal of Biological Chemistry*, **277**(31), 28340–28350.

Gong, D., Guo, Y., Schumaker, K. S. & Zhu, J. K. (2004) The SOS3 Family of Calcium Sensors and SOS2 Family of Protein Kinases in Arabidopsis. *Plant Physiology*, **134**(3), 919–926.

Gong, D., Zhang, C., Chen, X., Gong, Z. & Zhu, J. K. (2002b) Constitutive activation and transgenic evaluation of the function of an arabidopsis PKS protein kinase. *Journal of Biological Chemistry*, **277**(44), 42088–42096.

Gong, Z., Lee, H., Xiong, L., Jagendorf, A., Stevenson, B. & Zhu, J.-K. (2002) RNA helicase-like protein as an early regulator of transcription factors for plant chilling and freezing tolerance. *Proceedings of the National Academy of Sciences of the United States of America*, **99**(17), 11507–11512.

Grabov, A. & Blatt, M. R. (1998) Membrane voltage initiates Ca^{2+} waves and potentiates Ca^{2+} increases with abscisic acid in stomatal guard cells. *Proceedings of the National Academy of Sciences of the United States of America*, **95**(8), 4778–4783.

Guan, L. M., Zhao J. & Scandalios J. G. (2000) Cis-elements and trans-factors that regulate expression of maize *Cat1* antioxidant gene in response to ABA and osmotic stress: H_2O_2 is the likely intermediary signaling molecule for the response. *Plant Journal*, **22**(2), 87–95.

Guo, Y., Halfter, U., Ishitani, M. & Zhu, J. K. (2001) Molecular characterization of functional domains in the protein kinase SOS2 that is required for plant salt tolerance. *Plant Cell*, **13**(6), 1383–1400.

Guo, Y., Qiu, Q., Quintero, F. J., Pardo, J. M., Ohta, M., Zhang, C., Schumaker, K. S. & Zhu, J. K. (2004) Transgenic evaluation of activated mutant alleles of SOS2 reveals a critical requirement for its kinase activity and C-terminal regulatory domain for salt tolerance in *Arabidopsis thaliana*. *Plant Cell*, **16**(2), 435–449.

Guo, Y., Xiong, L., Ishitani, M. & Zhu, J.-K. (2002a) An Arabidopsis mutation in translation elongation factor 2 causes superinduction of CBF/DREB1 transcription factor genes but blocks the induction of their downstream targets under low temperature. *Proceedings of the National Academy of Sciences of the United States of America*, **99**(11), 7786–7791.

Guo, Y., Xiong, L., Song, C. P., Gong, D., Halfter, U. & Zhu, J. K. (2002b) A calcium sensor and its interacting protein kinase are global regulators of abscisic acid signaling in *Arabidopsis*. *Developmental Cell*, **3**(2), 233–244.

Haake, V., Cook, D., Riechmann, J. L., Pineda, O., Thomashow, M. F. & Zhang, J. Z. (2002) Transcription factor CBF4 is a regulator of drought adaptation in Arabidopsis. *Plant Physiology*, **130**(2), 639–648.

Halfter, U., Ishitani, M. & Zhu, J. K. (2000) The *Arabidopsis* SOS2 protein kinase physically interacts with and is activated by the calcium-binding protein SOS3. *Proceedings of the National Academy of Sciences of the United States of America*, **97**(7), 3730–3734.

Hamilton, D. W., Hills, A., Kohler, B. & Blatt, M. R. (2000) Ca^{2+} channels at the plasma membrane of stomatal guard cells are activated by hyperpolarization and abscisic acid. *Proceedings of the National Academy of Sciences of the United States of America*, **97**(9), 4967–4972.

Hirayama, T., Ohto, C., Mizoguchi, T. & Shinozaki, K. (1995) A gene encoding a phosphatidylinositol-specific phospholipase C is induced by dehydration and salt stress in *Arabidopsis thaliana*. *Proceedings of the National Academy of Sciences of the United States of America*, **92**(9), 3903–3907.

Hirschi, K. D. (1999) Expression of Arabidopsis CAX1 in tobacco: altered calcium homeostasis and increased stress sensitivity. *Plant Cell*, **11**(11), 2113–2122.

Hoyos, M. E. & Zhang, S. (2000) Calcium-independent activation of salicylic acid-induced protein kinase and a 40-kilodalton protein kinase by hyperosmotic stress. *Plant Physiology*, **122**(4), 1355–1363.

Huang, Y., Li, H., Gupta, R., Morris, P. C., Luan, S. & Kieber, J. J. (2000) ATMPK4, an Arabidopsis homolog of mitogen-activated protein kinase, is activated in vitro by AtMEK1 through threonine phosphorylation, *Plant Physiology*, **122**(4), 1301–1310.

Ichimura, K., Mizoguchi, T., Irie, K., Morris, P., Giraudat, J., Matsumoto, K. & Shinozaki, K. (1998) Isolation of ATMEKK1 (a MAP kinase kinase kinase)-interacting proteins and analysis of a MAP kinase cascade in Arabidopsis. *Biochemical and Biophysical Research Communications*, **253**(2), 532–543.

Ichimura, K., Mizoguchi, T., Yoshida, R., Yuasa, T. & Shinozaki, K. (2000) Various abiotic stresses rapidly activate Arabidopsis MAP kinases ATMPK4 and ATMPK6. *Plant Journal*, **24**(5), 655–665.

Inaba, M., Suzuki, I., Szalontai, B., Kanesaki, Y., Los, D.A., Hayashi, H. & Murata, N. (2003) Gene-engineered rigidification of membrane lipids enhances the cold inducibility of gene expression in synechocystis. *Journal of Biological Chemistry*, **278**(14), 12191–12198.

Inoue, T., Higuchi, M., Hashimoto, Y., Seki, M., Kobayashi, M., Kato, T., Tabata, S., Shinozaki, K. & Kakimoto, T. (2001) Identification of CRE1 as a cytokinin receptor from Arabidopsis. *Nature*, **409**(6823), 1060–1063.

Ishitani, M., Liu, J., Halfter, U., Kim, C. S., Shi, W. & Zhu, J. K. (2000) SOS3 function in plant salt tolerance requires myristoylation and calcium-binding. *Plant Cell*, **12**(9), 1667–1678.

Iuchi, S., Kobayashi, M., Taji, T., Naramoto, M., Seki, M., Kato, T., Tabata, S., Kakubari, Y., Yamaguchi-Shinosaki, K. & Shinosaki, K. (2001) Regulation of drought tolerance by gene manipulation of 9-cis-epoxycarotenoid dioxygenase, a key enzyme in abscisic acid biosynthesis in Arabidopsis. *Plant Journal*, **27**(4), 325–333.

Iwasaki, T., Yamaguchi-Shinozaki, K. & Shinozaki, K. (1995) Identification of a cis-regulatory region of a gene in *Arabidopsis thaliana* whose induction by dehydration is mediated by abscisic acid and requires protein synthesis. *Molecular and General Genetics*, **247**(4), 391–398.

Jacob, T., Ritchie, S., Assmann, S. M. and Gilroy, S. (1999) Abscisic acid signal transduction in guard cells is mediated by phospholipase D activity. *Proceedings of the National Academy of Sciences of the United States of America*, **96**(21), 12192–12197.

Katagiri, T., Takahashi, S. and Shinozaki, K. (2001) Involvement of a novel *Arabidopsis* phospholipase D, AtPLD delta, in dehydration-inducible accumulation of phosphatidic acid in stress signaling. *Plant Journal*, **26**(6), 595–605.

Kawasaki, S., Borchert, C., Deyholos, M., Wang, H., Brazille, S., Kawai, K., Galbraith, D. & Bohnert, H. J. (2001) Gene expression profiles during the initial phase of salt stress in rice. *Plant Cell*, **13**(4), 889–905.

Kim, J. C., Lee, S. H., Cheong, Y. H., Yoo, C. M., Lee, S. I., Chun, H. J., Yun, D. J., Hong, J. C., Lee, S. Y., Lim, C. O. & Cho, M. J. (2001) A novel cold-inducible zinc finger protein from soybean, SCOF-1, enhances cold tolerance in transgenic plants. *Plant Journal*, **25**(3), 247–259.

Kim, K. N., Cheong, Y. H., Grant, J. J., Pandey, G. K. & Luan, S. (2003) CIPK3, a calcium sensor-associated protein kinase that regulates abscisic acid and cold signal transduction in Arabidopsis. *Plant Cell*, **15**(2), 411–423.

Knight, H. (2000) Calcium signaling during abiotic stress in plants. *International Review on Cytology*, **195**, 269–324.

Knight, H. & Knight, M. R. (2001) Abiotic stress signalling pathways: specificity and cross-talk. *Trends in Plant Science*, **6**(6), 262–267.

Knight, H., Trewavas, A. J. & Knight, M. R. (1996) Cold calcium signaling in Arabidopsis involves two cellular pools and a change in calcium signature after acclimation. *Plant Cell*, **8**(3), 489–503.

Knight, H., Trewavas, A. J. & Knight, M. R. (1997) Calcium signalling in *Arabidopsis thaliana* responding to drought and salinity. *Plant Journal*, **12**(5), 1067–1078.

Koiwa, H., Barb, A. W., Xiong, L., Li, F., McCully, M. G., Lee, B.-H., Sokolchik, I., Zhu, J., Gong, Z., Reddy, M., Sharkhuu, A., Manabe, Y., Yokoi, S., Zhu, J.-K., Bressan, R. A. & Hasegawa, P. M. (2002) C-terminal domain phosphatase-like family members (AtCPLs) differentially regulate *Arabidopsis thaliana* abiotic stress signaling, growth, and development. *Proceedings of the National Academy of Sciences of the United States of America*, **99**(16), 10893–10898.

Kopka, J., Pical, C., Gray, J. E. & Muller-Rober, B. (1998) Molecular and enzymatic characterization of three phosphoinositide-specific phospholipase C isoforms from potato. *Plant Physiology*, **116**(1), 239–250.

Kovtun, Y., Chiu, W. L., Tena, G. & Sheen, J. (2000) Functional analysis of oxidative stress-activated mitogen-activated protein kinase cascade in plants. *Proceedings of the National Academy of Sciences of the United States of America*, **97**(6), 2940–2945.

Kreps, J. A., Wu, Y., Chang, H. S., Zhu, T., Wang, X. & Harper, J. F. (2002) Transcriptome changes for Arabidopsis in response to salt, osmotic, and cold stress. *Plant Physiology*, **130**(4), 2129–2141.

Krochko, J. E., Abrams, G. D., Loewen, M. K., Abrams, S. R. & Culter, A. J. (1998) (+)-abscisic acid 8'-hydroxylase is a cytochrome P450 monooxygenase. *Plant Physiology*, **118**(3), 849–860.

Kudla, J., Xu, Q., Harter, K., Gruissem, W. & Luan, S. (1999) Genes for calcineurin B-like proteins in Arabidopsis are differentially regulated by stress signals. *Proceedings of the National Academy of Sciences of the United States of America*, **96**(8), 4718–4723.

Kushiro, T., Okamoto, M., Nakabayashi, K., Yamagishi, K., Kitamura, S., Asami, T., Hirai, N., Koshiba, T., Kamiya, Y. & Nambara, E. (2004) The Arabidopsis cytochrome P450 CYP707A encodes ABA 8'-ydroxylases: key enzymes in ABA catabolism. *EMBO Journal*, **23**(7), 1647–1656.

Lamb, C. & Dixon, R. A. (1997) The oxidative burst in plant disease resistance. *Annual Review on Plant Physiology amd Plant Molecular Biology*, **48**, 251–275.

Laurie, S., Feeney K. A., Maathuis F. J., Heard, P. J., Brown, S. J. & Leigh, R. A. (2002) A role for HKT1 in sodium uptake by wheat roots. *Plant Journal*, **32**(2), 139–149.

Lechleiter, J. D., John, L. M. & Camacho, P. (1998) Ca^{2+} wave dispersion and spiral wave entrainment in Xenopus laevis oocytes overexpressing Ca^{2+} ATPases. *Biophysical Chemistry*, **72**(1–2), 123–129.

Lee, H., Xiong, L., Gong, Z., Ishitani, M., Stevenson, B. & Zhu, J. K. (2001) The Arabidopsis *HOS1* gene negatively regulates cold signal transduction and encodes a RING-finger protein that displays cold-regulated nucleo-cytoplasmic partitioning. *Genes & Development*, **15**(7), 912–924.

Lee, H., Guo, Y., Ohta, M., Xiong, L., Stevenson, B. & Zhu, J.-K. (2002) LOS2, a genetic locus required for cold-responsive gene transcription encodes a bifunctional enolase. *EMBO Journal*, **21**(11), 2692–2702.

Lee, Y., Choi, Y. B., Suh, S., Lee, J., Assmann S. M., Joe, C. O., Kelleher, J. F. & Crain, R. C. (1996) Abscisic acid-induced phosphoinositide turnover in guard cell protoplasts of *Vicia faba*. *Plant Physiology*, **110**(3), 987–996.

Ligterink, W. & Hirt, H. (2001) Mitogen-activated protein (MAP) kinase pathways in plants: versatile signaling tools. *International Review of Cytology*, **201**, 209–275.

Liotenberg, S., North, H. & Marion-Poll, A. (1999) Molecular biology and regulation of abscisic acid biosynthesis in plants. *Plant Physiology and Biochemistry*, **37**(5), 341–350.

Liu, J. & Zhu, J. K. (1998) A calcium sensor homolog required for plant salt tolerance. *Science*, **280**(5371), 1943–1945.

Liu, J., Ishitani, M., Halfter, U., Kim, C. S. & Zhu, J. K. (2000) The *Arabidopsis thaliana SOS2* gene encodes a protein kinase that is required for salt tolerance. *Proceedings of the National Academy of Sciences of the United States of America*, **97**(7), 3735–3740.

Liu, Q., Kasuga, M., Sakuma, Y., Abe, H., Miura, S., Yamaguchi-Shinozaki, K. & Shinozaki, K. (1998) Two transcription factors, DREB1 and DREB2, with an EREBP/AP2 DNA binding domain separate two cellular signal transduction pathways in drought- and low temperature-responsive gene expression, respectively, in Arabidopsis. *Plant Cell*, **10**(8), 1391–1406.

Ludwig, A. A., Romeis, T. & Jones, J. D. (2004) CDPK-mediated signalling pathways: specificity and cross-talk. *Journal of Experimental Botany*, **55**(395), 181–188.

Maeda, T., Takekawa, M. & Saito, H. (1995) Activation of yeast PBS2 MAPKK by MAPKKKs or by binding of an SH3-containing osmosensor. *Science*, **269**(5223), 554–558.

Maeda, T., Wurgler-Murphy, S. M. & Saito, H. (1994) A two-component system that regulates an osmosen-sing MAP kinase cascade in yeast. *Nature*, **369**(6477), 242–245.

Marin, K., Suzuki, I., Yamaguchi, K., Ribbeck, K., Yamamoto, H., Kanesaki, Y., Hagemann, M. & Murata, N. (2003) Identification of histidine kinases that act as sensors in the perception of salt stress in Synecho-cystis sp. PCC 6803. *Proceedings of the National Academy of Sciences of the United States of America*, **100**(15), 9061–9066.

Martinec, J., Feltl, T., Scanlon, C. H., Lumsden, P. J. & Machackova, I. (2000) Subcellular localization of a high affinity binding site for D-myo-inositol 1,4,5-trisphosphate from *Chenopodium rubrum*. *Plant Physiology*, **124**(1), 475–483.

McAinsh, M. R. & Hetherington, A. M. (1998) Encoding specificity in Ca^{2+} signaling systems. *Trends in Plant Science*, **3**(1), 32–36.

Meijer, H. J. & Munnik T. (2003) Phospholipid-based signaling in plants. *Annual Review on Plant Biology*, **54**, 265–306.

Meijer, H. J., Divecha, N., van den Ende, H., Musgrave, A. & Munnik, T. (1999) Hyperosmotic stress induces rapid synthesis of phosphatidyl-D-inositol 3,5-bisphosphate in plant cells. *Planta*, **208**(2), 294–298.

Mikami, K., Kanesaki, Y., Suzuki, I. & Murata, N. (2002) The histidine kinase Hik33 perceives osmotic stress and cold stress in Synechocystis sp PCC 6803. *Molecular Microbiology*, **46**(4), 905–915.

Mikami, K., Katagiri, T., Luchi, S., Yamaguchi-Shinozaki, K. & Shinozaki, K. (1998) A gene encoding phosphatidylinositol 4-phosphate 5-kinase is induced by water stress and abscisic acid in *Arabidopsis thaliana*. *Plant Journal*, **15**(4), 563–568.

Mikolajczyk, M., Awotunde, O. S., Muszynska, G., Klessig, D. F. & Dobrowolska, G. (2000) Osmotic stress induces rapid activation of a salicylic acid-induced protein kinase and a homolog of protein kinase ASK1 in tobacco cells. *Plant Cell*, **12**(1), 165–178.

Mizoguchi, T., Irie, K., Hirayama, T., Hayashida, N., Yamaguchi-Shinozaki, K., Matsumoto, K. & Shinozaki, K. (1996) A gene encoding a mitogen-activated protein kinase kinase kinase is induced simultaneously with genes for a mitogen-activated protein kinase and an S6 ribosomal protein kinase by touch, cold, and water stress in *Arabidopsis thaliana*. *Proceedings of the National Academy of Sciences of the United States of America*, **93**(2), 765–769.

Monroy, A. F. & Dhindsa, R. S. (1995) Low-temperature signal transduction: induction of cold acclimation-specific genes of alfalfa by calcium at 25 degrees C. *Plant Cell*, **7**(3), 321–331.

Moon, H., Lee, B., Choi, G., Shin, D., Prasad, D. T., Lee, O., Kwak, S. S., Kim, D. H., Nam, J., Bahk, J., Hong, J. C., Lee, S. Y., Cho, M. J., Lim, C. O. & Yun, D. J. (2003) NDP kinase 2 interacts with two oxidative stress-activated MAPKs to regulate cellular redox state and enhances multiple stress tolerance in transgenic plants. *Proceedings of the National Academy of Sciences of the United States of America*, **100**(1), 358–363.

Munnik, T. (2001) Phosphatidic acid: an emerging plant lipid second messenger. *Trends in Plant Science*, **6**(5), 227–233.

Munnik, T. & Meijer, H. J. (2001) Osmotic stress activates distinct lipid and MAPK signaling pathways in plants. *FEBS Letters*, **498**(2–3), 172–178.

Munnik, T., Meijer, H. J., ter Riet, B., Frank, W., Bartels, D. & Musgrave, A. (2000) Hyperosmotic stress stimulates phospholipase D activity and elevates the levels of phosphatidic acid and diacylglycerol pyrophosphate. *Plant Journal*, **22**(2), 147–154.

Murata, Y., Pei, Z. M., Mori, I. C. & Schroeder, J. (2001) Abscisic acid activation of plasma membrane Ca^{2+} channels in guard cells requires cytosolic NAD(P)H and is differentially disrupted upstream and downstream of reactive oxygen species production in *abi1-1* and *abi2-1* protein phosphatase 2C mutants. *Plant Cell*, **13**(11), 2513–2523.

Mustilli, A. C., Merlot, S., Vavasseur, A., Fenzi, F. & Giraudat, J. (2002) Arabidopsis OST1 protein kinase mediates the regulation of stomatal aperture by abscisic acid and acts upstream of reactive oxygen species production. *Plant Cell*, **14**(12), 3089–3099.

Narusaka, Y., Nakashima, K., Shinwari, Z. K., Sakuma, Y., Furihata, T., Abe, H., Narusaka, M., Shinozaki, K. & Yamaguchi-Shinozaki, K. (2003) Interaction between two cis-acting elements, ABRE and DRE, in ABA-dependent expression of Arabidopsis rd29A gene in response to dehydration and high-salinity stresses. *Plant Journal*, **34**(2), 137–148.

Navazio, L., Bewell, M. A., Siddiqua, A., Dickinson, G. D., Galione, A. & Sanders, D. (2000) Calcium release from the endoplasmic reticulum of higher plants elicited by the NADP metabolite nicotinic acid adenine dinucleotide phosphate. *Proceedings of the National Academy of Sciences of the United States of America*, **97**(15), 8693–8698.

O'Rourke, S. M., Herskowitz, I. & O'Shea, E. K. (2002) Yeast go the whole HOG for the hyperosmotic response. *Trends in Genetics*, **18**(8), 405–412.

Orvar, B. L., Sangwan, V., Omann, F. & Dhindsa, R. (2000) Early steps in cold sensing by plant cells: the role of actin cytoskeleton and membrane fluidity. *Plant Journal*, **23**(6), 785–794.

Patharkar, O. R. & Cushman, J. C. (2000) A stress-induced calcium-dependent protein kinase from *Mesembryanthemum crystallium* phosphorylates a two-component pseudo-response regulator. *Plant Journal*, **24**(5), 679–691.

Pei, Z. M., Murata, Y., Benning, G., Thomine, S., Klusener, B., Allen, G. J., Grill, E. & Schroeder, J. I. (2000) Calcium channels activated by hydrogen peroxide mediate abscisic acid signaling in guard cells. *Nature*, **406**(6797), 731–734.

Pical, C., Westergren, T., Dove, S. K., Larsson, C. & Sommarin, M. (1999) Salinity and hyperosmotic stress induce rapid increases in phosphatidylinositol 4,5-bisphosphate, diacylglycerol pyrophosphate, and phosphatidyl-choline in *Arabidopsis thaliana* cells. *Journal of Biological Chemistry*, **274**(53), 38232–38240.

Posas, F., Wurgler-Murphy, S. M., Maeda, T., Witten, E. A., Thai, T. C. & Saito, H. (1996) Yeast HOG1 MAP kinase cascade is regulated by a multistep phosphorelay mechanism in the SLN1-YPD1-SSK1 'two-component' osmosensor. *Cell*, **86**(6), 865–875.

Prasad, T. K., Anderson, M. D., Martin, B. A. & Stewart, C. R. (1994) Evidence for chilling-induced oxidative stress in maize seedlings and a regulatory role for hydrogen peroxide. *Plant Cell*, **6**(1), 65–74.

Qin, X. & Zeevaart, J. A. (1999) The 9-*cis*-epoxycarotenoid cleavage reaction is the key regulatory step of abscisic acid biosynthesis in water-stressed bean. *Proceedings of the National Academy of Sciences of the United States of America*, **96**(26), 15354–15361.

Qiu, Q. S., Guo, Y., Dietrich, M. A., Schumaker, K. S. & Zhu, J. K. (2002) Regulation of SOS1, a plasma membrane Na+/H+ exchanger in *Arabidopsis thaliana*, by SOS2 and SOS3. *Proceedings of the National Academy of Sciences of the United States of America*, **99**(12), 8436–8441.

Qiu, Q. S., Guo, Y., Quintero, F. J., Pardo, J. M., Schumaker, K. S. & Zhu, J. K. (2004) Regulation of vacuolar Na$^+$/H$^+$ exchange in *Arabidopsis thaliana* by the salt-overly-sensitive (SOS) pathway. *Journal of Biological Chemistry*, **279**(1), 207–215.

Quintero, F. J., Garciadeblas, B. & Rodriguez-Navarro, A. (1996) The *SAL1* gene of Arabidopsis, encoding an enzyme with 3'(2'), 5'-bisphosphate nucleotide and inositol polyphosphate 1-phosphatase activities, increases salt tolerance in yeast. *Plant Cell*, **8**(3), 529–537.

Quintero, F. J., Ohta, M., Shi, H., Zhu, J. K. & Pardo, J. M. (2002) Reconstitution in yeast of the Arabidopsis SOS signaling pathway for Na+ homeostasis. *Proceedings of the National Academy of Sciences of the United States of America*, **99**(13), 9061–9066.

Raitt, D. C., Posas, F. & Saito, H. (2000) Yeast Cdc42 GTPase and Ste20 PAK-like kinase regulate Sho1-dependent activation of the Hog1 MAPK pathway. *EMBO Journal*, **19**(17), 4623–4631.

Rao, M. V., Paliyath, G. & Ormrod, D. P. (1996) Ultraviolet-B- and ozone-induced biochemical changes in antioxidant enzymes of *Arabidopsis thaliana*. *Plant Physiology*, **110**(1), 125–136.

Reddy, A. S. (2001) Calcium: silver bullet in signaling. *Plant Science*, **160**(3), 381–404.

Reiser, V., Raitt, D. C. & Saito, H. (2003) Yeast osmosensor Sln1 and plant cytokinin receptor Cre1 respond to changes in turgor pressure. *Journal of Cell Biology*, **161**(6), 1035–1040.

Rentel, M. C., Lecourieux, D., Ouaked, F., Usher, S. L., Petersen, L., Okamoto, H., Knight, H., Peck, S. C., Grierson, C. S., Hirt, H. & Knight, M. R. (2004) OXI1 kinase is necessary for oxidative burst-mediated signalling in Arabidopsis. *Nature*, **427**(6977), 858–861.

Richmond, T. & Somerville, S. (2000) Chasing the dream: plant EST microarrays. *Current Opinion in Plant Biology*, **3**(2), 108–116.

Ritchie, S. & Gilroy, S. (1998) Abscisic acid signal transduction in the barley aleurone is mediated by phospholipase D activity. *Proceedings of the National Academy of Sciences of the United States of America*, **95**(5), 2697–2702.

Roderick, H. L., Lechleiter, J. D. & Camacho, P. (2000) Cytosolic phosphorylation of calnexin controls intracellular Ca(2+) oscillations via an interaction with SERCA2b. *Journal of Cell Biology*, **149**(6), 1235–1248.

Romeis, T., Piedras, P. & Jones, J. D. (2000) Resistance gene-dependent activation of a calcium-dependent protein kinase in the plant defense response. *Plant Cell*, **12**(5), 803–816.

Rus, A., Yokoi, S., Sharkhuu, A., Reddy, M., Lee, B. H., Matsumoto, T. K., Koiwa, H., Zhu, J. K., Bressan, R. A. & Hasegawa, P. M. (2001) AtHKT1 is a salt tolerance determinant that controls Na(+) entry into plant roots. *Proceedings of the National Academy of Sciences of the United States of America*, **98**(24), 14150–14155.

Saijo, Y., Hata, S., Kyozuka, J., Shimamoto, K. & Izui, K. (2000) Overexpression of a single Ca^{2+} dependent protein kinase confers both cold and salt/drought tolerance on rice plants. *Plant Journal*, **23**(3), 319–327.

Samuel, M. A., Miles, G. P. & Ellis, B. E. (2000) Ozone treatment rapidly activates MAP kinase signalling in plants. *Plant Journal*, **22**(4), 367–376.

Sanchez, J. P. & Chua, N. H. (2001) Arabidopsis PLC1 is required for secondary responses to abscisic acid signals. *Plant Cell*, **13**(5), 1143–1154.

Sanders, D., Brownlee, C. & Harper, J. F. (1999) Communicating with calcium. *Plant Cell*, **11**(4), 691–706.

Sanders, D., Pelloux, J., Brownlee, C. & Harper, J. F. (2002) Calcium at the crossroads of signaling. *Plant Cell*, **14**(Suppl.), S401–S417.

Sangwan, V., Foulds, I., Singh, J. & Dhindsa, R. J. (2001) Cold-activation of *Brassica napus* BN115 promoter is mediated by structural changes in membranes and cytoskeleton, and requires Ca^{2+} influx. *Plant Journal*, **27**(1), 1–12.

Schuler, M. A. & Werck-Reichhart, D. (2003) Functional Genomics Of P450s. *Annual Review on Plant Physiology and Plant Molecular Biology*, **54**, 629–667.

Seki, M., Narusaka, M., Abe, H., Kasuga, M., Yamaguchi-Shinozaki, K., Carninci, P., Hayashizaki, Y. & Shinozaki, K. (2001) Monitoring the expression pattern of 1300 Arabidopsis genes under drought and cold stresses by using a full-length cDNA microarray. *Plant Cell*, **13**(1), 61–72.

Seki, M., Narusaka, M., Ishida, J., Nanjo, T., Fujita, M., Oono, Y., Kamiya, A., Nakajima, M., Enju, A., Sakurai, T., Satou, M., Akiyama, K., Taji, T., Yamaguchi-Shinozaki, K., Carninci, P., Kawai, J., Hayashizaki, Y. & Shinozaki, K. (2002a) Monitoring the expression profiles of 7000 Arabidopsis genes under drought, cold and high-salinity stresses using a full-length cDNA microarray. *Plant Journal*, **31**(3), 279–292.

Seki, M., Narusaka, M., Kamiya, A., Ishida, J., Satou, M., Sakurai, T., Nakajima, M., Enju, A., Akiyama, K., Oono, Y., Muramatsu, M., Hayashizaki, Y., Kawai, J., Carninci, P., Itoh, M., Ishii, Y., Arakawa, T., Shibata, K., Shinagawa, A. & Shinozaki, K. (2002b) Functional annotation of a full-length Arabidopsis cDNA collection. *Science*, **296**(5565), 141–145.

Seo, M., Peeters, A. J., Koiwai, H., Oritani, T., Marion-Poll, A., Zeevaart, J. A., Koornneef., Kamiya, Y. & Koshiba, T. (2000) The *Arabidopsis* aldehyde oxidase 3 (*AAO3*) gene product catalyzes the final step in abscisic acid biosynthesis in leaves. *Proceedings of the National Academy of Sciences of the United States of America*, **97**(23), 12908–12913.

Sheen, J. (1996) Ca^{2+}-dependent protein kinases and stress signal transduction in plants. *Science*, **274**(5294), 1900–1902.

Shi, H., Ishitani, M., Kim, C. & Zhu, J. K. (2000) The *Arabidopsis thaliana* salt tolerance gene SOS1 encodes a putative Na+/H+ antiporter. *Proceedings of the National Academy of Sciences of the United States of America*, **97**(12), 6896–6901.

Shinozaki, K. & Yamaguchi-Shinozaki, K. (2000). Molecular response to dehydration and low temperature: differences and cross-talk between two stress signaling pathways. *Current Opinion in Plant Biology*, **3**, 217–223.

Srivastava, A., Pines, M. & Jacoby, B. (1989) Enhanced potassium uptake and phosphatidylinositol 4,5-biphosphate turnover by hypertonic mannitol shock. *Physiologia Plantarum*, **77**(3), 320–325.

Stockinger, E. J., Gilmour, S. J. & Thomashow, M. F. (1997) *Arabidopsis thaliana CBF1* encodes an AP2 domain-containing transcriptional activator that binds to the C-repeat/DRE, a cis-acting DNA regulatory element that stimulates transcription in response to low temperature and water deficit. *Proceedings of the National Academy of Sciences of the United States of America*, **94**(3), 1035–1040.

Suzuki, I., Los, D. A., Kanesaki, Y., Mikami, K. & Murata, N. (2000) The pathway for perception and transcription of low-temperature signals in *Synechocystis*. *EMBO Journal*, **19**(6), 1327–1334.

Takahashi, S., Katagiri, T., Hirayama, T., Yamaguchi-Shinozaki, K. & Shinozaki, K. (2001) Hyperosmotic stress induced a rapid and transient increase in inositol 1,4,5-trisphosphate independent of abscisic acid in *Arabidopsis* cell culture. *Plant and Cell Physiology*, **42**(4), 214–222.

Takatsuji, H. & Matsumoto, T. (1996) Target-sequence recognition by separate-type Cys2/His2 zinc finger proteins in plants. *Journal of Biological Chemistry*, **271**(38), 23368–23373.

Tamura, T., Hara, K., Yamaguchi, Y., Koizumi, N. & Sano, H. (2003) Osmotic stress tolerance of transgenic tobacco expressing a gene encoding a membrane-located receptor-like protein from tobacco plants. *Plant Physiology*, **131**(2), 454–462.

Taylor, I. B., Burbidge, A. & Thompson, A. J. (2000) Control of abscisic acid synthesis. *Journal of Experimental Botany*, **51**(350), 1563–1574.

Thomashow, M. F. (1999) Plant cold acclimation: freezing tolerance genes and regulatory mechanisms. *Annual Review on Plant Physiology and Plant Molecular Biology*, **50**, 571–599.

Thompson, A. J., Jackson, A. C., Symonds, R. C., Mulholland, B. J., Dadswell, A. R., Blake, P. S., Burbidge, A. & Taylor, I. B. (2000) Ectopic expression of a tomato 9-cis-epoxycarotenoid dioxygenase gene causes over-production of abscisic acid. *Plant Journal*, **23**(3), 363–374.

Ulm, R., Ichimura, K., Mizoguchi, T., Peck, S. C., Zhu, T., Wang, X., Shinozaki, K. & Paszkowski, J. (2002) Distinct regulation of salinity and genotoxic stress responses by Arabidopsis MAP kinase phosphatase 1. *EMBO Journal*, **21**(23), 6483–6493.

Urao, T., Katagiri, T., Mizoguchi, T., Yamaguchi-Shinozaki, K., Hayashida, N. & Shinozaki, K. (1994) Two genes that encode Ca^{2+}-dependent protein kinases are induced by drought and high salt stresses in *Arabidopsis thaliana*. *Molecular and General Genetics*, **224**(4), 331–340.

Urao, T., Yakubov, B., Satoh, R., Yamaguchi-Shinozaki, K., Seki, M., Hirayama, T. & Shinozaki, K. (1999) A transmembrane hybrid-type histidine kinase in Arabidopsis functions as an osmosensor. *Plant Cell*, **11**(9), 1743–1754.

Viswanathan, C. & Zhu, J. K. (2002) Molecular genetic analysis of cold-regulated gene transcription. *Philosophical Transactions of the Royal Society of London Series B – Biological Sciences*, **357**(1423), 877–886.

Wang, X. (1999) The role of phospholipase D in signaling cascades. *Plant Physiology*, **120**(3), 645–651.

Wohlgemuth, H., Mittelstrass, K., Kschieschan, S., Bender, J., Weigel, H.-J., Overmeyer, K., Kangasarvi, J., Sandermann, H. & Langebartels, C. (2002) Activation of an oxidative burst is a general feature of sensitive plants exposed to the air pollutant ozone. *Plant Cell and Environment*, 25, 717–726.

Wood, N. T., Allan, A. C., Haley, A., Viry-Moussaïd, M. & Trewavas, A. J. (2000) The characterization of differential calcium signalling in tobacco guard cells. *Plant Journal*, **24**(3), 335–344.

Wu, Y., Kuzma, J., Marechal, E., Graeff, R., Lee, H. C., Foster, R. & Chua, N. H. (1997) Abscisic acid signaling through cyclic ADP-ribose in plants. *Science*, **278**(5346), 2126–2130.

Xiong, L. & Zhu, J. K. (2001) Abiotic stress signal transduction in plants: molecular and genetic perspectives. *Physiologia Plantarum*, **112**(2), 152–166.

Xiong, L., Gong, Z., Rock C., Subramanian, S., Guo, Y., Xu, W., Galbraith, D. & Zhu J. K. (2001a) Modulation of abscisic acid signal transduction and biosynthesis by an Sm-like protein in *Arabidopsis*. *Developmental Cell*, **1**(6), 771–781.

Xiong, L., Ishitani, M., Lee, H. & Zhu, J. K. (2001b) The Arabidopsis LOS5/ABA3 locus encodes a molybdenum cofactor sulfurase and modulates cold stress- and osmotic stress-responsive gene expression. *Plant Cell*, **13**(9), 2063–2083.

Xiong, L., Lee, B. H., Ishitani, M., Lee, H., Zhang, C. & Zhu, J. K. (2001c) *FIERY1* encoding an inositol polyphosphate 1-phosphatase is a negative regulator of abscisic acid and stress signaling in Arabidopsis. *Genes & Development*, **15**(15), 1971–1984.

Xiong, L., Lee, H., Ishitani, M., Tanaka, Y., Stevenson, B., Koiwa, H., Bressan, R. A., Hasegawa, P. M. & Zhu, J.-K. (2002a) Repression of stress-responsive genes by FIERY2, a novel transcription regulator unique to plants. *Proceedings of the National Academy of Sciences of the United States of America*, **99**(16), 10899–10904.

Xiong, L., Lee, H., Ishitani, M. & Zhu, J. K. (2002b) Regulation of osmotic stress-responsive gene expression by the LOS6/ABA1 locus in Arabidopsis. *Journal of Biological Chemistry*, **277**(10), 8588–8596.

Xiong, L., Schumaker, K. S. & Zhu, J. K. (2002c) Cell signaling during cold, drought, and salt stress. *Plant Cell*, **14**(Suppl.), S165–S183.

Xu, Q., Fu, H. H., Gupta, R. & Luan, S. (1998) Molecular characterization of a tyrosine-specific protein phosphatase encoded by a stress-responsive gene in Arabidopsis. *Plant Cell*, **10**(5), 849–857.

Yamada, H., Suzuki, T., Terada, K., Takei, K., Ishikawa, K., Miwa, K., Yamashino, T. & Mizuno, T. (2001) The Arabidopsis AHK4 histidine kinase is a cytokinin-binding receptor that transduces cytokinin signals across the membrane. *Plant and Cell Physiology*, **42**(9), 1017–1023.

Yamaguchi-Shinozaki, K. & Shinozaki, K. (1993a) Characterization of the expression of a desiccation-responsive *rd29* gene of *Arabidopsis thaliana* and analysis of its promoter in transgenic plants. *Molecular and General Genetics*, **236**(2–3), 331–340.

Yamaguchi-Shinozaki, K. & Shinozaki, K. (1993b) The plant hormone abscisic acid mediates the drought-induced expression but not the seed-specific expression of rd22, a gene responsive to dehydration stress in *Arabidopsis thaliana*. *Molecular and General Genetics*, **238**(1–2), 17–25.

Yamaguchi-Shinozaki, K. & Shinozaki, K. (1994) A novel *cis*-acting element in an Arabidopsis gene is involved in responsiveness to drought, low-temperature, or high-salt stress. *Plant Cell*, **6**(2), 251–264.

Yang, Y., Fang, S., Jensen, J. P., Weissman, A. M. & Ashwell, J. D. (2000) Ubiquitin protein ligase activity of IAPs and their degradation in proteasomes in response to apoptotic stimuli. *Science*, **288**(5467), 874–877.

Zhao, Z., Chen, G. & Zhang, C. (2001) Interaction between reactive oxygen species and nitric oxide in drought-induced abscisic acid synthesis in root tips of wheat seedlings. *Australian Journal of Plant Physiology*, **28**(10), 1055–1061.

Zheng, M., Aslund, F. & Storz, G. (1998) Activation of the OxyR transcription factor by reversible disulfide bond formation. *Science*, **279**(5357), 1718–1721.

Zhu, J. K. (2001) Cell signaling under salt, water and cold stresses. *Current Opinion in Plant Biology*, **4**(5), 401–406.

Zhu, J. K. (2002) Salt and drought stress signal transduction in plants. *Annual Review of Plant Biology*, **53**, 247–273.

9   Genomic analysis of stress response

Motoaki Seki, Junko Ishida, Maiko Nakajima, Akiko Enju,
Kei Iida, Masakazu Satou, Miki Fujita, Yoshihiro Narusaka,
Mari Narusaka, Tetsuya Sakurai, Kenji Akiyama, Youko Oono,
Ayako Kamei, Taishi Umezawa, Saho Mizukado, Kyonoshin
Maruyama, Kazuko Yamaguchi-Shinozaki and Kazuo Shinozaki

9.1   Introduction

Plant growth is greatly affected by environmental abiotic stresses, such as
drought, high salinity and low temperature, and biotic stresses, such as pathogen infection. Plants respond and adapt to these stresses in order to survive.
These stresses induce various biochemical and physiological responses in
plants. Several genes that respond to these stresses at the transcriptional
level have been studied (Thomashow, 1999; Hasegawa *et al.*, 2000; Shinozaki
& Yamaguchi-Shinozaki, 2000; Wan *et al.*, 2002; Zhu, 2002; Narusaka *et al.*,
2003; Seki *et al.*, 2003; Shinozaki *et al.*, 2003). The products of the stress-
inducible genes have been classified into two groups: those that directly
protect against environmental stresses and those that regulate gene expression
and signal transduction in the stress response. Stress-inducible genes have
been used to improve the stress tolerance of plants by gene transfer (Salmeron
& Vernooij, 1998; Thomashow, 1999; Hasegawa *et al.*, 2000; Shinozaki &
Yamaguchi-Shinozaki, 2000; Hammond-Kosack & Parker, 2003; Seki *et al.*,
2003; Shinozaki *et al.*, 2003; Zhang, 2003). It is important to analyze the
functions of stress-inducible genes not only to understand the molecular
mechanisms of stress tolerance and the responses of higher plants but also to
improve the stress tolerance of crops by gene manipulation. Hundreds of genes
are thought to be involved in the responses to abiotic (Shinozaki & Yamaguchi-Shinozaki, 1999, 2000; Xiong & Zhu, 2001, 2002; Xiong *et al.*, 2002; Zhu,
2002; Seki *et al.*, 2003; Shinozaki *et al.*, 2003) and biotic (Kazan *et al.*, 2001;
Wan *et al.*, 2002; Narusaka *et al.*, 2003) stresses.

In this chapter, we mainly highlight recent progress in the genomic analysis
of abiotic stress responses.

9.2   Expression profiling under stress conditions by cDNA microarray analysis

Microarray technology has recently become a powerful tool for the systematic
analysis of expression profiles of large numbers of genes (Eisen & Brown,

1999; Richmond & Somerville, 2000; Seki *et al.*, 2001b). Several groups have reported the application of microarray technology to the analysis of expression profiles in response to abiotic stresses, such as drought, cold and high-salinity (Kawasaki *et al.*, 2001; Seki *et al.*, 2001a, 2002b, 2002c; Chen *et al.*, 2002; Fowler & Thomashow, 2002), and biotic stresses (Narusaka *et al.*, 2003). Several reviews on the transcriptome analysis in abiotic (Bray, 2002; Ramanjulu & Bartels, 2002; Hazen *et al.*, 2003; Seki *et al.*, 2003, 2004) and biotic (Kazan *et al.*, 2001; Wan *et al.*, 2002) stress conditions in higher plants have been published recently. In this chapter, we mainly summarize the recent progress in the transcriptome analysis under abiotic stress conditions.

9.3 DNA microarrays are an excellent tool for identifying genes regulated by various stresses

A number of genes that respond to drought, cold and high-salinity stresses at the transcriptional level have been described. However, many unidentified genes were thought to be involved in the responses to drought, cold and high-salinity stress responses. Therefore, we applied 7K RAFL (RIKEN *Arabidopsis* full-length) cDNA microarray containing ~7000 *Arabidopsis* full-length cDNA groups to identify new drought-, cold-, high-salinity- or abscisic acid (ABA)–inducible genes (Seki *et al.*, 2002b, 2002c). In this study, we identified 299 drought-inducible genes, 54 cold-inducible genes, 213 high-salinity-stress-inducible genes and 245 ABA-inducible genes (Seki *et al.*, 2002b, 2002c). Venn diagram analysis indicated the existence of significant crosstalk between drought and high-salinity stress signaling processes (Seki *et al.*, 2002b). Many ABA-inducible genes are induced after drought- and high-salinity stress treatments, which indicates the existence of significant crosstalk between drought and ABA responses (Seki *et al.*, 2002c). These results support our previous model on strong overlap of gene expression in response to drought, high-salinity and ABA (Shinozaki & Yamaguchi-Shinozaki, 2000) and partial overlap of gene expression in response to cold and osmotic stress.

Drought-, high-salinity- or cold-stress-inducible gene products can be classified into two groups (Shinozaki & Yamaguchi-Shinozaki, 1999, 2000; Seki *et al.*, 2002b, 2003). The first group includes functional proteins, or proteins that probably function in stress tolerance. The second group contains regulatory proteins, that is, protein factors involved in further regulation of signal transduction and gene expression that probably function in response to stress (Shinozaki & Yamaguchi-Shinozaki, 1999, 2000; Seki *et al.*, 2002b, 2003). They are various transcription factors, protein kinases, protein phosphatases, enzymes involved in phospholipid metabolism and other signaling molecules, such as calmodulin-binding protein (Seki *et al.*, 2002b).

Kreps *et al.* (2002) studied the expression profiles in leaves and roots from *Arabidopsis* subjected to salt (100 mM NaCl), hyperosmotic (200 mM mannitol) and cold (4°C) stress treatments using *Arabidopsis* 8K Affymetrix GeneChip. RNA samples were collected separately from leaves and roots after 3- and 27-h stress treatments. Kreps *et al.* (2002) identified a total of 2409 unique stress-regulated genes that displayed a greater than twofold change in expression compared with the control. The results suggested the majority of changes were stress-specific. At the 3-h time point, less than 5% (118 genes) of the changes were observed as shared by all three stress responses, and by 27 h, the number of shared responses was reduced more than tenfold (< 0.5%). Roots and leaves displayed very different changes. For example, less than 14% of the cold-specific changes were shared between roots and leaves at both 3 and 27 h. Kreps *et al.* (2002) also identified 306 stress-regulated genes among the 453 known circadian-controlled genes (Harmer *et al.*, 2000). These results suggested that ~68% of the circadian-controlled genes are linked to a stress response pathway and supported the hypothesis that an important function of the circadian clock is to 'anticipate' predictable stresses such as cold nights.

Lee and Lee (2003) have characterized the global gene expression patterns of *Arabidopsis* pollen using serial analysis of gene expression (SAGE) and cDNA microarray analysis. The expression level of the majority of transcripts was unaffected by cold treatment at 0° for 72 h, whereas pollen tube growth and seed production were substantially reduced. Many genes thought to be responsible for cold acclimation, such as *cor*, lipid transfer protein, and β-amylase, which are highly induced in *Arabidopsis* leaves, were only expressed at their normal level or weakly induced in the pollen, suggesting that poor accumulation of proteins that play a role in stress tolerance may be why *Arabidopsis* pollen is cold-sensitive.

9.4 DNA microarrays are a useful tool for identifying the target genes of the stress-related transcription factors

We reported that overexpression of the *DREB1A/CBF3* cDNA under the control of the cauliflower mosaic virus (CaMV) 35S promoter or the stress-inducible *rd29A* promoter in transgenic plants gave rise to strong constitutive expression of the stress-inducible DREB1A target genes and increased tolerance to freezing, drought and salt stresses (Liu *et al.*, 1998; Kasuga *et al.*, 1999). Kasuga *et al.* (1999) identified six DREB1A target genes. However, it was not well understood how overexpression of the *DREB1A* cDNA in transgenic plants increases stress tolerance to freezing, drought and high-salinity stresses. To study the molecular mechanisms of drought and freezing tolerance, it is important to identify and analyze more genes that are controlled by DREB1A. Therefore, we applied the full-length cDNA microarray containing ~1300 *Arabidopsis* full-length cDNAs to identify new target genes of DREB1A (Seki *et al.*, 2001a). Twelve stress-inducible genes were identified as target stress-inducible genes of DREB1A, and six of them were novel. All DREB1A target genes identified contained DRE- or DRE-related CCGAC core motif sequences in their promoter regions (Seki *et al.*, 2001a). These results show that cDNA microarray is useful for identifying target genes of stress-related transcription factors and for identifying potential *cis*-acting DNA elements by combining expression data with genomic sequence data.

Recently, Maruyama *et al.* (2004) searched for new DREB1A target genes using the 7K RAFL cDNA microarray and 8K Affymetrix GeneChip. Maruyama *et al.* (2004) identified 38 genes as the DREB1A target genes, including 20 unreported new target genes. Many of the products of these genes were proteins known to function against stress and were probably responsible for the stress tolerance of the transgenic plants. The target genes also included genes for protein factors involved in further regulation of signal transduction and gene expression in response to stress. The identified genes were classified into direct target genes of DREB1A and the others based on their expression patterns in response to cold stress. Maruyama *et al.* (2004) also searched for conserved sequences in the promoter regions of the direct target genes and found A/GCCGACNT in their promoter regions from −51 to −450 as a consensus DRE. The recombinant DREB1A protein bound to A/GCCGACNT more efficiently than to A/GCCGACNA/G/C.

Fowler and Thomashow (2002) identified 306 cold-regulated genes and 41 *DREB/CBF*-regulated genes using the Affymetrix GeneChip (Fowler & Thomashow, 2002). Twenty-eight percent of the cold-responsive genes were not regulated by the CBF transcription factors, suggesting the existence of the unidentified cold stress response pathways. However, several differences between our results (Seki *et al.*, 2001a, 2002b; Maruyama *et al.*, 2004) and those of Fowler and Thomashow exist. This difference may be due to differences in gene annotation, expression-profiling methods, ecotypes used and plant growth conditions.

The *INDUCER OF CBF EXPRESSION1 (ICE1)* gene was identified through map-based cloning of the *Arabidopsis ice1* mutation, which affected the expression of the *CBF3/DREB1A* promoter::LUC transgenic plants (Chinnusamy *et al.*, 2003). ICE1 encodes a MYC-type bHLH transcription factor that regulates the expression of *CBF3/DREB1A* genes. Microarray analysis using Affymetrix Arabidopsis ATH1 genome array GeneChips showed that out of 306 genes induced threefold or more in the wild type by a 6-h cold treatment, 217 are either not induced in the *ice1* mutant or their induction is

50% or less of that in the wild type (Chinnusamy *et al.*, 2003). Thirty-two of these encode putative transcription factors, suggesting that ICE1 may control many cold-responsive regulons. Overexpression of ICE1 in transgenics resulted in improved freezing tolerance, supporting an important role for ICE1 in the cold stress response (Chinnusamy *et al.*, 2003).

The transgenic plants overexpressing *AtMYC2* and *AtMYB2* cDNAs have higher sensitivity to ABA (Abe *et al.*, 2003). Abe *et al.* (2003) studied the expression profiles in the transgenic plants overexpressing *AtMYC2* and *AtMYB2* cDNAs using the 7K RAFL cDNA microarray. mRNAs prepared from 35S:*AtMYC2*/*AtMYB2* and wild-type plants were used for the generation of Cy3-labeled and Cy5-labeled cDNA probes, respectively. Microarray analysis of the transgenic plants revealed that several ABA-inducible genes were upregulated in the 35S:*AtMYC2*/*AtMYB2* transgenic plants. Abe *et al.* (2003) searched for the MYC recognition sequence (CANNTG) and the MYB recognition sequences (A/TAACCA and C/TAACG/TG) located within the 10- to 600-bp upstream region from each putative TATA box in the promoter regions of the 32 upregulated genes identified. Abe *et al.* (2003) found that 29 genes have the MYC recognition sequence, 29 genes have the MYB recognition sequence, and 26 genes have both MYC and MYB recognition sequences in their promoter regions. *Ds* insertion mutant of the *AtMYC2* gene was less sensitive to ABA and showed significantly decreased ABA-induced gene expression of *rd22* and *AtADH1*. These results indicated that both AtMYC2 and AtMYB2 function as transcriptional activators in ABA-inducible gene expression under drought stress conditions in plants.

In rice, Dubouzet *et al.* (2003) isolated five cDNAs for DREB homologues: *OsDREB1A*, *OsDREB1B*, *OsDREB1C*, *OsDREB1D* and *OsDREB2A*. Expression of *OsDREB1A* and *OsDREB1B* was induced by cold stress, whereas expression of *OsDREB2A* was induced by dehydration and high-salinity stresses. The OsDREB1A and OsDREB2A proteins specifically bound to DRE and activated the transcription of the GUS reporter gene driven by DRE in rice protoplasts. Overexpression of *OsDREB1A* in transgenic *Arabidopsis* plants resulted in improved tolerance to drought, high-salinity and freezing stresses, indicating that OsDREB1A has functional similarity to DREB1A (Dubouzet *et al.*, 2003). Several OsDREB1A target genes were identified by the cDNA microarray and RNA gel blot analyses. Computer analysis showed that the seven OsDREB1A target genes have at least one core GCCGAC sequence as the DRE core motif in their promoter regions. Some of the DREB1A target genes such as *kin1*, *kin2* and *erd10*, containing ACCGAC but not GCCGAC as the DRE core motifs in their promoter regions, were not overexpressed in the 35S:*OsDREB1A* plants. These results indicated that the OsDREB1A protein binds more preferentially to GCCGAC than to ACCGAC in the promoter regions, whereas the DREB1A protein binds to both GCCGAC and ACCGAC efficiently (Dubouzet *et al.*, 2003).

9.5 Expression profiling in various stress-related mutants

Hugouvieux *et al.* (2001) isolated a recessive ABA-hypersensitive *Arabidopsis* mutant, *abh1*. *ABH1* encodes a functional mRNA cap-binding protein. DNA chip experiments showed that 18 genes including *RD20, KIN2* and *COR15b* had significant and threefold reduced transcript levels in the *abh1* mutant, and seven of these genes are ABA-regulated in the wild type. Consistent with these results, *abh1* plants showed ABA-hypersensitive stomatal closing and reduced wilting during drought. Hugouvieux *et al.* (2001) showed ABA-hypersensitive cytosolic calcium increases in *abh1* guard cells. These results indicate a functional link between mRNA processing and modulation of early ABA signal transduction.

Hoth *et al.* (2002) applied massively parallel signature sequencing (MPSS) to samples from *Arabidopsis* wild type and *abi1-1* mutant seedlings. Hoth *et al.* (2002) identified 1354 genes that are either up- or downregulated by ABA treatment of wild-type seedlings. Among these ABA-responsive genes, many encode signal transduction components. In addition, Hoth *et al.* (2002) identified novel ABA-responsive gene families including those encoding ribosomal proteins and proteins involved in regulated proteolysis. In the ABA-insensitive mutant *abi1-1*, ABA regulation of about 84.5% and 6.9% of the identified genes was impaired and strongly diminished, respectively; however, 8.6% of the genes remained appropriately regulated. Regulation of the majority of the genes by ABA was impaired in the ABA-insensitive mutant *abi1-1*. However, a subset of genes continued to be appropriately regulated by ABA, which suggests the presence of at least two ABA signaling pathways, only one of which is blocked in *abi1-1*.

The *sfr6* mutant of *Arabidopsis* displays a deficit in freezing tolerance after cold acclimation. Boyce *et al.* (2003) reported that the effects of *sfr6* mutant upon transcript levels were reflected in the levels of the encoded proteins, confirming that the cold-inducible protein expression was affected by the *sfr6* mutation. Using microarray analysis, Boyce *et al.* (2003) found not only that this effect may be general to cold-inducible genes with DRE/CRT promoter elements, but also that it extends to some other genes whose promoter lacks a DRE/CRT element.

Provart *et al.* (2003) have analyzed the molecular phenotypes of 12 chilling-sensitive (*chs*) mutants exposed to chilling (13°C) conditions by GeneChip experiments. The number and pattern of expression of chilling-responsive genes in the mutants were consistent with their final degree of chilling injury. The mRNA accumulation profiles for the chilling-lethal mutants *chs1, chs2* and *chs3* were highly similar and included extensive chilling-induced and mutant-specific alterations in gene expression. Provart *et al.* (2003) have identified 634 chilling-responsive genes with aberrant expression in all the chilling-lethal mutants. Among the genes identified, genes related to lipid metabolism, chloroplast function, carbohydrate metabolism and free radical detoxification existed.

9.6 Rehydration- or proline-inducible genes and functions of their gene products identified by RAFL cDNA microarray

Proline (Pro) is one of the most widely distributed osmolytes in water-stressed plants. L-Pro is metabolized to L-Glu via Δ^1-pyrroline-5-carboxylate (P5C) by two enzymes, Pro dehydrogenase (ProDH) and P5C dehydrogenase (Strizhov *et al.*, 1997; Yoshiba *et al.*, 1997). The ProDH gene in *Arabidopsis* is upregulated not only by rehydration after dehydration, but also by L-Pro and hypoosmolarity (Kiyosue *et al.*, 1996; Nakashima *et al.*, 1998). Satoh *et al.* (2002) analyzed the promoter regions of ProDH to identify *cis*-acting elements involved in L-Pro-induced and hypoosmolarity-induced expression in transgenic tobacco and *Arabidopsis* plants. Satoh *et al.* (2002) found that a 9-bp sequence, ACTCATCCT, in the ProDH promoter is necessary for the efficient expression of ProDH in response to L-Pro and hypoosmolarity and that the ACTCAT sequence is a core *cis*-acting element. To elucidate whether the promoter region of the other L-Pro-inducible genes have the ACTCAT sequence, Satoh *et al.* (2002) used the 7K RAFL cDNA microarray and found that 27 L-Pro-inducible genes identified have the ACTCAT sequence in their promoter regions. Twenty-one genes among the 27 genes showed L-Pro-inducible expression based on RNA gel blot analysis. These results suggest that the ACTCAT sequence is conserved in many L-Pro-inducible promoters and plays a key role in L-Pro-inducible gene expression.

Few studies on genes involved in the recovery process from dehydration have been published until recently. Analysis of rehydration-inducible genes should help not only to understand the molecular mechanisms of stress responses of higher plants, but also to improve the stress tolerance of crops by gene manipulation. Oono *et al.* (2003) applied the 7K RAFL cDNA microarray to identify genes that are induced during the rehydration process after dehydration stress and to analyze expression profiles of gene expression in the recovery process from drought stress. Oono *et al.* (2003) identified 152 rehydration-inducible genes. These genes can be classified into the following three major groups:

(1) regulatory proteins involved in further regulation of signal transduction and gene expression,
(2) functional proteins involved in the recovery process from dehydration-stress-induced damages and
(3) functional proteins involved in plant growth.

Venn diagram analysis also showed that among the rehydration-inducible genes, at least two gene groups, that is, genes functioning in adjustment of cellular osmotic conditions and those functioning in the repair of drought-stress-induced damages existed and that most of the rehydration-downregulated genes are dehydration-inducible. Furthermore, promoter analysis suggested that the ACTCAT sequence is a major *cis*-acting element involved in rehydration-inducible gene expression and that some novel *cis*-acting elements involved in rehydration-inducible gene expression existed.

9.7 Abiotic stress-inducible genes identified using microarrays in monocots

Studies on gene expression profile under abiotic stress have been conducted in other plant species, such as rice (Bohnert *et al.*, 2001; Kawasaki *et al.*, 2001; Rabbani *et al.*, 2003), barley (Ozturk *et al.*, 2002) and maize (Wang *et al.*, 2003).

Kawasaki *et al.* (2001) compared the gene expression profiles between a salinity-sensitive rice (var IR29) and a salinity-tolerant one (var Pokkali) after salinity stress (150 mM NaCl) treatment. The sensitive IR29 had a similar response to high-salinity stress as the tolerant Pokkali. However, a difference between the two accessions existed with respect to the onset of the initial response to salinity stress, that is, IR29 responded more slowly than Pokkali. In this case, it appears that more rapid transcriptional regulation under the salt stress may be responsible for salt tolerance.

Rabbani *et al.* (2003) prepared a rice cDNA microarray including about 1700 independent cDNAs derived from cDNA libraries prepared from drought-, cold-, and high-salinity-treated rice plants. Rabbani *et al.* (2003) confirmed stress-inducible expression of the candidate genes selected by microarray analysis using RNA gel blot analysis and finally identified a total of 73 genes as stress-inducible including 58 novel unreported genes in rice. Among them, 36, 62, 57 and 43 genes were induced by cold, drought, high salinity and ABA, respectively. Venn diagram analysis revealed greater cross-talk between the signaling processes for drought stress and high-salinity stress or for drought stress and ABA application than between the signaling processes for cold stress and high-salinity stress or for cold stress and ABA application, which is consistent with our observation on the overlap of the stress-responsive gene expression in *Arabidopsis* (Seki *et al.*, 2002b, 2002c). Comparative analysis of *Arabidopsis* and rice showed that among the 73 stress-inducible rice genes, 51 already have been reported in *Arabidopsis* with a similar function or gene name, suggesting that there are similar molecular mechanisms of the stress tolerance and responses between dicots and monocots.

9.8 Many stress- or hormone-inducible transcription factor genes have been identified by transcriptome analysis

Transcription factors play important roles in plant response to environmental stresses and its development. Transcription factors are sequence-specific DNA-binding proteins that are capable of activating and repressing transcription. The *Arabidopsis* genome encodes more than 1500 transcription factors (Riechmann *et al.*, 2000) and a number of transcription factor families have been implicated in plant stress responses (Shinozaki & Yamaguchi-Shinozaki, 2000; Shinozaki *et al.*, 2003). Several studies have demonstrated overlap among various stress- or hormone-signaling pathways (Seki *et al.*, 2001a, 2002b, 2002c; Chen *et al.*, 2002; Cheong *et al.*, 2002; Kimura *et al.*, 2003; Narusaka *et al.*, 2003; Oono *et al.*, 2003; Shinozaki *et al.*, 2003). Microarray analyses have also identified many stress- or hormone-inducible transcription factor genes (Seki *et al.*, 2001a, 2002b, 2002c; Chen *et al.*, 2002; Cheong *et al.*, 2002; Kimura *et al.*, 2003; Narusaka *et al.*, 2003; Oono *et al.*, 2003). These transcription factors probably function in stress- or hormone-inducible gene expression, although most of their target genes have not yet been identified. In this chapter, we summarize two transcriptome analysis, that is, RAFL cDNA microarray analysis and GeneChip analysis of Chen *et al.* (2002), for identification of the stress- or hormone-inducible transcription factors.

9.8.1   *7K RAFL cDNA microarray analysis*

We have applied the 7K RAFL cDNA microarray to study the expression profiles of *Arabidopsis* genes under various stress conditions, such as drought, cold and high-salinity stresses (Seki *et al.*, 2001a, 2002b) and high light stress (Kimura *et al.*, 2003), various treatment conditions, such as ABA (Seki *et al.*, 2002c), rehydration treatment after dehydration (Oono *et al.*, 2003), ethylene (Narusaka *et al.*, 2003), jasmonic acid (JA) (Narusaka *et al.*, 2003), salicylic acid (SA) (Narusaka *et al.*, 2003), reactive oxygen species (ROS)-inducing compounds such as paraquat and rose bengal (Narusaka *et al.*, 2003), UV-C (Narusaka *et al.*, 2003), Pro (Satoh *et al.*, 2002) and inoculation with pathogen (Narusaka *et al.*, 2003). Among the 7K RAFL cDNA microarrays, ~400 transcription factor genes are included. In this chapter, we summarize stress- or hormone-inducible transcription factor genes identified by 7K RAFL cDNA microarray analysis. Genes with expression ratios (treated/untreated) greater than three times that of the lambda control template DNA fragment in at least one time-course point are considered as stress- or hormone-inducible genes. The number of genes identified as transcription factor genes induced in Columbia wild type by drought, cold, high-salinity, ABA, water-soaked treatment, rehydration, pathogen inoculation, water-spray treatment, ethylene, JA, paraquat, rose bengal, SA, UV and high light were 67, 27, 54, 68, 39, 50, 51, 5, 60, 13, 20, 10, 24, 26 and 8, respectively (Table 9.1). Nine genes were also identified as pathogen-inoculation-inducible transcription factor genes in *pad3* mutants (Table 9.1). These results suggest that various transcriptional regulatory mechanisms function in the various stress or hormone signal transduction pathways. The list of genes is available at http://pfgweb.gsc.riken.-go.jp/index.html (as supplemental tables 1 and 2). These transcription factors probably regulate various stress- or hormone-inducible genes either cooperatively or separately. Functional analysis of these stress- or hormone-inducible transcription factors should provide more information on signal transduction in response to various stresses and hormones.

9.8.2   GeneChip analysis

Chen *et al.* (2002) used the expression profiles generated from the 8K Gene-Chip experiments to deduce the functions of genes encoding *Arabidopsis*

Table 9.1   Stress- or hormone-inducible transcription factor genes identified by the 7K RAFL cDNA microarray analysis

Treatment	Number of transcription factor genes[*]	Reference
Drought	67	Seki *et al.* (2002b)
Cold	27	Seki *et al.* (2002b)
High salinity	54	Seki *et al.* (2002b)
ABA	68	Seki *et al.* (2002c)
Water-soaked treatment	39	Seki *et al.* (2002b)
Rehydration after 2 h of dehydration	50	Oono *et al.* (2003)
Inoculation with *Alternaria brassicicola* in wild type	51	Narusaka *et al.* (2003)
Inoculation with *Alternaria brassicicola* in *pad3* mutants	9	Narusaka *et al.* (2003)
Water-spray treatment	5	Narusaka *et al.* (2003)
Ethylene	60	Narusaka *et al.* (2003)
Jasmonic acid	13	Narusaka *et al.* (2003)
Paraquat	20	Narusaka *et al.* (2003)
Rose bengal	10	Narusaka *et al.* (2003)
Salicylic acid	24	Narusaka *et al.* (2003)
UV	26	Narusaka *et al.* (2003)
High light	8	Kimura *et al.* (2003)

[*]In this study, we regarded the genes with expression ratios (treated/untreated) greater than three times that of lambda control template DNA fragment in at least one time-course point as stress- or hormone-inducible genes.

transcription factors. The expression levels of the 402 transcription factor genes were monitored in various organs, at different developmental stages, and under various biotic and abiotic stresses. A two-dimensional matrix (genes versus treatments or developmental stages/tissues) describing the changes in the mRNA levels of the 402 transcription factor genes was constructed for these experiments. The data represent 19 independent experiments, with samples derived from different organs such as roots, leaves, inflorescence stems, flowers and siliques and at different developmental stages and > 80 experiments representing 57 independent treatments with cold, salt, osmoticum, wounding, JA and different types of pathogens at different time points. The results showed that the transcription factors potentially controlling downstream gene expression in stress signal transduction pathways were identified by observing the activation and repression of the genes after certain stress treatments and that the mRNA levels of a number of previously characterized transcription factor genes were changed significantly in connection with other regulatory pathways, suggesting their multifunctional nature (Chen et al., 2002). Among the 43 transcription factor genes that are induced during senescence, 28 of them also are induced by stress treatment, suggesting that the signaling pathway activated by senescence may overlap substantially with the stress signaling pathways (Chen et al., 2002). The statistical analysis of the promoter regions of the genes responsive to cold stress indicated that two elements, the ABRE-like element and the DRE-like element (Shinozaki & Yamaguchi-Shinozaki, 2000), occur at significantly higher frequencies in the promoters from the late cold response cluster than their average frequency in all of the promoters of the genes on the *Arabidopsis* GeneChip (Chen et al., 2002). These results suggest that the ABRE-like element and the DRE-like element are two major elements important for the transcriptional regulation of genes in the late cold response cluster.

9.9   Application of full-length cDNAs to structural and functional analysis of plant proteins

Full-length cDNAs are useful resources for determining the three-dimensional structures of proteins by X-ray crystallography and nuclear magnetic resonance (NMR) spectroscopy. We have determined the three-dimensional structures of plant proteins using the RAFL cDNAs by NMR spectroscopy in collaboration with the protein research group (Principal Investigator: Shigeyuki Yokoyama) in the RIKEN Structural Genomics Initiative (Yokoyama et al., 2000). In the RIKEN Structural Genomics Initiative, we are mainly using cell-free protein synthesis systems for protein expression. Cell-free *in vitro* systems have three advantages over conventional *in vivo* expression systems:

(1) cell-free systems are suitable for automated, high-throughput expression, as proteins can be produced without the need for cloning genes into expression vectors;
(2) milligram quantities of proteins can be obtained in several hours; and
(3) proteins that are difficult for expression *in vivo* can be produced *in vitro*.

We have applied this system to plant protein expression for structure determination in the RIKEN Structural Genomics Initiative (Yokoyama *et al.*, 2000) and determined the domain structure of 29 proteins containing plant-specific-type transcription factors (Yamasaki *et al.*, 2004) as of March 2004. Determination of the three-dimensional structure of the DNA-binding domain of the stress-inducible transcription factors may be applied to alter the target genes for improvement of stress tolerance in the future.

Recently, Endo's group at Ehime University has modified the wheat germ cell-free protein synthesis system to produce milligram quantities of proteins (Madin *et al.*, 2000; Sawasaki *et al.*, 2002). We have applied this system for expression of plant proteins to study protein–protein and protein–DNA interactions (Sawasaki *et al.*, 2004).

Systematic approaches using transgenic plants, such as overexpression (Liu *et al.*, 1998; Kasuga *et al.*, 1999; Zhang, 2003), antisense suppression (Nanjo *et al.*, 1999) and double-stranded RNA interference (dsRNAi) (Chuang & Meyerowitz, 2000; Smith *et al.*, 2000), are also effective for functional analysis of plant proteins. Full-length cDNAs are also useful for these transgenic studies. The production of transgenic *Arabidopsis* plant lines that overexpress full-length cDNAs under the control of the CaMV 35S promoter or a chemical-inducible system is thought to be a useful alternative to activation tagging (Ichikawa *et al.*, unpublished).

9.10 Conclusions and perspectives

The microarray-based expression profiling method is useful for analyzing the expression pattern of plant genes under various stress and hormone treatments, identifying target genes of transcription factors involved in stress or hormone signal transduction pathways and identifying potential *cis*-acting DNA elements by combining expression data with genomic sequence data. By the expression profiling approach, many stress-inducible genes and stress-inducible transcription factor genes have been identified, suggesting that various transcriptional regulatory mechanisms function in each stress signal transduction pathway. Functional analysis of these stress-inducible genes should provide more information on the signal transduction in these stress responses.

By genetic approaches and biochemical analyses of signal transduction and stress tolerance of drought, cold and high-salinity stress, several mutants on the signal transduction and stress tolerance of these stresses have been identified (Browse & Xin, 2001; Knight & Knight, 2001; Finkelstein *et al.*, 2002; Xiong *et al.*, 2002; Zhu, 2002). Reverse genetic approaches, such as transgenic analyses, have also become useful for studying the function of the signaling components (Hasegawa *et al.*, 2000; Iuchi *et al.*, 2001; Apse & Blumwald, 2002; Finkelstein *et al.*, 2002; Gong *et al.*, 2002; Guo *et al.*, 2002; Xiong *et al.*, 2002). The availability of the *Arabidopsis* and rice genome sequences will not only greatly facilitate the isolation of mutations identified by the above genetic screening, but also will offer many other useful opportunities for study stress signal transduction. Genomewide expression profiling of the stress-resistant or stress-sensitive mutants, and mutants on the stress signal transduction should help identify more genes that are regulated at the transcriptional level by the signaling components.

Moreover, full-length cDNAs (Seki *et al.*, 2002a; Kikuchi *et al.*, 2003; Yamada *et al.*, 2003) are useful for transgenic analyses, such as overexpression, antisense suppression and dsRNAi, and biochemical analyses to study the function of the encoded proteins. T-DNA- or transposon-knockout mutants also offer the opportunity to study the function of the genes. Genomewide protein interaction studies help to identify the interactions among signaling components and to construct the signal networks 'dissected' by the above genetic analysis. Whole genome tilling array studies (Yamada *et al.*, 2003) will also become powerful tools for identification of stress-inducible microexons or noncoding RNAs. The information generated by focused studies of gene function in *Arabidopsis* is the springboard for a new wave of strategies to improve stress tolerance in agriculturally important crops.

9.11 Summary

Plants respond and adapt to abiotic and biotic stresses in order to survive. Molecular and genomic studies have shown that many genes with various functions are induced by abiotic and biotic stresses, and the various transcription factors are involved in the regulation of stress-inducible genes. The development of microarray-based expression profiling methods, together with the availability of genomic and cDNA sequence data, has allowed significant progress in the characterization of plant stress response.

Acknowledgments

This work was supported in part by a grant for Genome Research from RIKEN, the Program for Promotion of Basic Research Activities for Innovative Biosciences, the Special Coordination Fund of the Science and Technology Agency, and a Grant-in-Aid from the Ministry of Education, Culture, Sports, Science and Technology of Japan (MECSST) to K.S. It was also supported in part by a Grant-in-Aid for Scientific Research on Priority Areas (C) 'Genome Science' from MECSST to M.S.

References

Abe, H., Urao, T., Ito, T., Seki, M., Shinozaki, K. & Yamaguchi-Shinozaki, K. (2003) *Arabidopsis* AtMYC2 (bHLH) and AtMYB2 (MYB) function as transcriptional activators in abscisic acid signaling. *Plant Cell*, **15**, 63–78.

Apse, M. P. & Blumwald, E. (2002) Engineering salt tolerance in plants. *Current Opinion in Biotechnology*, **13**, 146–150.

Bohnert, H. J., Ayoubi, P., Borchert, C., Bressan, R. A., Burnap, R. L., Cushman, J. C., Cushman, M. A., Deyholos, M., Fischer, R., Galbraith, D. W., Hasegawa, P. M., Jenks, M., Kawasaki, S., Koiwa, H., Koreeda, S., Lee, B. H., Michalowski, C. B., Misawa, E., Nomura, M., Ozturk, N., Postier, B., Prade, R., Song, C. P., Tanaka, Y., Wang, H. & Zhu, J. K. (2001) A genomics approach towards salt stress tolerance. *Plant Physiology and Biochemistry*, **39**, 295–311.

Boyce, J. M., Knight, H., Deyholos, M., Openshaw, M. R., Galbraith, D. W., Warren, G. & Knight, M. R. (2003) The sfr6 mutant of Arabidopsis is defective in transcriptional activation via CBF/DREB1 and DREB2 and shows sensitivity to osmotic stress. *Plant Journal*, **34**, 395–406.

Bray, E. A. (2002) Classification of genes differentially expressed during water-deficit stress in *Arabidopsis thaliana*: an analysis using microarray and differentia; expression data. *Annals of Botany*, **89**, 803–811.

Browse, J. & Xin, Z. (2001) Temperature sensing and cold acclimation. *Current Opinion in Plant Biology*, **4**, 241–246.

Chen, W., Provart, N. J., Glazebrook, J., Katagiri, F., Chang, H. S., Eulgem, T., Mauch, F., Luan, S., Zou, G., Whitham, S. A., Budworth, P. R., Tao, Y., Xie, Z., Chen, X., Lam, S., Kreps, J. A., Harper, J. F., Si-Ammour, A., Mauch-Mani, B., Heinlein, M., Kobayashi, K., Hohn, T., Dangl, J. L., Wang, X. & Zhu, T. (2002) Expression profile matrix of *Arabidopsis* transcription factor genes suggests their putative functions in response to environmental stresses. *Plant Cell*, **14**, 559–574.

Cheong, Y. H., Chang, H. S., Gupta, R., Wang, X., Zhu, T. & Luan, S. (2002) Transcriptional profiling reveals novel interactions between wounding, pathogen, abiotic stress, and hormonal responses in *Arabidopsis*. *Plant Physiology*, **129**, 661–677.

Chinnusamy, V., Ohta, M., Kannar, S., Lee, B., Hong, Z. & Agarwal, A. & Zhu, J. K. (2003) ICE1: a regulator of cold-induced transcriptome and freezing tolerance in *Arabidopsis*. *Genes & Development*, **17**, 1043–1054.

Chuang, C. F. & Meyerowitz, E. (2000) Specific and heritable genetic interference by double-stranded RNA in *Arabidopsis thaliana*. *Proceedings of the National Academy of Sciences of the United States of America*, **97**, 4985–4990.

Dubouzet, J. G., Sakuma, Y., Ito, Y., Kasuga, M., Dubouzet, E. G., Miura, S., Seki, M., Shinozaki, K. & Yamaguchi-Shinozaki, K. (2003) OsDREB genes in rice, *Oryza sativa* L., encode transcription activators that function in drought-, high-salt- and cold-responsive gene expression. *Plant Journal*, **33**, 751–763.

Eisen, M. B. & Brown, P. O. (1999) DNA arrays for analysis of gene expression. *Methods in Enzymology*, **303**, 179–205.

Finkelstein, R. R., Gampala, S. S. L. & Rock, C. D. (2002) Abscisic acid signaling in seeds. *Plant Cell,* **14**, S15–S45.

Fowler, S. & Thomashow, M. F. (2002) *Arabidopsis* transcriptome profiling indicates that multiple regulatory pathways are activated during cold acclimation in addition to the CBF cold response pathway. *Plant Cell*, **14**, 1675–1690.

Gong, D., Zhang, C., Chen, X., Gong, Z. & Zhu, J. K. (2002) Constitutive activation and transgenic evaluation of the function of an *Arabidopsis* PKS protein kinase. *Journal of Biological Chemistry*, **277**, 42088–42096.

Guo, Y., Xiong, L., Song, C. P., Gong, D., Halfter, U. & Zhu, J. K. (2002) A calcium sensor and its interacting protein kinase are global regulators of abscisic acid signaling in *Arabidopsis*. *Developmental Cell*, **3**, 233–244.

Hammond-Kosack, K. E. & Parker, J. E. (2003) Deciphering plant-pathogen communication: fresh perspectives for molecular resistance breeding. *Current Opinion in Biotechnology*, **14**, 177–193.

Harmer, S. L., Hogenesch, J. B., Straume, M., Chang, H. S., Han, B., Zhu, T., Wang, X., Kreps, J. A. & Kay, S. A. (2000) Orchestrated transcription of key pathways in *Arabidopsis* by the circadian clock. *Science*, **290**, 2110–2113.

Hasegawa, P. M., Bressan, R. A., Zhu, J. K. & Bohnert, H. J. (2000) Plant cellular and molecular responses to high salinity. *Annual Review of Plant Physiology and Plant Molecular Biology*, **51**, 463–499.

Hazen, S. P., Wu, Y. & Kreps, J. A. (2003) Gene expression profiling of plant responses to abiotic stress. *Functional and Integrative Genomics*, **3**, 105–111.

Hoth, S., Morgante, M., Sanchez, J. P., Hanafey, M. K., Tingey, S. V. & Chua, N. H. (2002) Genome-wide gene expression profiling in *Arabidopsis thaliana* reveals new targets of abscisic acid and largely impaired gene regulation in the *abi1-1* mutant. *Journal of Cell Science*, **115**, 4891–4900.

Hugouvieux, V., Kwak, J. M. & Schroeder, J. I. (2001) An mRNA cap binding protein, ABH1, modulates early abscisic acid signal transduction in *Arabidopsis*. *Cell*, **106**, 477–487.

Iuchi, S., Kobayashi, M., Taji, T., Naramoto, M., Seki, M., Kato, T., Tabata, S., Kakubari, Y., Yamaguchi-Shinozaki, K. & Shinozaki, K. (2001) Regulation of drought tolerance by gene manipulation of 9-*cis*-epoxycarotenoid dioxygenase, a key enzyme in abscisic acid biosynthesis, in *Arabidopsis*. *Plant Journal*, **27**, 325–333.

Kasuga, M., Liu, Q., Miura, S., Yamaguchi-Shinozaki, K. & Shinozaki, K. (1999) Improving plant drought, salt, and freezing tolerance by gene transfer of a single stress-inducible transcription factor. *Nature Biotechnology*, **17**, 287–291.

Kawasaki, S., Borchert, C., Deyholos, M., Wang, H., Brazille, S., Kawai, K., Galbraith, D. & Bohnert, H. (2001) Gene expression profiles during the initial phase of salt stress in rice. *Plant Cell*, **13**, 889–905.

Kazan, K., Schenk, P. M., Wilson, I. & Manners, J. M. (2001) DNA microarrays: new tools in the analysis of plant defence responses. *Molecular Plant Pathology*, **2**, 177–185.

Kikuchi, S., Satoh, K., Nagata, T., Kawagashira, N., Doi, K., Kishimoto, N., Yazaki, J., Ishikawa, M., Yamada, H., Ooka, H., Hotta, I., Kojima, K., Namiki, T., Ohneda, E., Yahagi, W., Suzuki, K., Li, C. J., Ohtsuki, K., Shishiki, T., Otomo, Y., Murakami, K., Iida, Y., Sugano, S., Fujimura, T., Suzuki, Y., Tsunoda, Y., Kurosaki, T., Kodama, T., Masuda, H., Kobayashi, M., Xie, Q., Lu, M., Narikawa, R., Sugiyama, A., Mizuno, K., Yokomizo, S., Niikura, J., Ikeda, R., Ishibiki, J., Kawamata, M., Yoshimura, A., Miura, J., Kusumegi, T., Oka, M., Ryu, R., Ueda, M., Matsubara, K., Kawai, J., Carninci, P., Adachi, J., Aizawa, K., Arakawa, T., Fukuda, S., Hara, A., Hashidume, W., Hayatsu, N., Imotani, K., Ishii, Y., Itoh, M., Kagawa, I., Kondo, S., Konno, H., Miyazaki, A., Osato, N., Ota, Y., Saito, R., Sasaki, D., Sato, K., Shibata, K., Shinagawa, A., Shiraki, T., Yoshino, M. & Hayashizaki, Y. (2003) Collection, mapping, and annotation of over 28,000 cDNA clones from japonica rice. *Science*, **301**, 376–379.

Kimura, M., Yamamoto, Y. Y., Seki, M., Sakurai, T., Satou, M., Abe, T., Yoshida, S., Manabe, K., Shinozaki, K. & Matsui, M. (2003) Identification of *Arabidopsis* genes regulated by high light stress using cDNA microarray. *Photochemistry and Photobiology*, **77**, 226–233.

Kiyosue, T., Yoshiba, Y., Yamaguchi-Shinozaki, K. & Shinozaki, K. (1996) A nuclear gene encoding mitochondrial proline dehydrogenase, an enzyme involved in proline metabolism, is upregulated by proline but downregulated by dehydration in Arabidopsis. *Plant Cell*, **8**, 1323–1335.

Knight, H. & Knight, M. R. (2001) Abiotic stress signalling pathways: specificity and cross-talk, *Trends in Plant Science*, **6**, 262–267.

Kreps, J. A., Wu, Y., Chang, H. S., Zhu, T., Wang, X. & Harper, J. F. (2002) Transcriptome changes for Arabidopsis in response to salt, osmotic, and cold stress. *Plant Physiology*, **130**, 2129–2141.

Lee, J. Y. & Lee, D. H. (2003) Use of serial analysis of gene expression technology to reveal changes in gene expression in Arabidopsis pollen undergoing cold stress. *Plant Physiology*, **132**, 517–529.

Liu, Q., Kasuga, M., Sakuma, Y., Abe, H., Miura, S., Yamaguchi-Shinozaki, K. & Shinozaki, K. (1998) The transcription factors, DREB1 and DREB2, with an EREBP/AP2 DNA binding domain separate two cellular signal transduction pathways in drought- and low-temperature-responsive gene expression, respectively, in *Arabidopsis*. *Plant Cell*, **10**, 1391–1406.

Madin, K., Sawasaki, T., Ogasawara, T. & Endo, Y. (2000) A highly efficient and robust cell-free protein synthesis system prepared from wheat embryos: plants apparently contain a suicide system directed at ribosomes. *Proceedings of the National Academy of Sciences of the United States of America*, **97**, 559–564.

Maruyama, K., Sakuma, Y., Kasuga, M., Ito, Y., Seki, M., Goda, H., Shimada, Y., Yoshida, S., Shinozaki, K. & Yamaguchi-Shinozaki, K. (2004) Identification of cold-inducible downstream genes of the Arabidopsis DREB1A/CBF3 transcriptional factor using two microarray systems. *Plant Journal*, **38**, 982–993.

Nakashima, K., Satoh, R., Kiyosue, T., Yamaguchi-Shinozaki, K. & Shinozaki, K. (1998) A gene encoding proline dehydrogenase is not only induced by proline and hypoosmolarity, but is also developmentally regulated in the reproductive organs of *Arabidopsis*. *Plant Physiology*, **118**, 1233–1241.

Nanjo, T., Kobayashi, M., Yoshiba, Y., Sanada, Y., Wada, K., Tsukaya, H., Kakubari, Y., Yamaguchi-Shinozaki, K., & Shinozaki, K. (1999) Biological functions of proline in morphogenesis and osmotolerance revealed in antisense transgenic *Arabidopsis thaliana*. *Plant Journal*, **18**, 185–193.

Narusaka, Y., Narusaka, M., Seki, M., Ishida, J., Nakashima, M., Kamiya, A., Enju, A., Sakurai, T., Satoh, M., Kobayashi, M., Tosa, Y., Park, P. & Shinozaki, K. (2003) The cDNA microarray analysis using an *Arabidopsis pad3* mutant reveals the expression profiles and classification of genes induced by *Alternaria brassicicola* attack. *Plant and Cell Physiology*, **44**, 377–387.

Oono, Y., Seki, M., Nanjo, T., Narusaka, M., Fujita, M., Satoh, R., Satou, M., Sakurai, T., Ishida, J., Akiyama, K., Iida, K., Maruyama, K., Sato, S., Yamaguchi-Shinozaki, K. & Shinozaki, K. (2003) Monitoring expression profiles of *Arabidopsis* gene expression during rehydration process after dehydration using ca. 7000 full-length cDNA microarray. *Plant Journal*, **34**, 868–887.

Ozturk, Z. N., Talame, V., Deyholos, M., Michalowski, C. B., Galbraith, D. W., Gozukirmizi, N., Tuberosa, R. & Bohnert, H. J. (2002) Monitoring large-scale changes in transcript abundance in drought- and salt-stressed barley. *Plant Molecular Biology*, **48**, 551–573.

Provart, N. J., Gil, P., Chen, W., Han, W., Chang, H. S., Wang, X. & Zhu, T. (2003) Gene expression phenotypes of Arabidopsis associated with sensitivity to low temperatures. *Plant Physiology*, **132**, 893–906.

Rabbani, M. A., Maruyama, K., Abe, H., Khan, M. A., Katsura, K., Ito, Y., Yoshiwara, K., Seki, M., Shinozaki, K. & Yamaguchi-Shinozaki, K. (2003) Monitoring expression profiles of rice (*Oryza sativa* L.) genes under cold, drought and high-salinity stresses, and ABA application using both cDNA microarray and RNA gel blot analyses. *Plant Physiology*, **133**, 1755–1767.

Ramanjulu, S. & Bartels, D. (2002) Drought- and desiccation-induced modulation of gene expression in plants. *Plant Cell and Environment*, **25**, 141–151.

Richmond, T. & Somerville, S. (2000) Chasing the dream: plant EST microarrays. *Current Opinion in Plant Biology*, **3**, 108–116.

Riechmann, J. L., Heard, J., Martin, G., Reuber, L., Jiang, C. Z., Keddie, J., Adam, L., Pineda, O., Ratcliffe, O. J., Samaha, R. R., Creelman, R., Pilgrim, M., Broun, P., Zhang, J. Z., Ghandehari, D., Sherman, B. K. & Yu, G. L. (2000) *Arabidopsis* transcription factors: genome-wide comparative analysis among eukaryotes. *Science*, **290**, 2105–2110.

Salmeron, J. M. & Vernooij, B. (1998) Transgenic approaches to microbial disease resistance in crop plants. *Current Opinion in Plant Biology*, **1**, 347–352.

Satoh, R., Nakashima, K., Seki, M., Shinozaki, K. & Yamaguchi-Shinozaki, K. (2002) ACTCAT, a novel *cis*-acting element for proline- and hypoosmolarity-responsive expression of the *ProDH* gene encoding proline dehydrogenase in *Arabidopsis*. *Plant Physiology*, **130**, 709–719.

Sawasaki, T., Ogasawara, T., Morishita, R. & Endo, Y. (2002). A cell-free protein synthesis system for high-throughput proteomics. *Proceedings of the National Academy of Sciences of the United States of America*, **99**, 14652–14657.

Sawasaki, T., Hasegawa, Y., Morishita, R., Seki, M., Shinozaki, K. & Endo, Y. (2004) Genome-scale, biochemical annotation method based on the wheat germ cell-free protein synthesis system. *Phytochemistry*, **65**, 1549–1555.

Seki, M., Narusaka, M., Abe, H., Kasuga, M., Yamaguchi-Shinozaki, K., Carninci, P., Hayashizaki, Y. & Shinozaki, K. (2001a) Monitoring the expression pattern of 1300 *Arabidopsis* genes under drought and cold stresses using a full-length cDNA microarray. *Plant Cell*, **13**, 61–72.

Seki, M., Narusaka, M., Yamaguchi-Shinozaki, K., Carninci, P., Kawai, J., Hayashizaki, Y. & Shinozaki, K. (2001b) *Arabidopsis* encyclopedia using full-length cDNAs and its application. *Plant Physiology and Biochemistry*, **39**, 211–220.

Seki, M., Narusaka, M., Kamiya, A., Ishida, J., Satou, M., Sakurai, T., Nakajima, M., Enju, A., Akiyama, K., Oono, Y., Muramatsu, M., Hayashizaki, Y., Kawai, J., Carninci, P., Itoh, M., Ishii, Y., Arakawa, T., Shibata, K., Shinagawa, A. & Shinozaki, K. (2002a) Functional annotation of a full-length *Arabidopsis* cDNA collection. *Science*, **296**, 141–145.

Seki, M., Narusaka, M., Ishida, J., Nanjo, T., Fujita, M., Oono, Y., Kamiya, A., Nakajima, M., Enju, A., Sakurai, T., Satou, M., Akiyama, K., Taji, T., Yamaguchi-Shinozaki, K., Carninci, P., Kawai, J., Hayashizaki, Y. & Shinozaki, K. (2002b) Monitoring the expression profiles of 7000 *Arabidopsis* genes under drought, cold, and high-salinity stresses using a full-length cDNA microarray. *Plant Journal*, **31**, 279–292.

Seki, M., Ishida, J., Narusaka, M., Fujita, M., Nanjo, T., Umezawa, T., Kamiya, A., Nakajima, M., Enju, A., Sakurai, T., Satou, M., Akiyama, K., Yamaguchi-Shinozaki, K., Carninci, P., Kawai, J., Hayashizaki, Y. & Shinozaki, K. (2002c) Monitoring the expression pattern of ca. 7000 Arabidopsis genes under ABA treatments using a full-length cDNA microarray. *Functional and Integrative Genomics*, **2**, 282–291.

Seki, M., Kamei, A., Satou, M., Sakurai, T., Fujita, M., Oono, Y., Yamaguchi-Shinozaki, K. & Shinozaki, K. (2003) Transcriptome analysis in abiotic stress conditions in higher plants. *Topics in Current Genetics*, **4**, 271–295.

Seki, M., Satou, M., Sakurai, T., Akiyama, K., Iida, K., Ishida, J., Nakajima, M., Enju, A., Narusaka, M., Fujita, M., Oono, Y., Kamei, A., Yamaguchi-Shinozaki, K. & Shinozaki, K. (2004) RIKEN Arabidopsis full-length (RAFL) cDNA and its applications for expression profiling under abiotic stress conditions. *Journal of Experimental Botany*, **55**, 213–223.

Shinozaki, K. & Yamaguchi-Shinozaki, K. (1999) Molecular responses to drought stress. In K. Shinozaki & K. Yamaguchi-Shinozaki (eds) *Molecular Responses to Cold, Drought, Heat and Salt Stress in Higher Plants*. RG Landes, Austin, TX, pp. 11–28.

Shinozaki, K. & Yamaguchi-Shinozaki, K. (2000) Molecular responses to dehydration and low temperature: differences and cross-talk between two stress signaling pathways. *Current Opinion in Plant Biology*, **3**, 217–223.

Shinozaki, K., Yamaguchi-Shinozaki, K. & Seki, M. (2003) Regulatory network of gene expression in the drought and cold stress responses. *Current Opinion in Plant Biology*, **6**, 410–417.

Smith, N. A., Singh, S. P., Wang, M. B., Stoutjesdijk, P. A., Green, A. G. & Waterhouse, P. M. (2000) Total silencing by intron-spliced hairpin RNAs. *Nature*, **407**, 319–320.

Strizhov, N., Abraham, E., Okresz, L., Blickling, S., Zilberstein, A., Schell, J., Koncz, C. & Szabados, L. (1997) Differential expression of two P5CS genes controlling proline accumulation during salt-stress requires ABA and is regulated by *ABA1*, *ABI1* and *AXR2* in *Arabidopsis*. *Plant Journal*, **12**, 557–569.

Thomashow, M. F. (1999) Plant cold acclimation: freezing tolerance genes and regulatory mechanisms. *Annual Review of Plant Physiology and Plant Molecular Biology*, **50**, 571–599.

Wan, J., Dunning, F. M. & Bent, A. F. (2002) Probing plant-pathogen interactions and downstream defense signaling using DNA microarrays. *Functional and Integrative Genomics*, **2**, 259–273.

Wang, H., Miyazaki, S., Kawai, K., Deyholos, M., Galbraith, D. W. & Bohnert, H. J. (2003) Temporal progression of gene expression responses to salt shock in maize roots. *Plant Molecular Biology*, **52**, 873–891.

Xiong, L., Schumaker, K. S. & Zhu, J. K. (2002) Cell signaling during cold, drought, and salt stress. *Plant Cell Supplement*, **14**, S165–S183.

Xiong, L. & Zhu, J. K. (2001) Abiotic stress signal transduction in plants: molecular and genetic perspectives. *Physiologia Plantarum*, **112**, 152–166.

Xiong, L. & Zhu, J. K. (2002) Molecular and genetic aspects of plant responses to osmotic stress. *Plant Cell and Environment*, **25**, 131–139.

Yamada, K., Lim, J., Dale, J. M., Chen, H., Shinn, P., Palm, C. J., Southwick, A. M., Wu, H. C., Kim, C., Nguyen, M., Pham, P., Cheuk, R., Karlin-Neumann, G., Liu, S. X., Lam, B., Sakano, H., Wu, T., Yu, G., Miranda, M., Quach, H. L., Tripp, M., Chang, C. H., Lee, J. M., Toriumi, M., Chan, M. M. H., Tang, C. C., Onodera, C. S., Deng, J. M., Akiyama, K., Ansari, Y., Arakawa, T., Banh, J., Banno, F., Bowser, L., Brooks, S., Carninci, P., Chao, Q., Choy, N., Enju, A., Goldsmith, A. D., Gurjal, M., Hansen, N. F., Hayashizaki, Y., Johnson-Hopson, C., Hsuan, V. W., Iida, K., Karnes, M., Khan, S., Koesema, E., Ishida, J., Jiang, P. X., Jones, T., Kawai, J., Kamiya, A., Meyers, C., Nakajima, M., Narusaka, M., Seki, M., Sakurai, T., Satou, M., Tamse, R., Vaysberg, M., Wallender, E. K., Wong, C., Yamamura, Y., Yuan, S., Shinozaki, K., Davis, R. W., Theologis, A. & Ecker, J. R. (2003) Empirical analysis of transcriptional activity in the *Arabidopsis* genome. *Science*, **302**, 842–846.

Yamasaki, K., Kigawa, T., Inoue, M., Tateno, M., Yamasaki, T., Yabuki, T., Aoki, M., Seki, E., Matsuda, T., Nunokawa, E., Ishizuka, Y., Terada, T., Shirouzu, M., Osanai, T., Tanaka, A., Seki, M., Shinozaki, K. & Yokoyama, S. (2004) A novel zinc-binding motif revealed by solution structures of DNA-binding domains of *Arabidopsis* SBP-family transcription factors. *Journal of Molecular Biology*, **337**, 49–63.

Yokoyama, S., Hirota, H., Kigawa, T., Yabuki, T., Shirouzu, M., Terada, T., Ito, Y., Matsuo, Y., Kuroda, Y., Nishimura, Y., Kyogoku, Y., Miki, K., Masui, R. & Kuramitsu, S. (2000) Structural genomics projects in Japan. *Nature Structural Biology*, **7**, 943–945.

Yoshiba, Y., Kiyosue, T., Nakashima, K., Yamaguchi-Shinozaki, K. & Shinozaki, K. (1997) Regulation of levels of proline as an osmolyte in plants under water stress. *Plant and Cell Physiology*, **38**, 1095–1102.

Zhang, J. Z. (2003) Overexpression analysis of plant transcription factors. *Current Opinion in Plant Biology*, **6**, 430–440.

Zhu, J. K. (2002) Salt and drought stress signal transduction in plants. *Annual Review of Plant Biology*, **53**, 247–273.

Index